STATISTICAL
MECHANICS

Advanced Book Classics

Anderson: Basic Notions of Condensed Matter Physics, ABC ppbk,
ISBN 0-201-32830-5

Atiyah: K-Theory, ABC ppbk, ISBN 0-201-40792-2

Bethe: Intermediate Quantum Mechanics, ABC ppbk, ISBN 0-201-32831-3

Clemmow: Electrodynamics of Particles and Plasmas, ABC ppbk,
ISBN 0-20147986-9

Davidson: Physics of Nonneutral Plasmas, ABC ppbk
ISBN 0-201-57830-1

DeGennes: Superconductivity of Metals and Alloys, ABC ppbk,
ISBN 0-7382-0101-4

d'Espagnat: Conceptual Foundations Quantum Mechanics, ABC ppbk,
ISBN 0-7382-0104-9

Feynman: Photon-Hadron Interactions, ABC ppbk, ISBN 0-201-36074-8

Feynman: Quantum Electrodynamics, ABC ppbk, ISBN 0-201-36075-4

Feynman: Statistical Mechanics, ABC ppbk, ISBN 0-201-36076-4

Feynman: Theory of Fundamental Processes, ABC ppbk, ISBN 0-201-36077-2

Forster: Hydrodynamic Fluctuations, Broken Symmetry, and Correlation Functions,
ABC ppbk, ISBN 0-201-41049-4

Gell-Mann/Ne'eman: The Eightfold Way, ABC ppbk, ISBN 0-7382-0299-1

Gottfried: Quantum Mechanics, ABC ppbk, ISBN 0-201-40633-0

Kadanoff/Baym: Quantum Statistical Mechanics, ABC ppbk, ISBN 0-201-41046-X

Khalatnikov: An Intro to the Theory of Superfluidity, ABC ppbk,
ISBN 0-7382-0300-9

Ma: Modern Theory of Critical Phenomena, ABC ppbk, ISBN 0-7382-0301-7

Migdal: Qualitative Methods in Quantum Theory, ABC ppbk, ISBN 0-7382-0302-5

Negele/Orland: Quantum Many-Particle Systems, ABC ppbk, ISBN 0-7382-0052-2

Nozieres/Pines: Theory of Quantum Liquids, ABC ppbk, ISBN 0-7382-0229-0

Nozieres: Theory of Interacting Fermi Systems, ABC ppbk, ISBN 0-201-32824-0

Parisi: Statistical Field Theory, ABC ppbk, ISBN 0-7382-0051-4

Pines: Elementary Excitations in Solids, ABC ppbk, ISBN 0-7382-0115-4

Pines: The Many-Body Problem, ABC ppbk, ISBN 0-201-32834-8

Quigg: Gauge Theories of the Strong, Weak, and Electromagnetic Interactions,
     ABC ppbk, ISBN 0-201-32832-1

Richardson: Experimental Techniques in Condensed Matter Physics at Low
     Temperatures, ABC ppbk ISBN 0-201-36078-0

Rohrlich: Classical Charges Particles, ABC ppbk ISBN 0-201-48300-9

Schrieffer: Theory of Superconductivity, ABC ppbk ISBN 0-7382-0120-0

Schwinger: Particles, Sources, and Fields Vol. 1, ABC ppbk
     ISBN 0-7382-0053-0

Schwinger: Particles, Sources, and Fields Vol. 2, ABC ppbk
     ISBN 0-7382-0054-9

Schwinger: Particles, Sources, and Fields Vol. 3, ABC ppbk
     ISBN 0-7382-0055-7

Schwinger: Quantum Kinematics and Dynamics, ABC ppbk, ISBN 0-7382-0303-3

Thom: Structural Stability and Morphogenesis, ABC ppbk, ISBN 0-201-40685-3

Wyld: Mathematical Methods for Physics, ABC ppbk, ISBN 0-7382-0125-1

STATISTICAL MECHANICS

A Set of Lectures

RICHARD P. FEYNMAN
late, California Institute of Technology

Advanced Book Program

CRC Press
Taylor & Francis Group
Boca Raton London New York

CRC Press is an imprint of the
Taylor & Francis Group, an **informa** business

First published 1972 by Westview Press

Published 2018 by CRC Press
Taylor & Francis Group
6000 Broken Sound Parkway NW, Suite 300
Boca Raton, FL 33487-2742

CRC Press is an imprint of the Taylor & Francis Group, an informa business

Visit the Taylor & Francis Web site at
http://www.taylorandfrancis.com

and the CRC Press Web site at
http://www.crcpress.com

ISBN 13: 978-0-201-36076-9 (pbk)

Cover design by Suzanne Heiser

Editor's Foreword

Editor's Foreword

Addison-Wesley's *Frontiers in Physics* series has, since 1961, made it possible for leading physicists to communicate in coherent fashion their views of recent developments in the most exciting and active fields of physics—without having to devote the time and energy required to prepare a formal review or monograph. Indeed, throughout its nearly forty-year existence, the series has emphasized informality in both style and content, as well as pedagogical clarity. Over time, it was expected that these informal accounts would be replaced by more formal counterparts—textbooks or monographs—as the cutting-edge topics they treated gradually became integrated into the body of physics knowledge and reader interest dwindled. However, this has not proven to be the case for a number of the volumes in the series: Many works have remained in print on an on-demand basis, while others have such intrinsic value that the physics community has urged us to extend their life span.

The *Advanced Book Classics* series has been designed to meet this demand. It will keep in print those volumes in *Frontiers in Physics* or its sister series, *Lecture Notes and Supplements in Physics*, that continue to provide a unique account of a topic of lasting interest. And through a sizable printing, these classics will be made available at a comparatively modest cost to the reader.

These notes on Richard Feynman's lectures on Statistical Mechanics were first published some twenty-five years ago. As is the case with all of Feynman's lectures, the presentation in this work reflects his deep physical insight, the freshness and originality of his approach to understanding physics, and the overall pedagogical wizardry of Richard Feynman. This volume will be of interest to everyone concerned with teaching and learning statistical mechanics. In

addition to providing an elegant and concise introduction to the basic concepts
of statistical physics, the notes contain a description of some of the many original and profound contributions—ranging from polaron theory to the theory of
liquid helium—that Feynman made to the field of condensed matter physics.

David Pines
Urbana, Illinois
December 1997

CONTENTS

ACKNOWLEDGMENTS

This volume is based on a series of lectures sponsored by Hughes Research Laboratories in 1961. The notes for the majority of lectures were taken by R. Kikuchi and H. A. Feiveson.

Others who took notes for one or more of the lectures were F. L. Vernon, Jr., W. R. Graham, Jr., R. W. Hellwarth, D. P. Devor, J. R. Christman, R. N. Byrne, and J. L. Emmett.

The notes were edited by Dr. Jacob Shaham who also prepared the Index.

Statistical
Mechanics

CHAPTER 1

INTRODUCTION TO STATISTICAL MECHANICS

1.1 THE PARTITION FUNCTION

The key principle of statistical mechanics is as follows:

If a system in equilibrium can be in one of N states, then the probability of the system having energy E_n is $(1/Q)e^{-E_n/kT}$, where

$$Q = \sum_{n=1}^{N} e^{-E_n/kT},$$

k = Boltzmann's constant, and T = temperature. Q is called the *partition function.*

If we take $|i\rangle$ as a state with energy E_i and A as a quantum-mechanical operator for a physical observable, then the expected value of the observable is

$$\langle A \rangle = \frac{1}{Q} \sum_{|i\rangle} \langle i|A|i\rangle e^{-E_i/kT}.$$

This fundamental law is the summit of statistical mechanics, and the entire subject is either the slide-down from this summit, as the principle is applied to various cases, or the climb-up to where the fundamental law is derived and the concepts of thermal equilibrium and temperature T clarified. We will begin by embarking on the climb.

If a system is very weakly coupled to a heat bath at a given "temperature," if the coupling is indefinite or not known precisely, if the coupling has been on for a long time, and if all the "fast" things have happened and all the "slow" things not, the system is said to be in *thermal equilibrium.*

For instance, an enclosed gas placed in a heat bath will eventually erode its enclosure; but this erosion is a comparatively slow process, and sometime before the enclosure is appreciably eroded, the gas will be in thermal equilibrium.

Consider two different states of the system that have the same energy, $E_r = E_s$. The probabilities of the system being in states r and s are then equal. For if the system is in state r, any extremely small perturbation will cause the system to go into a different state of essentially the same energy, such as s. The

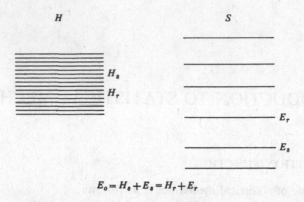

$$E_0 = H_s + E_s = H_r + E_r$$

Fig. 1.1 Energy levels in a system S and a heat bath H.

same is true if the system is in state s. Since the system remains in contact with the heat bath for a long time, one would expect states of equal energy to be equally likely. Also, states of different energies would be expected to have different probabilities.

Because two states of the same energy are equally probable, the probability of a state having energy E is a function only of the energy; $P = P(E)$.

Now consider a system, S, in equilibrium with a large heat bath, H (see Fig. 1.1). Since experience shows that the behavior of a system in equilibrium is independent of the nature of the heat bath, the bath may be assumed extremely large and its total energy E very great. Also, the possible energy levels of the heat bath may be assumed quasi-continuous.

Let the energy levels of the heat bath be denoted by H_i. These levels are distributed quasi-continuously. Let the energy levels of S be denoted by E_j. Then $H_i \gg E_j$ for all i, j. The bath plus the system can be thought of as a new system, T, which is also in thermal equilibrium.

T has a definite energy, but as that energy is not fixed exactly (the bath is in contact with the outside world), we can assume that the energy may be anywhere in the range $E_0 \pm \Delta$. If Δ is sufficiently small, we can assume that the states of the heat bath are equally likely in the range $H_i \pm \Delta$. Let $\eta(H_r)$ be the number of states per unit energy range in the heat bath H around energy H_r.

The probability, $P(E_r)$, that S is in a state with energy E_r is proportional to the number of ways S can have that energy. In other words, it is proportional to $\eta(E_0 - E_r)2\Delta$ = the number of states of H that allow T to have energy in the range $E_0 \pm \Delta$. Then

$$\frac{P(E_r)}{P(E_{r'})} = \frac{\eta(E_0 - E_r)}{\eta(E_0 - E_{r'})} = e^{\ln \eta(E_0 - E_r) - \ln \eta(E_0 - E_{r'})}.$$

Remember that $E_r \ll E_0$. If it is true that $(d/dE) \ln \eta(E) = \beta(E)$ is almost constant for E in the range under consideration,* then we can say

$$\frac{P(E_r)}{P(E_{r'})} = e^{-\beta(E_r - E_{r'})};$$

so

$$P(E_{r'}) \propto e^{-\beta E_{r'}}.$$

Normalization requires that $P(E_i) = (1/Q)e^{-\beta E_i}$ where $Q = \sum_i e^{-\beta E_i}$.

The fundamental law has just been shown to be quite plausible. But for those who doubt the constancy of $\beta(E)$, let us consider some examples.

First, assume that the heat bath consists of N independent harmonic oscillators. The energy of the bath is

$$F = \sum_{i=1}^{N} n_i \hbar w_i,$$

where we assume that n_i is very large, and we neglect the zero-point energy. How many states are there with energy less than F? If $N = 2$, we have the situation shown in Fig. 1.2.

For $N = 2$, the number of states is proportional to the area of the triangle. Clearly the number of states with energy less than F for large F is proportional to F^N; so the number of states per unit energy range is $\eta(F) \propto (d/dF)F^N \propto F^{N-1}$.

$$\ln \eta(F) = \text{constant} + (N - 1) \ln F,$$

$$\frac{d \ln \eta(F)}{dF} = \frac{N - 1}{F} \approx \frac{N - 1}{E_0} \qquad \text{since} \qquad E_r \ll E_0.$$

---

* The basic assumption made here is, that the system governing the probabilities has a quasi-continuous spectrum in the region considered and for which there is no particular characteristic energy. For such a system, if ε_1 and ε_2 are two energy values, then, since energy is only defined up to an additive constant ε, we should have

$$\frac{f(\varepsilon_1)}{f(\varepsilon_2)} = \frac{f(\varepsilon_1 + \varepsilon)}{f(\varepsilon_2 + \varepsilon)},$$

where $f(\varepsilon)$ is the probability. Defining

$$f(\varepsilon) = g(\varepsilon - \varepsilon_2),$$

we obtain

$$g(\varepsilon)g(\varepsilon_1 - \varepsilon_2) = g(0)g(\varepsilon_1 - \varepsilon_2 + \varepsilon),$$

which is (uniquely) solved by

$$g(\varepsilon) = g(0)e^{-\beta\varepsilon} \qquad (\beta \text{ constant}),$$

that is,

$$\frac{f(\varepsilon_1)}{f(\varepsilon_2)} = e^{-\beta(\varepsilon_1 - \varepsilon_2)}$$

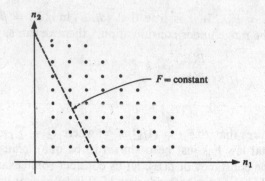

Fig. 1.2 States with energy less than F for two independent harmonic oscillators.

For N large,

$$\beta = \frac{1}{\text{energy per oscillator}} = \frac{1}{W},$$

where $W =$ energy per oscillator.

Alternatively, we can go back to the equation

$$\frac{P(E_r)}{P(E_{r'})} = \frac{\eta(E_0 - E_r)}{\eta(E_0 - E_{r'})}$$

and plug in N directly. We get

$$\frac{\eta(E_0 - E_r)}{\eta(E_0 - E_{r'})} = \frac{(E_0 - E_r)^{N-1}}{(E_0 - E_{r'})^{N-1}}$$

$$= \frac{(1 - E_r/NW)^{N-1}}{(1 - E_{r'}/NW)^{N-1}} \to \frac{e^{-E_r/W}}{e^{-E_{r'}/W}} \quad \text{as} \quad N \to \infty.^*$$

We then get the same value for β as by our previous method.

As a second example, consider the case of a heat bath consisting of N particles in a box.

$$F = \frac{P_{x_1}^2 + P_{y_1}^2 + P_{z_1}^2}{2m} + \frac{P_{x_2}^2 + P_{y_2}^2 + P_{z_2}^2}{2m} + \cdots + \frac{P_{x_N}^2 + P_{y_N}^2 + P_{z_N}^2}{2m}.$$

Assuming periodic boundary conditions,

$$P_x = \frac{2\pi\hbar(\text{integer})}{L} = \frac{2n_x\pi\hbar}{L},$$

---

* *Problem*: Why is W independent of N?

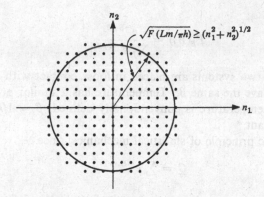

Fig. 1.3 Number of states with energy less than F is roughly equal to the area of the circle shown.

where L is the length of the sides of the box. Then

$$F = \frac{2\pi\hbar}{2Lm} \left[n_{x_1}^2 + n_{y_1}^2 + n_{z_1}^2 + n_{x_2}^2 + \cdots + n_{z_N}^2 \right].$$

If there were just two η's, we would calculate β with the help of Fig. 1.3.
The number of states with energy less than F is roughly equal to the area of the circle with radius $\sqrt{F(Lm/\pi\hbar)}$. For N particles, we must use a hypersphere in $3N$ dimensions, and the number of states with energy less than F is proportional to $(\sqrt{F})^{3N}$. It follows that β is roughly constant, and equals $1/(\frac{2}{3}W)$, where $W =$ energy per particle.

Consider two independent systems S_A and S_B with energy levels A_i and B_j. The probability of system S_A having energy A_i is

$$P_A(A_i) = \frac{e^{-\beta A_i}}{\sum e^{-\beta A_j}}.$$

Now place S_A and S_B in loose contact with each other and consider the combined system $S_T = S_A + S_B$, with combined energy $T_k = A_i + B_j$.

$$P_T(T_k) = P_T(A_i + B_j) = \frac{e^{-\beta_T(A_i + B_j)}}{\sum_i e^{-\beta_T A_i} \sum_j e^{-\beta_T B_j}}$$

$$= \frac{e^{-\beta_T A_i}}{\sum_i e^{-\beta_T A_i}} \frac{e^{-\beta_T B_j}}{\sum_j e^{-\beta_T B_j}}.$$

The probability of system S_T being such that system S_A has energy A_i is

$$P_T(A_i) = \sum_J \left[\frac{e^{-\beta_T A_i}}{\sum_i e^{-\beta_T A_i}} \right] \frac{e^{-\beta_T B_j}}{\sum_j e^{-\beta_T B_j}} = \frac{e^{-\beta_T A_i}}{\sum_i e^{-\beta_T A_i}} = P_A(A_i).$$

Similarly,

$$P_B(B_j) = \frac{e^{-\beta_T B_j}}{\sum_j e^{-\beta_T B_j}}.$$

We see that if two systems are placed in loose contact with each other, in equilibrium they have the same β. Temperature has a similar property, and in fact, by the way temperature is conventionally defined, $\beta = 1/kT$ where k is Boltzmann's constant.*

From the basic principle of statistical mechanics, once

$$Q = \sum_n e^{-E_n/kT}$$

is known, all thermodynamic properties can be found. We define F, the Helmholtz free energy, so that

$$Q = \sum_n e^{-E_n/kT} \equiv e^{-F/kT}, \tag{1.1}$$

$$F = -kT \ln Q = -kT \ln\left(\sum_n e^{-E_n/kT}\right), \tag{1.2}$$

$$S = \text{entropy, is defined as } -k \sum_n P_n \ln P_n, \tag{1.3}$$

---

* Furthermore, suppose our spectrum has a density of states $n(\varepsilon)$ which is relatively constant when $\varepsilon \to \beta\varepsilon$, that is

$$n(\beta\varepsilon) = n(\varepsilon)q(\beta) \tag{*}$$

Then $U(\beta)$ (see 1.7) $= C/\beta$ and we recognize in C the usual (total) heat capacity, so that

$$\frac{C_1}{\beta_1} + \frac{C_2}{\beta_2} = \frac{C_1 + C_2}{\beta},$$

which is the well-known (experimental) formula.

Equation (*) has the unique solution

$$n(\varepsilon) = A\varepsilon^n$$

which is, indeed, the form of $n(\varepsilon)$ for microscopic bodies (see pp. 3, 5).

Also, it can easily be seen (from 1.7 below) that

$$\frac{dU}{d\beta} = -\frac{1}{2Q^2} \sum_{n,m} (\varepsilon_n - \varepsilon_m)^2 e^{-\beta(\varepsilon_n + \varepsilon_m)} < 0,$$

so that U is a decreasing function of β. Suppose we mix two systems having different β's, $\beta_1 < \beta_2$. After mixing there will be a common β, according to

$$U_1(\beta_1) + U_2(\beta_2) = U_1(\beta) + U_2(\beta),$$

whence, since the U's decrease, $\beta_1 < \beta < \beta_2$, and energy flows from the low-β body to the high-β body. This also fits into our intuitive perception of temperature.

where

$$P_n = \frac{1}{Q} e^{-E_n/kT} \tag{1.4}$$

From Eq. (1.2) it can be seen that

$$-\left(\frac{\partial F}{\partial T}\right)_V = -k \sum_n \left[\frac{1}{Q} e^{-E_n/kT} \left\{\frac{-E_n}{kT} - \ln Q\right\}\right] = S \tag{1.5}$$

$$P = \text{pressure} = \sum_n - P_n \frac{\partial E_n}{\partial \text{Vol}} = -\left(\frac{\partial F}{\partial V}\right)_T \tag{1.6}$$

$$U = \text{average energy} = \frac{1}{Q} \sum_n E_n e^{-E_n/kT} \tag{1.7}$$

But

$$\frac{\partial Q}{\partial T} = \frac{1}{kT^2} \sum_n E_n e^{-E_n/kT},$$

$$U = \frac{kT^2}{Q} \frac{\partial Q}{\partial T} = -T^2 \frac{\partial}{\partial T}\left(\frac{F}{T}\right) = F - T \frac{\partial F}{\partial T} = \frac{\partial}{\partial(1/T)} \frac{F}{T}, \tag{1.8}$$

since $S = -\partial F/\partial T$, $U = F + TS$.

$$C_v = \left(\frac{\partial U}{\partial T}\right)_V = \frac{\partial F}{\partial T} - \frac{\partial}{\partial T}\left(T \frac{\partial F}{\partial T}\right) = \frac{\partial F}{\partial T} - \frac{\partial F}{\partial T} - T \frac{\partial^2 F}{\partial T^2}$$

$$= -T \left(\frac{\partial^2 F}{\partial T^2}\right)_V \tag{1.9}$$

To get a clearer idea of the nature of pressure, consider a possible alternative definition. The Hamiltonian operator for the system is dependent on the volume. Set $H = \text{Hamiltonian} = H(V)$. We can take as a pressure operator $P_{OP} = -\partial H/\partial V$.

$$P = \frac{1}{Q} \sum_i \langle i|P_{OP}|i\rangle e^{-E_i/kT}.$$

Our alternative definition is equivalent to the first one if

$$\langle i| \frac{\partial H}{\partial V} |i\rangle = \frac{\partial E_i}{\partial V}.$$

But

$$\frac{\partial E_i}{\partial V} = \lim_{V' \to V} \frac{E_{i'}(V') - E_i(V)}{V' - V} = \lim_{V' \to V} \frac{\langle i'|H'|i'\rangle - \langle i|H|i\rangle}{V' - V},$$

where $|i'\rangle$ is the eigenvector of H' corresponding to $|i\rangle$, the ith eigenvector of H. $H' = H(V')$.

$H' = H + (H' - H)$; so we can apply first-order perturbation theory to say that $\langle i'| H' |i'\rangle \approx \langle i| H |i\rangle + \langle i| H' - H |i\rangle$ for $H' - H \approx 0$. Then

$$\frac{\partial E_i}{\partial V} = \lim_{V' \to V} \frac{\langle i|H' - H|i\rangle}{V' - V} = \langle i| \frac{\partial H}{\partial V} |i\rangle.$$

Our two definitions are equivalent.

Because H may be a function of the shape of the system, as well as its volume, our definition of pressure might depend on how the volume is changed. In general, for any parameter α, there is a force that can be computed by

$$(\text{Force})_\alpha = \frac{1}{Q} \sum_i \left(-\frac{\partial E_i}{\partial \alpha} \right) e^{- E_i(\alpha)/kT},$$

and we can write

$$\frac{\partial E_i}{\partial \alpha} = \langle i| \frac{\partial H}{\partial \alpha} |i\rangle.$$

A third definition of pressure is

$$P = - \frac{\partial U}{\partial V} (V, S).$$

That this definition is equivalent to the other two may be verified without difficulty from the equation $U = F + TS$.

The equation $S = -(\partial F/\partial T)_V$ holds only at equilibrium, where F is defined. Away from equilibrium, S always increases with time. To see this, note that time-dependent perturbation theory gives

$$\frac{dP_m}{dt} = \sum_n (|V_{mn}|^2 P_n - |V_{mn}|^2 P_m),$$

where $|V_{nm}|^2$ is the probability per unit time of transition from state n to state m, and $|V_{nm}|^2 = |V_{mn}|^2$. Then

$$\frac{dS}{dt} = -k \sum_i \left(\frac{dP_i}{dt} + \ln P_i \frac{dP_i}{dt} \right) = -k \sum_i \frac{dP_i}{dt} \ln P_i,$$

since

$$\sum_i \frac{dP_i}{dt} = \frac{d}{dt} \sum_i P_i = \frac{dP}{dt} = 0.$$

Then

$$\frac{dS}{dt} = -k \sum_{mn} |V_{nm}|^2 (P_n - P_m) \ln P_m$$

$$= - \frac{k}{2} \sum_{mn} |V_{nm}|^2 (P_n - P_m)(\ln P_m - \ln P_n).$$

But each term in the sum is negative, for the sign of $P_n - P_m$ is opposite to the sign of ($\ln P_m - \ln P_n$). So $dS/dt > 0$.

Before we start to make calculations, note that if we have a system that is a combination of several independent subsystems, with $E_{total} = E_n = \sum_i E_{n_i} = \sum_i$ energies of the subsystems,

$$F = -kT \ln Q = -kT \ln \sum_n e^{-\beta E_n} = -kT \ln \sum_{n_1, n_2, \dots} e^{-\beta \sum_i E_{n_i}}$$

$$= -kT \ln \prod_i \left(\sum_{n_i} e^{-\beta E_{n_i}} \right) = -kT \sum_i \left[\ln \sum_{n_i} e^{-\beta E_{n_i}} \right] = \sum_i F_i.$$

The free energy of the whole system is the sum of the free energies of its non-interacting parts.

1.2 LINEAR HARMONIC OSCILLATORS

Consider a system of harmonic oscillators in thermal equilibrium. The partition function Q, free energy F, and average energy of the system of oscillators can be found as follows: The oscillators do not interact with each other, but only with the heat bath. Since each oscillator is independent, one can find F_i of the ith oscillator and then

$$F = \sum_{i=1}^{M} F_i$$

(M oscillators).

$$Q_i = \sum_n e^{-E_n^i/kT} \tag{1.10}$$

$$E_n^i = \hbar \omega_i (n + \tfrac{1}{2}) \qquad \text{from quantum mechanics } (n = 0, 1, 2, \dots) \tag{1.11}$$

$$Q_i = \sum_n e^{-\hbar \omega_i (n + 1/2)/kT}$$

$$Q_i = \frac{e^{-\hbar \omega_i / 2kT}}{1 - e^{-\hbar \omega_i / kT}} \left[\frac{1}{1 - e^{-\hbar \omega_i / kT}} = 1 + e^{-\hbar \omega_i / kT} + e^{-2\hbar \omega_i / kT} + \cdots \right] \tag{1.12}$$

$$F_i = -kT \ln Q_i = \frac{\hbar \omega_i}{2} + kT \ln \left(1 - e^{-\hbar \omega_i / kT} \right) \tag{1.13}$$

U_i = average energy of a single oscillator in thermal equilibrium

$$= \frac{1}{Q_i} \sum_n E_n^i e^{-E_n^i/kT} = \frac{\partial}{\partial (1/T)} \frac{F_i}{T} \tag{1.14}$$

$$= \frac{\hbar \omega_i}{2} + \frac{\hbar \omega_i e^{-\hbar \omega_i / kT}}{1 - e^{-\hbar \omega_i / kT}} = \frac{\hbar \omega_i}{2} + \frac{\hbar \omega_i}{e^{\hbar \omega_i / kT} - 1}. \tag{1.15}$$

$$F = \sum_i F_i = \sum_i \left[\frac{\hbar\omega_i}{2} + kT \ln\left(1 - e^{-\hbar\omega_i/kT}\right) \right] \tag{1.16}$$

$$U = \sum_i U_i = \sum_i \left[\frac{\hbar\omega_i}{2} + \frac{\hbar\omega_i}{e^{\hbar\omega_i/kT} - 1} \right].$$

It is customary to define an average n_i, n_i, according to

$$U_i = (n_i + \tfrac{1}{2})\hbar\omega_i.$$

Thus

$$n_i = \frac{1}{e^{\hbar\omega_i/kT} - 1}.$$

Returning to the example on p. 3 we see that for very high temperature (and U) we indeed have

$$W \approx U_i \approx kT = 1/\beta,$$

which is independent of U.

Note that the contribution to F of the ith oscillator is negligible if $\hbar\omega_i \gg kT$ (except for the $\hbar\omega_i/2$ term). At low temperatures the high-frequency modes are "frozen out" and do not contribute to the specific heat.

1.3 BLACKBODY RADIATION

In dealing with blackbody radiation, our point of view will be as follows: In a cavity (blackbody), there are a great number of modes of oscillation. The number of modes per unit volume per frequency bandwidth is given by classical considerations. Each mode, however, behaves as an independent quantum harmonic oscillator, except that the $\hbar\omega/2$ term is neglected. That is, $E_n = n\hbar\omega$.

We want to get rid of the $\hbar\omega/2$ because it leads to infinite energy when there are an infinite number of modes. A Hamiltonian that eliminates the $\hbar\omega/2$ is

$$H_i = \tfrac{1}{2}(P_i^2 + \omega_i^2 q_i^2) - \frac{\hbar\omega_i}{2}.$$

With the above assumptions, we can obtain an expression for the energy per unit volume per unit frequency.

First we will find the number of modes per unit volume per frequency (or wave number).

Assume a gigantic box of dimensions a, b, and c. The demand is made that the waves be periodic at the walls of the box.

$$\frac{1}{\lambda_x} = \text{number of waves/cm in } x\text{-direction}$$

$$\frac{a}{\lambda_x} = \text{number of waves in box } (x\text{-direction}).$$

Because of the periodic boundary condition,

$$\frac{a}{\lambda_x} = n_x = \text{an integer.}$$

Let $k_x = 2\pi/\lambda_x$. Then

$$\frac{ak_x}{2\pi} = n_x,$$

and similarly

$$\frac{bk_y}{2\pi} = n_y, \qquad \frac{ck_z}{2\pi} = n_z.$$

$$\frac{a\,dk_x}{2\pi} = dn_x, \qquad \frac{b\,dk_y}{2\pi} = dn_y, \qquad \frac{c\,dk_z}{2\pi} = dn_z$$

$$d^3n = dn_x\,dn_y\,dn_z = abc\,\frac{d^3k}{(2\pi)^3} = (abc)\,\frac{dk_x\,dk_y\,dk_z}{(2\pi)^3}.$$

For each k, there are two possible polarizations. Thus, the number of modes per unit volume with wave number between k and $k + dk$ is

$$\frac{2\,d^3n}{(abc)} = 2\,\frac{d^3k}{(2\pi)^3}.$$

Now $\omega = kc$ where c is the velocity of light. Also, the number of modes is so large that the sum over the modes can be replaced by an integral.

$$\frac{F}{V} = \iiint k_b T \ln\left(1 - \exp\left[\frac{-\hbar\omega(k)}{k_b T}\right]\right)\frac{2\,d^3k}{(2\pi)^3}. \tag{1.17}$$

(Remember that the $\hbar\omega/2$ term has been omitted.)

From symmetry,

$$\frac{F}{V} = 2\int k_b T \ln\left(1 - \exp\left[\frac{-\hbar\omega(k)}{k_b T}\right]\right)\frac{4\pi k^2\,dk}{(2\pi)^3}.$$

$$\frac{U}{V} = 2\int \frac{\hbar\omega(k)\exp\left[-\hbar\omega(k)/k_b T\right]}{1 - \exp\left[-\hbar\omega(k)/k_b T\right]}\frac{4\pi k^2\,dk}{(2\pi)^3}. \tag{1.18}$$

Let $x = \hbar\omega/k_b T = \hbar kc/k_b T$. Then

$$\frac{U}{V} = \frac{8\pi}{(2\pi)^3}\frac{(k_b T)^4}{\hbar^3 c^3}\int_0^\infty \frac{e^{-x}}{1 - e^{-x}}x^3\,dx$$

$$= \frac{\pi^2 (k_b T)^4}{15\hbar^3 c^3} = \sigma T^4;$$

$$\sigma = \frac{\pi^2 k_b^4}{15\hbar^3 c^3}. \tag{1.19}$$

The preceding results can be summarized and put in a more familiar form by replacing $\hbar\omega$ with $h\nu$.

The number of modes/unit volume between k and $k + dk$ is

$$2 \cdot \frac{4\pi k^2 \, dk}{(2\pi)^3} = \frac{k^2 \, dk}{\pi^2} = \frac{8\pi\nu^2 \, d\nu}{c^3},$$

since $k = 2\pi\nu/c$. The average energy of an oscillator of frequency ν is $h\nu/e^{h\nu/kT}-1$. Thus, the energy per unit volume between ν and $\nu + d\nu$ is

$$U_\nu \, d\nu = \frac{dU}{V} = \frac{8\pi\nu^2}{c^3} \frac{h\nu \, d\nu}{e^{h\nu/kT} - 1}. \tag{1.20}$$

This is the *Planck Radiation Law*, and is the same as Eq. (1.18).

$$\int U_\nu \, d\nu = \sigma T^4$$

is the *Stefan Boltzmann Law*.

For T very large ($kT \gg h\nu$),

$$U_\nu \, d\nu = \frac{8\pi\nu^2}{c^3} \frac{h\nu \, d\nu}{1 + (h\nu/kT) - 1} = \frac{8\pi\nu^2}{c^3} kT \, d\nu. \tag{1.21}$$

This is the *Rayleigh-Jeans Law*.

Let $F^* = F/V$ and $U^* = U/V$.

$$C_v = \frac{\partial U^*}{\partial T} = 4\sigma T^3$$

is the heat capacity of a gas of photons in equilibrium with the container.

If an oscillator (or mode) is excited to the Nth level, that is, if $E = \hbar\omega_i N$, one says that there are N photons with energy $\hbar\omega_i$. Since photons are defined as the degree of excitation of a mode, one cannot consider a permutation of photons as a new state. That is, photons are indistinguishable. This point is the basis of quantum statistics and will be dealt with in a later section.

1.4 VIBRATIONS IN A SOLID

We want to find the specific heat of a solid. However, en route to the specific heats, we will derive important results that will prove useful in many other cases; chief among these results is the calculation of the normal modes of a crystal.

The program will be as follows:

1. Consider the solid to be a crystal lattice of atoms, each atom behaving as an harmonic oscillator. These oscillators are, of course, coupled.

2. Find the normal modes of the system. There are as many normal modes as there are degrees of freedom; namely, $3(AN)$, where (AN) is the number of atoms in the crystal. The modes behave as independent quantum oscillators.

3. Given the modes, calculate F, the free energy.

4. From F, calculate C_v and any other thermodynamic quantities of interest. Once F is found, we can quickly calculate U, the total energy of the system. In the following derivation, we will calculate C_v directly from U without always writing down F first.

Method of Labeling

Consider a crystal with A atoms per unit cell. For convenience, assume that the unit cell is a rectangular solid of dimensions a, b, c, along three mutually per-pendicular axes, x, y, z. Let the origin be at the "center" of a cell. This cell can be denoted by the triplet $(0, 0, 0)$. The cell to its right along the x-axis is denoted by $(1, 0, 0)$, and so on. Thus, any cell can be denoted by a vector $N = n_x a + n_y b + n_z c$ where $a = a\hat{i}, b = b\hat{j}, c = c\hat{k}$. If there are A atoms/cell, $3A$ additional coordinates must be given to locate each atom. Let α denote one of these $3A$ coordinates.

Call the displacement from equilibrium of the coordinate in the Nth cell $Z_{\alpha,N}$. (We either consider $m = $ mass equal to 1 or absorb it in Z.)

$Z_{\alpha,N+M}$ is the displacement of an atom in a cell close to N; if, for example, there are two atoms A_1, A_2 per cell, the displacement of A_1 in the x direction is denoted by $Z_{1,N}$, that of A_2 (in same cell) by $Z_{4,N}$, and that of A_1 in an adjacent cell by $Z_{1,N+1}$. $1 = (1, 0, 0)$ or $(0, 1, 0)$, or $(0, 0, 1)$.

Normal Modes

$$T = (1/2) \sum_{\substack{\text{particles} \\ \text{and directions}}} \dot{Z}_{\alpha,N}^2 = (1/2) \sum_{\alpha,N} \dot{Z}_{\alpha,N}^2 \qquad (1.22)$$

$$V = V(0) + \sum_{\alpha,N} \left(\frac{\partial V}{\partial Z_{\alpha,N}} \right)_{Z=0} Z_{\alpha,N}$$

$$+ \frac{1}{2} \sum_{\alpha,\beta,N,M} \left(\frac{\partial^2 V}{\partial Z_{\alpha,N} \, \partial Z_{\beta,N+M}} \right)_{Z=0} Z_{\alpha,N} Z_{\beta,N+M} + \cdots \qquad (1.23)$$

Assume that the electrons in the crystal always have time to adjust them-selves to the configuration with lowest energy, even when the crystal is vibrating. In this configuration, there is no net force on the nuclei when the $Z_{\alpha,N}$ are zero, so

$$\left(\frac{\partial V}{\partial Z_{\alpha,N}} \right)_{Z=0} = 0.$$

The additive constant, $V(0)$, will not affect our answers, so we might as well drop it. Let

$$\left(\frac{\partial^2 V}{\partial Z_{\alpha,N}\,\partial Z_{\beta,N+M}}\right)_{Z=0} = C_{\alpha\beta}^{M},$$

a number that depends on the relative positions of the cells of the two atoms, and not on their absolute positions. Note that $C_{\alpha\beta}^{M} = C_{\beta\alpha}^{-M}$.

Neglecting higher orders, take

$$V = \tfrac{1}{2} \sum_{\alpha,\beta,N,M} C_{\alpha\beta}^{M} Z_{\alpha,N} Z_{\beta,N+M}. \tag{1.24}$$

For low temperatures it is not too unreasonable to neglect higher orders, because the separation between atoms is of the order of 1 Å and at room temperature the vibrations have amplitude of the order of 0.1 Å. But we should not be too surprised if experiment shows our idealization to be false. In matters such as the one under consideration, the general approach is to make idealizations and then try to find corrections to our assumptions that will give better results.

In order to motivate the procedure that we will use for finding the vibrations of a solid, let us consider the classical problem of vibrations of coupled oscillators. Let the Hamiltonian be

$$H = \sum_{i} \frac{P_i'^2}{2M_i} + \sum_{ij} \tfrac{1}{2} C_{ij}' q_i' q_j',$$

where the q_i' are the coordinates of the amount of displacement from equilibrium, $P_i' = M_i \dot{q}_i'$ is the momentum, and $C_{ij}' = C_{ji}'$ are constants. To eliminate the constants M_i, let

$$q_i = q_i'\sqrt{M_i} \quad \text{and} \quad C_{ij} = \frac{C_{ij}'}{\sqrt{M_i M_j}}.$$

$$P_i = \frac{\partial \text{Lagrangian}}{\partial \dot{q}_i} = \frac{1}{\sqrt{M_i}} P_i'.$$

Then we get

$$H = \sum_{i} \frac{P_i^2}{2} + \frac{1}{2} \sum_{ij} C_{ij} q_i q_j. \tag{1.25}$$

The equations of motion are

$$\dot{q}_i = \frac{\partial H}{\partial P_i} = P_i, \qquad \dot{P}_i = \frac{-\partial H}{\partial q_i} = -\sum_{j} C_{ij} q_j.$$

We now break the motion of the system into modes, each of which has its own frequency. The total motion of the system is a sum of the motions of the modes.

Let the αth mode have frequency ω_α so that

$$q_i^{(\alpha)} = e^{-i\omega_\alpha t}\, a_i^{(\alpha)}$$

for the motion of the αth mode, with $a_i^{(\alpha)}$ independent of time. Then

$$\omega_\alpha^2 a_i^{(\alpha)} = \sum_j C_{ij} a_j^{(\alpha)}.$$

The classical problem of vibrations of coupled oscillators has just been reduced to the problem of finding the eigenvalues and eigenvectors of the real, symmetric matrix, $\|C_{ij}\|$. In order to get the ω_α we must solve the equation

$$\det \|C_{ij} - \omega^2 \delta_{ij}\| = 0. \tag{1.26}$$

Then the $a_i^{(\alpha)}$ (the eigenvectors) can be found. It is possible to choose the $a_i^{(\alpha)}$ so that

$$\sum_i a_i^{(\alpha)} a_i^{(\beta)} = \delta_{\alpha\beta}.$$

The general solution for q_i is

$$q_i = \sum_\alpha C_\alpha q_i^{(\alpha)},$$

where the C_α are arbitrary constants. If we take $Q_\alpha = C_\alpha e^{-i\omega_\alpha t}$, we get $q_i = \sum_\alpha a_i^{(\alpha)} Q_\alpha$. From this it follows that

$$\sum_i a_i^{(j)} q_i = \sum_{\alpha,i} a_i^{(j)} a_i^{(\alpha)} Q_\alpha = \sum_\alpha \delta_{\alpha j} Q_\alpha = Q_j.$$

Making the change of variables, $Q_j = \sum_i a_i^{(j)} q_i$, we get $H = \sum_\alpha H_\alpha$, where

$$H_\alpha = \tfrac{1}{2} P_\alpha^2 + \tfrac{1}{2}\omega_\alpha^2 Q_\alpha^2.$$

This has the expected solutions $Q_\alpha = C_\alpha e^{-i\omega_\alpha t}$.

Now suppose we wish to solve the quantum-mechanical problem of coupled oscillators. Again we have

$$H = \sum_i \frac{P_i'^2}{2M_i} + \sum_{ij} C_{ij} q_i' q_j',$$

where this time

$$P_i' = \frac{\hbar}{i} \frac{\partial}{\partial q_i'}.$$

Making the same change of variables as before, we get

$$Q_\alpha = \sum_i a_i^{(\alpha)} q_i = \sum_i a_i^{(\alpha)} \sqrt{M_i}\, q_i'$$

$$H = \sum_\alpha H_\alpha,$$

where

$$H_\alpha = \frac{-\hbar^2}{2} \frac{\partial^2}{\partial Q_\alpha^2} + \tfrac{1}{2}\omega_\alpha^2 Q_\alpha^2.$$

It follows immediately that the eigenvalues of our original Hamiltonian are $E = \sum_\alpha (N_\alpha + \tfrac{1}{2})\hbar\omega_\alpha$. The solution of a quantum-mechanical system of coupled oscillators is trivial once we have solved the equation

$$0 = \det \|C_{ij} - \omega^2 \delta_{ij}\| = \det \|C'_{ij}/\sqrt{M_i M_j} - \omega^2 \delta_{ij}\|.$$

If we have a solid with $\tfrac{1}{3}(10^{23})$ atoms we must apparently find the eigenvalues of a 10^{23} by 10^{23} matrix. But if the solid is a crystal, the problem is enormously simplified. The classical Hamiltonian for a crystal is

$$H = \tfrac{1}{2} \sum_{\alpha,N} \dot{Z}_{\alpha,N}^2 + \tfrac{1}{2} \sum_{\alpha,\beta,N,M} C_{\alpha\beta}^M Z_{\alpha,N} Z_{\beta,N+M},$$

and the classical equation of motion is (using $C_{\alpha\beta}^M = C_{\beta\alpha}^{-M}$)

$$\ddot{Z}_{\alpha,N} = - \sum_{M,\beta} C_{\alpha,\beta}^M Z_{\beta,N+M}.$$

In a given mode, if one cell of the crystal is vibrating in a certain manner, it is reasonable to expect all cells to vibrate the same way, but with different phases. So we try

$$Z_{\alpha,N} = a_\alpha(K)e^{-i\omega t}e^{iK \cdot N},$$

where K expresses the relative phase between cells. The $e^{iK \cdot N}$ factor allows for wave motion. We now want to find the dispersion relations, or $\omega = \omega(K)$.

$$\omega^2 a_\alpha e^{iK \cdot N} = \sum_{M,\beta} (C_{\alpha,\beta}^M a_\beta e^{iK \cdot M})e^{iK \cdot N}.$$

Let

$$\gamma_{\alpha\beta}(K) = \sum_M C_{\alpha,\beta}^M e^{iK \cdot M}.$$

(Note that $\gamma_{\alpha\beta}(K)$ is Hermitian. See end of p. 18).

Then $\omega^2 a_\alpha = \sum_\beta \gamma_{\alpha\beta} a_\beta$, and we must solve the characteristic equation of a $3A$-by-$3A$ matrix:

$$\det \|\gamma_{\alpha\beta} - \omega^2 \delta_{\alpha\beta}\| = 0. \tag{1.27}$$

The solutions of the characteristic equation are

$$\omega^{(r)}(K) \equiv \omega_K^{(r)},$$

where r runs from 1 to $3A$. The motion for a particular mode can be written

$$Z_{\alpha,N}^{(r)}(K) = a_\alpha^r(K)e^{-i\omega^{(r)}(K)^r}e^{iK \cdot N}$$

where

$$\sum_\alpha a_\alpha^r a_\alpha^{*r'} = \delta_{rr'}.$$

Then the general motion can be described by

$$Z_{\alpha,N} = \sum_{K,r} \frac{C_r(K)}{\sqrt{\eta}} a_\alpha^r(K) e^{-i\omega^{(r)}(K)r} e^{iK \cdot N}$$

where the $C_r(K)/\sqrt{\eta}$ are arbitrary constants, and η is the total number of unit cells.

The factor $1/\sqrt{\eta}$ is inserted to make things look nicer later, but it is not strictly necessary.

Let $Q_r(K) = C_r(K)e^{-i\omega^{(r)}(K)r}$. $Q_r(K)$ describes the motion of a particular mode.

$$Z_{\alpha,N} = \sum_{K,r} Q_r(K) a_\alpha^r(K) e^{iK \cdot N} (1/\sqrt{\eta}). \qquad (1.28)$$

It follows that

$$Q_r(K) \propto \sum_{\alpha,N} Z_{\alpha,N} a_\alpha^{*r}(K) e^{-iK \cdot N}, \qquad (1.29)$$

and the Hamiltonian for the system is

$$H = \frac{1}{2} \sum_{\alpha,N} \left[\dot{Z}_{\alpha,N}^2 + \sum_{\beta,M} C_{\alpha\beta}^M Z_{\alpha,N} Z_{\beta,N+M} \right]$$

$$= \frac{1}{2} \sum_{K,r} \left[|\dot{Q}_r(K)|^2 + \omega^{2(r)}(K)|Q_r(K)|^2 \right]. \qquad (1.30)$$

If we consider $Q_r(K)$ and its complex conjugate to be independent variables, we get the same equations of motion from

$$H = \sum_{K,r} \left[|\dot{Q}_r(K)|^2 + \omega^{2(r)}(K)|Q_r(K)|^2 \right]. \qquad (1.31)$$

1.5 SPECIFIC HEAT OF A CRYSTAL

We now want to find F, so we must sum over all possible modes.

For one mode, $F = kT \ln (1 - e^{-\hbar\omega_K(p)/kT})$. There is a quasi-continuum of K values, and for each K value there are $3A$ $\omega_K^{(p)}$'s. Assuming the crystal has volume V, and assuming periodic boundary conditions, we have in the range K to $K + dK$ approximately $d3K/(2\pi)^3 V$ modes with a given p. Thus,

$$\frac{F}{V} = kT \int_K \sum_p \frac{\hbar\omega_K(p)}{2} + kT \ln \left[1 + e^{-\hbar\omega_K(p)/kT} \right] \frac{d^3K}{(2\pi)^3}$$

$$d^3K = dK_x \, dK_y \, dK_z. \qquad (1.32)$$

$\int_K$ is shorthand for

$$\int_{-\pi/a}^{\pi/a} \int_{-\pi/b}^{\pi/b} \int_{-\pi/c}^{\pi/c} .$$

The reason for the limits $-\pi/a$, π/a is as follows: the factor K was introduced in $e^{iK \cdot N}$. Then $e^{iK \cdot N} = e^{iK_x n_x a} e^{iK_y n_y b} e^{iK_z n_z c}$; n_x, n_y, n_z are integers. But,

$$e^{iK_x n_x a} = e^{i(K_x + 2\pi/a)n_x a} = e^{iK_x n_x a} e^{2\pi i}.$$

No new modes are introduced if K ranges beyond the prescribed limits.

We now demonstrate that the specific heat of a crystal determined by our present method agrees with experiment; that is, at high T, $C_v = 3R$. At low T, $C_v \to 0$ as T^3.

We choose to look at U rather than at F. First we note that there is a maximum frequency, ω_M. Two adjacent atoms can be no more than 180° out of phase, and thus the minimum wavelength must be of the order of twice the atomic spacing (call it $2a$). Thus $K_{max} \approx 2\pi/2a = \pi/a$, and $\omega_M = \omega_M^{(p)}(K_M)$.

First, we consider the high-temperature limit. For $kT \gg \hbar\omega_M$, all modes are excited to approximately the same energy; that is, if there are N atoms,

$$U = \sum_i \frac{\hbar\omega_i}{2} + \sum_i \frac{\hbar\omega_i}{\exp(\hbar\omega_i/kT) - 1} \approx \frac{\hbar}{2} \sum_i \omega_i + \sum_i kT$$

$$= U_0 + 3NkT = U_0 + 3RT, \tag{1.33}$$

if N is Avogadro's number. $C_v = 3R$ for large T. Note that we might as well neglect the zero-point energy, U_0.

For very small values of T ($kT \ll \hbar\omega_M$), the behavior of C_v can also be approximately determined.

Ignoring the zero-point energy, we see that the contribution to U of the high-frequency modes is very small, because

$$\frac{\hbar\omega}{\exp(\hbar\omega/kT) - 1} \approx 0$$

when $\hbar\omega \gg kT$. We also know that when K is zero, there are three zero-frequency modes arising from translation of the entire crystal. For small K, there should be three very low-frequency modes. To see how those modes vary with K, consider $\gamma_{\alpha\beta}(K)$ for low K. Since

$$\gamma_{\alpha\beta}(K) = \sum_M C_{\alpha,\beta}^M e^{iK \cdot M},$$

where the $C_{\alpha\beta}^M$ are real,

$$\gamma_{\beta\alpha}^*(K) = \sum_M C_{\beta,\alpha}^M e^{-iK \cdot M} = \sum_M C_{\beta,\alpha}^{-M} e^{iK \cdot M} = \gamma_{\alpha\beta}(K).$$

Also, for real (K), $\gamma_{\beta\alpha}^*(K) = \gamma_{\beta\alpha}(-K)$, so $\gamma_{\alpha\beta}(K) = \gamma_{\beta\alpha}(-K)$, and

$$f(\omega, K) = \det \|\gamma_{\alpha\beta}(K) - \omega^2 \delta_{\alpha\beta}\| = \det \|\gamma_{\beta\alpha}(K) - \omega^2 \delta_{\beta\alpha}\|$$

$$= \det \|\gamma_{\alpha\beta}(-K) - \omega^2 \delta_{\alpha\beta}\| = f(\omega^2, -K).$$

Then, because f is an even function of K,

$$\frac{\partial f}{\partial K}(\omega^2, 0) = 0$$

Since the characteristic equation must have a solution with $\omega = 0$ for $K = 0$, $f(0, 0) = 0$. The characteristic equation becomes

$$0 = \omega^2 \frac{\partial f}{\partial \omega^2} - \frac{1}{2!} \sum_{i,j} K_i K_j \frac{\partial^2 f}{\partial K_i \, \partial K_j}\bigg|_0 + \text{higher order terms,}$$

so that, for low ω and $|K|$ we have

$$\omega = |K| \left(\frac{\cos \alpha_i \cos \alpha_j \dfrac{\partial^2 f}{\partial K_i \, \partial K_j}\bigg|_0}{2 \partial f / \partial \omega^2} \right)^{1/2},$$

where $\cos \alpha_i$ are the direction cosines of K.

In other words, $\omega = |K|/V$, where V is the velocity of sound, which may depend on the direction of K when the eigenvalues of $\partial^2 f / \partial K_i dK_j|_0$ are not all equal. But, for sufficiently low frequency, V does not depend on the frequency. This is the *Debye approximation*.

For convenience, assume that the three sound velocities are equal; that is, $\omega = V_0 K$.

$$U = \sum_i \left[\frac{\hbar \omega_i}{2} + \frac{\hbar \omega_i}{e^{\hbar \omega_i / kT} - 1} \right]$$

$$= \sum_i \frac{\hbar \omega_i}{2} + 3 \int_0^{K_M} \frac{\hbar \omega(K)}{e^{\hbar \omega / kT} - 1} \frac{4\pi K^2}{(2\pi)^3} \, dK; \quad K = \frac{\omega}{V_0}.$$

The factor 3 before the integral is there because for each K there are three modes with low ω.

$$U = U_0 + \frac{3\hbar}{2\pi^2 V_0^3} \int_0^{\omega_M} \frac{\omega^3}{e^{\hbar \omega / kT} - 1} \, d\omega, \quad \text{where} \quad \omega_M = V_0 K_{\max}. \quad (1.34)$$

$K_{\max}$ can be found by setting the total number of modes equal to $3(AN)$, where (AN) is the number of atoms in the crystal.

$$3(AN) = \int_0^{K_{\max}} \frac{3 \, d^3 K V}{(2\pi)^3} = \int_0^{K_{\max}} \frac{3V}{(2\pi)^3} 4\pi K^2 \, dK = \frac{K_M^3 V}{2\pi^2}.$$

Then

$$K_M = (6\pi^2 \rho)^{1/3} \text{ where } \rho \text{ is the number of atoms per unit volume.}$$

Let

$$x = \frac{\hbar\omega}{kT} = \frac{\Theta}{T} ; \qquad \omega = \frac{kTx}{\hbar} ; \qquad \omega^3 \, d\omega = \frac{k^3 T^3 x^3}{\hbar^3} \frac{kT \, dx}{\hbar} ; \qquad \Theta = \frac{\hbar K V}{k} .$$

$$\frac{U}{V} = \frac{U_0}{V} + \frac{3k^4 T^4}{2\pi^2 \hbar^3 V_0^3} \int_0^{\Theta_M/T} \frac{x^3 \, dx}{e^x - 1} = U_0 + AT^4. \tag{1.35}$$

Here Θ is the *Debye temperature*.

$$C_v = 4AT^3 \tag{1.36}$$

where $A \to \pi^2 k^4 / 10 h^3 V_0^3$ as $T/\Theta \to 0$.
So C_v is proportional to T^3 at low temperatures.

The T^3 dependence of the specific heat at low T can be made more plausible as follows: We assume that for $\hbar\omega_i > kT$, the contribution to the energy of the ω_i mode is negligible and for $\hbar\omega_i < kT$, the contribution is kT.

The number of modes with wave number less than K is

$$3V \int_0^K \frac{4\pi k^2}{(2\pi)^3} \, dk = \frac{K^3 V}{2\pi^2} .$$

At low T, $K = \omega/V_0$. Therefore the number of modes with frequency less than ω_c is $\omega_c^3 V / 2\pi^2 V_0^3$. But $\hbar\omega_c = kT$ or $\omega_c = kT/\hbar$ and

$$n = \text{(number of modes with frequency less than } \omega_c) = k^3 T^3 V / 2\pi^2 \bar{V}_0^3 h^3 .$$

The energy varies as

$$\frac{U}{V} = \frac{k^3 T^3 V}{2\pi^2 \hbar^3 V_0^3} (kT) = \frac{k^4 T^4 V}{2\pi^2 \hbar^3 V_0^3} .$$

And thus C_v is proportional to T^3. The numerical factor is, of course, not correct.

If the mode with frequency ω_i is excited to the nth level, $E_i = \hbar\omega_i(n + \frac{1}{2})$, we say that there are n phonons of frequency ω_i and energy $h\omega_i$ in the crystal.

Figure 1.4 shows the general form of the specific heat of a solid as a function of temperature.

Example: Assume the unit cell to be a cubic lattice with one atom per cell (Fig. 1.5). Each atom behaves as an harmonic oscillator, with spring constants k_A (nearest neighbors), and k_B (next-nearest neighbors). This case is fairly simple, and we can simplify the notation: $\alpha = 1, 2, 3$.

$$Z_{1,N} = X_N, \qquad Z_{2,N} = Y_N, \qquad Z_{3,N} = Z_N.$$

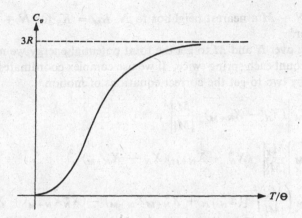

Fig. 1.4 Specific Heat of a Solid.

We wish to find the three natural frequencies associated with each K of the crystal. To do this, we must find $C_{\alpha,\beta}^M$ and then $\gamma_{\alpha,\beta}$. In complex coordinates

$$V = \sum_{\alpha,\beta} V_{\alpha\beta}$$

where

$$V_{\alpha\beta} = \sum_{N,M} C_{\alpha\beta}^M Z_{\alpha,N}^* Z_{\beta,N+M}. \tag{1.37}$$

For example,

$$V_{11} = \sum_{N,M} C_{11}^M X_N^* X_{N+M}. \tag{1.37'}$$

If we express the displacement of atom N from its normal position as X_N, then the potential energy from the distortion of the spring between atoms N and M is

$$\tfrac{1}{2} K_M \left[(X_N - X_{N+M}) \cdot \frac{M}{|M|} \right]^2 ;$$

Fig. 1.5 Cubic lattice with one atom per cell.

$K_M = K_A$ for $N + M$ a nearest neighbor to N, $K_M = K_B$ for $N + M$ a next-nearest neighbor.

In summing over N and M to get the total potential energy we must divide by two, for we count each spring twice. If we use complex coordinates, however, we multiply V by two to get the correct equations of motion.

$$V = \frac{1}{2} \sum_{N,M} K_M \left| (X_N - X_{N+M}) \cdot \frac{M}{|M|} \right|^2, \tag{1.38}$$

$$V_{11} = \frac{1}{2} \sum_{N,M} K_M \left| \frac{M_X}{|M|} \right|^2 (X_N^* - X_{N+M})(X_N - X_{N+M})$$

$$= \frac{1}{2} \sum_{N,M} K_M \left(\frac{M_X}{|M|} \right)^2 [(X_N^* X_N + X_{N+M}^* X_{N+M}) - (X_N^* X_{N+M} + X_{N+M}^* X_N)]$$

$$= \sum_{N,M} K_M \left(\frac{M_X}{|M|} \right)^2 [X_N^* X_N - X_N^* X_{N+M}]. \tag{1.38'}$$

Comparing Eqs. (1.37') and (1.38'), we see that

$$C_{11}^0 = \sum_M K_M \left(\frac{M_X}{|M|} \right)^2; \qquad C_{11}^{M \neq 0} = -K_M \left(\frac{M_X}{|M|} \right)^2. \tag{1.39}$$

Here $(M_X/|M|)^2 = 1$ for $M = (\pm 1, 0, 0)$, $\frac{1}{2}$ for $M = (\pm 1, \pm 1, 0)$ and $M = (\pm 1, 0, \pm 1)$, and zero for the other nearest and next nearest neighbors. So

$$C_{11}^0 = 2K_A + 4K_B, \qquad C_{11}^{\pm(1,0,0)} = -K_A, \quad \text{and so on.}$$

In this way, all the $C_{\alpha,\beta}^M$ can be found. We can then calculate

$$\gamma_{\alpha,\beta}(K) = \sum_M C_{\alpha,\beta}^M [e^{iK \cdot M}].$$

We wish to solve $\det |\gamma_{\alpha,\beta} - \omega^2 \delta_{\alpha,\beta}| = 0$.

For each K, there are three solutions for ω. Thus we obtain $3N$ values of ω, $\omega^{(p)}(K)$.

1.6 THE MÖSSBAUER EFFECT

If a free, excited nucleus goes to its ground state by emission of a photon, the energy of the photon will be less than the excitation energy because the nucleus recoils. But if the excited nucleus is in a crystal, there is a finite probability that it will emit a photon with the full excitation energy. In other words, there is a finite probability that the state of the crystal after radiation will be the same as the initial state. Similarly, there is a finite probability that the crystal state will be unchanged by absorption of a photon. This effect, called the *Mössbauer effect*, can be discussed in terms of modes of oscillation of a crystal. For example, suppose the temperature is absolute zero. Then the crystal must be in its vibrational ground state before emission of a photon. We will find an expression for the probability that the crystal is in its ground state after emission.

Let R be the position of the excited atom, and let P be the momentum of the emitted photon. We will assume, without proof, that the amplitude for the crystal being in a given final state is

$$a = \langle \text{final} \,|Ae^{iP \cdot R/\hbar}|\, \text{initial} \rangle.$$

Furthermore, we will consider only the vibrational states of the crystal, and will neglect any effect due to the fact that the nucleus changes state.

Let the Mth atom be the one that emits the photon; let $R_{0,M}$ be its mean position, and let Z_M be its displacement from that position.

$$R = R_{0,M} + Z_M.$$

We can take the origin of our coordinate system at the mean position of the excited atom. Then $R_{0,M} = 0$.

A single one-dimensional harmonic oscillator in the ground state is described by the wave function

$$\psi = \left(\frac{\omega}{\pi h} \right)^{1/4} e^{-\omega(Q^2/2\hbar)},$$

where Q is the position measured in units such that the mass can be set equal to unity. The wave function of a crystal in its ground state is a product of the wave functions for each mode of vibration.

$$\psi_{\text{crystal}} = \prod_{K,r} \left(\frac{\omega^{(r)}(K)}{\pi \hbar} \right)^{1/4} e^{-\omega^{(r)}(K)|Q_r^2(K)|/2\hbar}$$

We wish to calculate:

$$F = C \left| \iiint \psi_{\text{crystal}}^* e^{iP \cdot Z_M/\hbar} \psi_{\text{crystal}} \right|^2 \tag{1.40}$$

where F is the probability that the crystal will remain in its ground state and C is a constant independent of P. As $P \to 0$, F must tend to unity. So, if we can get a formula for F without the correct constant in front of it, the constant can be determined easily. We will ignore such factors as "A" and "$(\omega/\pi\hbar)^{1/4}$."

We take $Z_{\alpha,M} = Z_{\alpha,0}$ as the displacement of the nucleus that emits the photon. For simplicity, assume the momentum of the photon is in the α direction. Then we wish to compute

$$F^{1/2} \propto \iiint \left(\prod \exp\left[\frac{-\omega^{(r)}(K)Q_r^2(K)}{2\hbar} \right] \right) (\exp[iPZ_{\alpha,0}/\hbar])$$
$$\times \left(\prod \exp\left[\frac{-\omega^{(r')}(K')Q_r^2(K')}{2\hbar} \right] \right).$$

But

$$Z_{\alpha,0} = \sum_{K,r} Q_r(K) a_\alpha^r(K) \frac{1}{\sqrt{\eta}},$$

so

$$F^{1/2} \propto \prod_{K,r} \left[\int \exp\left[\frac{-\omega^{(r)}(K)|Q_r^2(K)|}{\hbar} \right] \exp\left[i\left(\frac{P}{\hbar\sqrt{\eta}}\right) Q_r(K) a_\alpha^r(K) \right] dQ_r(K) \right]$$

$$\propto \prod_{K,r} \exp\left[\frac{-P^2|a_\alpha^r(K)|^2}{4\hbar\omega^{(r)}(K)\eta} \right] = \exp\left[-\sum_{K,r} \frac{P^2|a_\alpha^r(K)|^2}{4\hbar\omega^{(r)}(K)\eta} \right].$$

Notice that

$$\langle (Z_{\alpha N})^2 \rangle = \left\langle \sum_{\substack{K,r \\ K',r'}} a_\alpha^{(r)} [a_\alpha^{(r')}]^* Q_r(K) Q_{r'}^*(K') e^{i(K-K')\cdot N} \frac{1}{\eta} \right\rangle$$

$$= \sum_{K,r} |a_\alpha^r(K)|^2 \langle Q_r^2(K) \rangle \frac{1}{\eta} = \sum_{K,r} \frac{\hbar|a_\alpha^r(K)|^2}{2\eta\omega^{(r)}(K)}.$$

So we can write

$$F \propto \exp\left[-\left(\frac{P^2}{\hbar^2}\right) \langle Z_{\alpha,M}^2 \rangle \right].$$

Because $F = 1$ at $P = 0$, we can write

$$F = \exp\left[-\left(\frac{P^2}{\hbar^2}\right) \langle Z_{\alpha,M}^2 \rangle \right]. \tag{1.41}$$

The value of F can also be found for nonzero temperature. If "i" is the number describing the state of the crystal, we must compute

$$F_i = |\langle i| e^{iPZ_{\alpha,0}/\hbar} |i\rangle|^2$$

and then

$$F = \frac{1}{Q} \sum_i F_i e^{-E_i/kT},$$

$$F_i e^{-E_i/kt} = \prod_{K,r} \left| \langle n| \exp\left[\frac{iP}{\hbar\sqrt{\eta}} Q_r(K) a_\alpha^r(K) \right] |n\rangle \right|^2 \exp\left[\frac{-(n+\frac{1}{2})\hbar\omega^{(r)}(K)}{kT} \right],$$

where $|n\rangle$ denotes the nth state of a single one-dimensional harmonic oscillator with frequency $\omega^{(r)}(K)$. As η is very large, we can write

$$F_i \exp\left(\frac{-E_i}{kT}\right) = \prod_{K,r} \left| 1 - \frac{1}{2!} \frac{P^2}{\eta\hbar^2} \langle Q_r^2(K) \rangle_n |a_\alpha^r(K)|^2 + \cdots \right|^2$$

$$\times \exp\left[-(n + \tfrac{1}{2}) \frac{\hbar\omega^r(K)}{kT} \right]$$

$$\approx \prod_{K,r} \left| 1 - \frac{1}{2} \frac{P^2(2n+1)\hbar}{\eta\hbar^2 2\omega^{(r)}(K)} |a_\alpha^r(K)|^2 \right|^2 \exp\left[\frac{-(2n+1)\hbar\omega^{(r)}(K)}{2kT} \right],$$

$$F \propto \sum \prod \left[1 - \frac{P^2(2n+1)}{2\omega^{(r)}(K)\eta\hbar} |a_\alpha^r(K)|^2 \right] \exp\left[\frac{-(2n+1)\hbar\omega^{(r)}(K)}{2kT} \right]$$

$$= \prod_{K,r} \sum_{n=0}^{\infty} \left[1 - \frac{P^2(2n+1)}{2\omega\eta\hbar} |a_\alpha^r|^2 \right] \exp\left[\frac{-(2n+1)\hbar\omega}{2kT} \right]$$

$$\propto \prod_{K,r} \left[1 - \frac{P^2}{2\omega\eta\hbar} |a_\alpha^r|^2 \frac{1 + e^{-\hbar\omega/kT}}{1 - e^{-\hbar\omega/kT}} \right]$$

$$\approx - \sum_{K,r} \left[\frac{P^2}{2\omega\eta\hbar} |a_\alpha^r(K)|^2 \frac{1 + e^{-\hbar\omega/kT}}{1 - e^{-\hbar\omega/kT}} \right]$$

As before, we get for our answer

$$F = \exp\left[\frac{-P^2}{\hbar^2} \langle Z_{\alpha,M}^2 \rangle \right]. \tag{1.41}$$

1.7 QUANTUM STATISTICS FOR A MANY-PARTICLE SYSTEM

Consider a system of N identical particles, and assume that there is no interaction among them. Any two configurations of the system that differ only by an interchange of two or more identical particles are regarded as one and the same state. Thus the state of the system is determined by giving the number of particles n_a with energy ε_a.

Our problem is to calculate the partition function Q with the condition $\sum n_a = $ constant. The values admitted for every n_a may be

(a) $n_a = 0, 1, 2, 3, 4, \ldots$ Bose-Einstein case

(b) $n_a = 0, 1$ Fermi-Dirac case

Bose-Einstein statistics must be used for particles with integral spin (for example, He^4) and Fermi-Dirac statistics for particles with half-integral spin (such as, electrons). For Bose particles, any number of particles may occupy a given state. For Fermi particles, however, there can only be one particle at most in each state (Pauli exclusion principle).

A state of the system is described by the set of numbers n_a, which can take on any set of values allowed both by the statistics and by the condition $\sum_a n_a = N$.

$$Q = \sum_{n_1, n_2, \ldots} \exp\left(-\beta \sum_a n_a \varepsilon_a\right). \tag{1.42}$$

If there were no restriction on the number of particles, we could write

$$Q = \prod_a \left(\sum_{n_a} e^{-\beta n_a \varepsilon_a}\right), \tag{1.43}$$

and we would have in the Bose-Einstein case

$$Q = \prod_a \left(\frac{1}{1 - e^{-\beta \varepsilon_a}}\right).$$

In the Fermi-Dirac case we would have

$$Q = \prod_a \left(1 + e^{-\beta \varepsilon_a}\right).$$

Unfortunately, Eq. (1.43) is incorrect, because we have an auxiliary condition that $\sum_a n_a = N = $ constant. With this auxiliary condition, the problem of finding Q becomes much more difficult. It is possible, however, to get around the restriction by considering the system of particles to be a box connected to a large reservoir of particles (Fig. 1.6), and by assuming that it is possible for particles to pass to and from the reservoir.

We further assume that the statistical mechanics of the total system acts as if it took energy μ to remove a particle from the box to the reservoir. μ can be adjusted by, say, a voltage regulator. In the case of electrons in a metal, for example, μ is the work function of the metal. In general, increasing μ will increase the expected number, $\langle N \rangle$, of particles in the box. If we can find $\langle N \rangle$ as

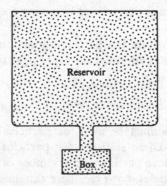

Fig. 1.6 System of particles considered as a box connected to a large reservoir.

a function of μ, we can in principle select μ so that any desired number of particles are in the box.

For a given μ, the energy levels of a particle in the box become $\varepsilon_a - \mu$, rather than ε_a.

By the key principle of statistical mechanics (as applied to the box), the probability of the gas having energy $E = \sum_a n_a(\varepsilon_a - \mu)$ is proportional to $e^{-\beta E}$.

$$Q^{(\mu)} = \sum_{n_1, n_2, \ldots} \exp -\beta[\sum n_a(\varepsilon_a - \mu)] . \tag{1.44}$$

The summation can be made without restriction. We will now show how μ is determined by N and how thermodynamic quantities depend on $Q^{(\mu)}$ or $g \equiv -1/\beta \ln Q^{(\mu)}$.

$$e^{-\beta g} = Q^{(\mu)} = \sum_{n_1, n_2, \ldots} \exp\left(-\beta\left[\sum_a n_a(\varepsilon_a - \mu)\right]\right). \tag{1.45}$$

Let $N = \sum_a n_a$.

$$\frac{\partial Q^{(\mu)}}{\partial \mu} = \sum \beta \left(\sum_a n_a\right) \exp\left(-\beta\left[\sum_a n_a(\varepsilon_a - \mu)\right]\right)$$

$$= \sum \beta N \exp\left(-\beta\left[\sum_a n_a(\varepsilon_a - \mu)\right]\right). \tag{1.46}$$

But

$$\langle N \rangle = \frac{1}{Q^{(\mu)}} \sum N \exp\left(-\beta\left[\sum_a n_a(\varepsilon_a - \mu)\right]\right). \tag{1.47}$$

$$\langle N \rangle = \frac{1}{\beta Q^{(\mu)}} \frac{\partial Q^{(\mu)}}{\partial \mu} = \frac{1}{\beta} \frac{\partial}{\partial \mu} \ln Q^{(\mu)} = \frac{-\partial g}{\partial \mu} .$$

This equation gives $\langle N \rangle = \langle N(\mu) \rangle$. Inverting it gives μ as a function of $\langle N \rangle$.* Similarly, let $\langle n_a \rangle$ be the average value of n_a for a given μ (or a given $\langle N \rangle$). Then $\langle n_a \rangle = \partial g/\partial f_a$ and $\langle N \rangle = \sum_a \langle n_a \rangle$.

For the purpose of computing the probability of a state, we assumed that the system acts as if it had energy $E = \sum_a n_a(\varepsilon_a - \mu)$. But the μ was inserted only in order to provide for the weighting factor that accounts for different probabilities of different numbers of particles in the system. For such purposes

---

* *Problem*: The function

$$f(\mu) = \sum_v \frac{1}{e^{\beta(\varepsilon_v - \mu)} - 1}$$

has many plus-minus infinity points, and therefore the equation $f(\mu) = N$ has many solutions. Why do we still talk about a unique chemical potential for bosons?

as computing the pressure, we do not consider μ to be part of the energy of the system.

$$U = \frac{1}{Q} \sum_{n_1, n_2, \ldots} \left(\sum_a n_a \varepsilon_a \right) \exp\left\{ -\beta \left[\sum_a n_a(\varepsilon_a - \mu) \right] \right\} = \frac{\partial \beta g}{\partial \beta} + \mu \langle N \rangle. \quad (1.48)$$

$$\text{Pressure} = P = \frac{1}{Q} \sum_{n_1, n_2, \ldots} -\left(\frac{\partial}{\partial V} \sum_a n_a \varepsilon_a \right) \exp\left\{ -\beta \left[\sum_a n_a(\varepsilon_a - \mu) \right] \right\}.$$

μ can be treated as a variable or as a function of V and $\langle N \rangle$. If we treat it as an independent variable, so that $g = g(V, \mu)$, we get

$$P = \left. \frac{-\partial g}{\partial V} \right|_{\mu = \text{const.}}$$

If we treat μ (and g) as functions of V and $\langle N \rangle$, that is, with $g = g[V, \mu(V, \langle N \rangle)]$, then

$$\left. \frac{-\partial g}{\partial V} \right|_{\langle N \rangle = \text{const.}} = P + \langle N \rangle \frac{\partial \mu}{\partial V}$$

is easily shown. Also,

$$\left. \frac{\partial g}{\partial N} \right|_V = -\langle N \rangle \left. \frac{\partial \mu}{\partial \langle N \rangle} \right|_V.$$

If $F = F(V, \langle N \rangle) = g(V, \mu(V, \langle N \rangle)) + \langle N \rangle \mu(V, \langle N \rangle)$, then

$$P = \left. \frac{-\partial F}{\partial V} \right|_{\langle N \rangle = \text{const.}} \quad \text{and} \quad \mu = \left. \frac{\partial F}{\partial \langle N \rangle} \right|_{V = \text{const.}}.$$

$$\text{Entropy} = S = -k \sum_{n_1, n_2, \ldots} P_{n_1, n_2, \ldots} \ln P_{n_1, n_2, \ldots}$$

where

$$P_{n_1, n_2, \ldots} = \frac{1}{Q^{(\mu)}} \exp\left[-\beta \sum_a n_a(\varepsilon_a - \mu) \right].$$

It is easily shown that

$$S = -\left. \frac{\partial g}{\partial T} \right|_{\mu, V = \text{const.}}.$$

In summary:

$$g = \frac{-1}{\beta} \ln Q^{(\mu)} = \frac{-1}{\beta} \ln \sum_{n_1, n_2, \ldots} \exp\left\{ -\beta \left[\sum_a n_a(\varepsilon_a - \mu) \right] \right\}, \quad (1.49)$$

$$\langle n_a \rangle = \frac{\partial g}{\partial \varepsilon_a}, \quad (1.50)$$

$$\langle N \rangle = \frac{-\partial g}{\partial \mu} = \sum_a \langle n_a \rangle, \quad (1.51)$$

$$U = \frac{1}{Q} \sum_{n_1, n_2, \ldots} \left(\sum_a n_a \varepsilon_a \right) \exp\left\{ -\beta \left[\sum_a n_a(\varepsilon_a - \mu) \right] \right\} = \frac{\partial \beta g}{\partial \beta} + \mu \langle N \rangle, \quad (1.52)$$

$$P = \frac{1}{Q} \sum_{n_1, n_2, \ldots} -\left(\frac{\partial}{\partial V} \sum_a n_a \varepsilon_a \right) \exp\left\{ -\beta \left[\sum_a n_a(\varepsilon_a - \mu) \right] \right\}$$

$$= \left. \frac{-\partial g}{\partial V} \right|_\mu, \quad (1.53)$$

$$S = \left. \frac{-\partial g}{\partial T} \right|_{\mu, V}. \quad (1.54)$$

Soon we will be able to find g for an ideal Bose gas, and then for an ideal Fermi gas. But first we must evaluate some integrals.

1.8 EVALUATION OF INTEGRALS

We will soon be dealing with integrals of the form $\int_{-\infty}^{\infty} e^{-ax^2} \, dx$ and $\int_{-\infty}^{\infty} x^2 e^{-ax^2} \, dx$. Let us pause for a second to calculate these integrals:

$$\left[\int_{-\infty}^{\infty} e^{-x^2} \, dx \right]^2 = \int_{-\infty}^{\infty} e^{-x^2} \, dx \int_{-\infty}^{\infty} e^{-y^2} \, dy$$

$$= \int_{-\infty}^{\infty} \int_{-\infty}^{\infty} e^{-(x^2 + y^2)} \, dx \, dy = \int_0^{2\pi} \int_0^{\infty} e^{-r^2} r \, dr \, d\theta$$

$$= -\frac{1}{2} \int_0^{2\pi} \int_0^{\infty} e^{-r^2} (-2r \, dr) \, d\theta = [\sqrt{\pi}]^2. \quad (1.55)$$

$$\int_{-\infty}^{\infty} e^{-x^2} \, dx = \sqrt{\pi}. \quad (1.56)$$

$$\int_{-\infty}^{\infty} e^{-ax^2} \, dx = \frac{1}{\sqrt{a}} \int_{-\infty}^{\infty} e^{-y^2} \, dy = \frac{\sqrt{\pi}}{\sqrt{a}} \quad (y = \sqrt{a} \, x). \quad (1.57)$$

$$\int_{-\infty}^{\infty} x^2 e^{-ax^2} \, dx = -\int_{-\infty}^{\infty} \frac{d}{da} e^{-ax^2} \, dx = -\frac{d}{da} \int_{-\infty}^{\infty} e^{-ax^2} \, dx$$

$$= -\frac{d}{da} \frac{\sqrt{\pi}}{\sqrt{a}} = \frac{1}{2} a^{-3/2} \sqrt{\pi}. \quad (1.58)$$

$$\int_{-\infty}^{\infty} x^3 e^{-ax^2} \, dx = 0 \quad \text{(odd function)} \quad (1.59)$$

$$\int_{-\infty}^{\infty} x^4 e^{-ax^2} \, dx = \frac{d^2}{da^2} \int_{-\infty}^{\infty} x^2 e^{-ax^2} \, dx = \frac{3}{4} a^{-5/2} \sqrt{\pi}, \quad (1.60)$$

and so forth.

1.9 THE IDEAL BOSE-EINSTEIN GAS

From Eq. (1.45) we have

$$e^{-\beta g} = \sum_{n_1, n_2, \ldots} \exp\{-\beta[n_1(\varepsilon_1 - \mu) + n_2(\varepsilon_2 - \mu) + \cdots]\}$$

$$= \sum_{n_1} \exp[-\beta n_1(\varepsilon_1 - \mu)] \sum_{n_2} \exp[-\beta n_2(\varepsilon_2 - \mu)] \cdots \qquad (1.61)$$

For a Bose-Einstein gas, $n_a = 0, 1, 2, \ldots$

$$\sum_{n_i} e^{-\beta n_i(\varepsilon_i - \mu)} = \frac{1}{1 - e^{-\beta(\varepsilon_i - \mu)}} \text{ (Bose Einstein)}, \qquad (1.62)$$

$$e^{-\beta g} = \prod_i \frac{1}{1 - e^{-\beta(\varepsilon_i - \mu)}}, \qquad (1.63)$$

$$g = \frac{1}{\beta} \sum_i \ln(1 - e^{-\beta(\varepsilon_i - \mu)}). \qquad (1.64)$$

Consider a particle contained in a box but otherwise free. The number of modes in the box with momentum in the three-dimensional region d^3p is $s(d^3k/(2\pi)^3)V$, where $V = $ volume and s is the number of possible spin states (e.g., $s = 3$ for spin 1 when the rest mass does not vanish; $s = 2$ for spin 1 when the rest mass vanishes, as for photons). The energy is

$$\varepsilon = \frac{p^2}{2m} = \frac{\hbar^2 k^2}{2m}.$$

Here we have approximated the discrete set of modes by a continuum (which is the case for a completely free particle). The sum for g can then be replaced by an integration:

$$g = s \frac{1}{\beta} \int \ln(1 - e^{-\beta p^2/2m} e^{\beta \mu}) \frac{d^3p}{(2\pi\hbar)^3} V \qquad (1.65)$$

$$\rho = \frac{\langle N \rangle}{V} = -\frac{1}{V} \frac{\partial g}{\partial \mu} = s \int \frac{e^{-\beta p^2/2m} e^{\beta \mu}}{1 - e^{-\beta p^2/2m} e^{\beta \mu}} \frac{d^3p}{(2\pi\hbar)^3}. \qquad (1.66)$$

For computational purposes, set $\alpha = e^{\beta \mu}$ and $x^2 = \beta p^2/2m$. Then $p^2 = 2mx^2/\beta$ and $dp = (2m/\beta)^{1/2} dx$.

$$\frac{1}{(2\pi\hbar)^3} d^3p \to \frac{4\pi p^2 \, dp}{(2\pi\hbar)^3} = \frac{x^2}{2\pi^2\hbar^3} \left(\frac{2m}{\beta}\right)^{3/2} dx.$$

$$\rho = s \int_0^\infty \frac{e^{-x^2}\alpha}{1 - e^{-x^2}\alpha} \left[\frac{x^2}{2\pi^2\hbar^3} \left(\frac{2m}{\beta}\right)^{3/2}\right] dx$$

$$= s \frac{1}{4\pi^2\hbar^3} \left(\frac{2m}{\beta}\right)^{3/2} \left[\int x^2 \{\alpha e^{-x^2} \, dx + \alpha^2 e^{-2x^2} \, dx + \cdots\}\right]$$

$$= s \left(\frac{mkT}{2\pi\hbar^2}\right)^{3/2} \left[\alpha + \frac{\alpha^2}{2^{3/2}} + \frac{\alpha^3}{3^{3/2}} + \cdots\right], \qquad (1.67)$$

where we have used $\int_{-\infty}^{\infty} x^2 e^{-ax^2}\, dx = (\sqrt{\pi}/2a)(1/\sqrt{a})$, from Eq. (1.58). Letting

$$\zeta_r(\alpha) = \sum_{n=1}^{\infty} \frac{\alpha^n}{n^r},$$

we obtain finally

$$\rho = s \left(\frac{mkT}{2\pi\hbar^2}\right)^{3/2} \zeta_{3/2}(\alpha). \tag{1.68}$$

Thus, given ρ, we can find $\alpha = e^{\beta\mu}$ (in principle) by solving the equation

$$\zeta_{3/2}(\alpha) = \frac{1}{s}\left(\frac{2\pi\hbar^2}{mk}\right)^{3/2} \frac{\rho}{T^{3/2}}.$$

For the total energy of the system we have

$$U = \frac{\partial(\beta g)}{\partial\beta} + \mu\langle N\rangle$$

$$= s \int \frac{(e^{-\beta p^2/2m} e^{\beta\mu}) p^2/2m}{1 - e^{-\beta p^2/2m} e^{\beta\mu}} V \frac{d^3 p}{(2\pi\hbar)^3_3}$$

$$= s\tfrac{3}{2}kT \left(\frac{mkT}{2\pi\hbar^2}\right)^{3/2} V\zeta_{5/2}(\alpha). \tag{1.69}$$

For very low ρ, high T, or both, $\rho/T^{3/2}$ is very small; then $\zeta_{3/2}(\alpha)$ is very small, and thus α is very small. In this case $\zeta_{3/2}(\alpha) \approx \alpha$, $\zeta_{5/2}(\alpha) \approx \alpha$, so that

$$\frac{\zeta_{5/2}(\alpha)}{\zeta_{3/2}(\alpha)} \approx 1$$

and $U/V \approx (3/2)kT\rho$. This condition therefore represents the classical limit.

Now let us look at the other limit. As T becomes lower and lower, with ρ fixed, $\zeta_{3/2}(\alpha)$ becomes larger and α approaches 1. For $\alpha > 1$, ζ diverges.

$$\zeta_{3/2}(1) = 1 + \frac{1}{2^{3/2}} + \frac{1}{3^{3/2}} + \cdots = 2.612,$$

$$\zeta_{5/2}(1) = 1 = \frac{1}{2^{5/2}} + \frac{1}{3^{5/2}} + \cdots = 1.341.$$

The temperature T_c at which $\alpha = 1$ is called the *critical temperature* for Bose-Einstein condensation.

$$\frac{\langle N\rangle}{V} = \rho = s\left(\frac{mkT_c}{2\pi\hbar^2}\right)^{3/2} (2.612), \tag{1.70}$$

$$T_c = \frac{2\pi\hbar^2}{mk}\left(\frac{\rho/s}{2.612}\right)^{2/3}. \tag{1.71}$$

The question arises: Why does our analysis break down at T_c? The answer is that for such low temperatures ($T < T_c$), we cannot replace the sum, Eq. (1.64), by the integral, Eq. (1.65).

Looking back at g and $\langle N \rangle$ before we approximated all summations by integrals, we recall that

$$g = \frac{1}{\beta} \sum_i \ln (1 - e^{-\beta \varepsilon_i} e^{\beta \mu}) = \frac{1}{\beta} \sum_i \ln (1 - e^{-\beta \varepsilon_i} \alpha);$$

$$N = \frac{-\partial g}{\partial \mu} = \sum_i \frac{e^{-\beta \varepsilon_i} \alpha}{1 - e^{-\beta \varepsilon_i} \alpha} = \sum_i \frac{1}{\alpha^{-1} e^{\beta \varepsilon_i} - 1} = \sum_i \frac{1}{e^{-\beta \mu} e^{\beta \varepsilon_i} - 1}.$$

For small α, $e^{-\beta \mu}$ is large; the terms with the lowest ε_i do not contribute much to the sum, and we can replace the sum with an integral. When $e^{-\beta \mu}$ is small, we cannot replace the sum with an integral because the first few discrete terms in the sum are important. Now

$$\langle n_a \rangle = \frac{\partial g}{\partial \varepsilon_a} = \frac{1}{(e^{-\beta \mu} e^{\beta \varepsilon_a}) - 1}.$$

Because $\langle n_a \rangle$ is positive, $(\varepsilon_a - \mu)$ must be greater than zero. For $\langle N \rangle$ to be larger than its value at $T = T_c$, $(\varepsilon_a - \mu)$ must be positive, but very small, in order that the low-energy terms in $\langle N \rangle$ be nonnegligible. Assuming no accidental degeneracy in the lowest level (that is, assuming $\varepsilon_0 \neq \varepsilon_1$) we can conclude that, for sufficiently low temperature,

$$\langle n_1 \rangle = \frac{1}{e^{\beta(\varepsilon_1 - \mu)} - 1} \ll \frac{1}{e^{\beta(\varepsilon_0 - \mu)} - 1} = \langle n_0 \rangle.$$

Without any loss of generality, we may take ε_0 as the zero of the energy. Then

$$\langle n_0 \rangle = \frac{1}{e^{-\beta \mu} - 1} \Rightarrow \mu = -kT \ln \left(1 + \frac{1}{\langle n_0 \rangle} \right) \approx -\frac{kT}{\langle n_0 \rangle} \quad \text{for large } \langle n_0 \rangle.$$

For low temperatures, μ is very close to zero, so for the energy states above ε_0, we can neglect μ. The sum for $\langle N - n_0 \rangle$ can be approximated by an integral:

$$N_{\text{exc}} = \langle N - n_0 \rangle = \sum_{i=1}^{\infty} \frac{1}{e^{\beta \varepsilon_i} - 1} \approx sV \int \frac{d^3 p}{(2\pi \hbar)^3} \frac{1}{e^{p^2/2mkT} - 1}$$

$$= 2.612 \left(\frac{mkT}{2\pi \hbar^2} \right)^{3/2} sV = \langle N \rangle \left(\frac{T}{T_c} \right)^{3/2}.$$

We see, then, that our definition of the critical temperature is such that when $T < T_c$, there must be a nonnegligible fraction of the particles in the ground state:

$$\langle n_0 \rangle = \langle N \rangle - N_{\text{exc}}$$

$$= \langle N \rangle [1 - (T/T_c)^{3/2}].$$

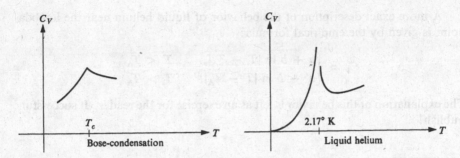

Fig. 1.7 A Bose-Einstein gas around T_c compared with liquid helium around the lambda point.

For $T < T_c$ the total energy is

$$U = \tfrac{3}{2}kT\, s \left(\frac{mkT}{2\pi\hbar^2}\right)^{3/2} V\zeta_{5/2}(1) = \tfrac{3}{2}kTN_{\text{exc}} \frac{\zeta_{5/2}(1)}{\zeta_{3/2}(1)}$$

$$= \tfrac{3}{2}kTN_{\text{exc}}(0.5134)$$

$$= \tfrac{3}{2}kT(0.5134)\left(\frac{T}{T_c}\right)^{3/2} \langle N\rangle. \tag{1.72}$$

Then $C_V = \partial U/\partial T \propto T^{3/2}$ at sufficiently low temperature.

For $T > T_c$, we have $\langle N\rangle = (mkT/2\pi\hbar^2)^{3/2}\zeta_{3/2}(\alpha)sV$. Using the definition of T_c, we have $\zeta_{3/2}(\alpha) = 2.612(T_c/T)^{3/2}$. This equation determines α from $\langle N\rangle$ and T. Then

$$U = \tfrac{3}{2}kT\langle N\rangle \frac{\zeta_{5/2}(\alpha)}{\zeta_{3/2}(\alpha)}. \tag{1.73}$$

The behavior of the specific heat of a Bose-Einstein gas around T_c is rather similar to the behavior of liquid helium near the so-called "lambda point" (Λ point). See Fig. 1.7.* For a given ρ, T_c and Λ are very close to one another (the lambda point is a few degrees K). Because of the mutual forces between particles, liquid helium is certainly not an ideal Bose gas, but perhaps part of the explanation of the lambda transition involves Bose condensation.

---

* For a calculation of the discontinuity in $(\partial C_v/\partial T)_v$ at the transition point, see L. D. Landau & E. M. Lifshitz, *Statistical Physics*, problem of §59, Pergamon Press, 1959.

A more exact description of the behavior of liquid helium near the lambda point is given by the empirical formula:

$$C_V \approx \begin{cases} a + b \ln |T - T_\Lambda|, & T < T_\Lambda, \\ a' + b \ln |T - T_\Lambda|, & . \quad T > T_\Lambda. \end{cases}$$

The explanation of this behavior is left as an exercise for the reader. If successful, publish!

1.10 THE FERMI-DIRAC GAS

For a Fermi gas, we proceed exactly as for the Bose case up to the point

$$e^{-\beta g} = \sum_{n_1} e^{-\beta n_1(\varepsilon_1 - \mu)} \sum_{n_2} e^{-\beta n_2(\varepsilon_2 - \mu)} \cdots. \tag{1.74}$$

Here $n_1 = 0, 1; n_2 = 0, 1;$ and so on. Thus

$$e^{-\beta g} = \prod_i (1 + e^{-\beta(\varepsilon_i - \mu)}) \tag{1.75}$$

$$g = -\left(\frac{1}{\beta}\right) \sum_i \ln (1 + e^{-\beta(\varepsilon_i - \mu)}). \tag{1.76}$$

Once again, we can approximate the sum by an integral. This time there is no danger that a sizable fraction of the particles will be in the lowest state. Assume that the gas is of electrons, so that $s = 2$.

$$g = -\left(\frac{1}{\beta}\right) \int \ln (1 + e^{-\beta(p^2/2m - \mu)}) \frac{2 \, d^3 p}{(2\pi\hbar)^3} V, \tag{1.77}$$

$$\rho = \frac{\langle N \rangle}{V} = -\frac{1}{V} \frac{\partial g}{\partial \mu} = \int \frac{e^{-\beta(p^2/2m)} e^{\beta\mu}}{1 + e^{-\beta p^2/2m} e^{\beta\mu}} \frac{2 \, d^3 p}{(2\pi\hbar)^3}, \tag{1.78}$$

$$\langle n_a \rangle = \frac{\partial g}{\partial \varepsilon_a} = \frac{e^{-\beta(\varepsilon_a - \mu)}}{1 + e^{-\beta(\varepsilon_a - \mu)}} = \frac{1}{e^{\beta(\varepsilon_a - \mu)} + 1}. \tag{1.79}$$

The form of $\langle n_a \rangle$ for $T = 0$ and for any $T > 0$ is shown in Fig. 1.8. Now at $T = 0$, $\langle n_a \rangle = 1$ if $\varepsilon_a < \mu$; and $\langle n_a \rangle = 0$ if $\varepsilon_a > \mu$. Let $\mu = \mu_0$ at $T = 0$. In other words, at $T = 0$ all the states with energy less than μ_0 are occupied, and all with energy greater than μ_0 are empty. μ_0 is called the *Fermi energy* and μ_0/k the *Fermi temperature*.

$$\rho = \int_0^{p_0} 2 \left(\frac{4\pi p^2}{(2\pi\hbar)^3}\right) dp = 2 \left(\frac{4\pi p_0^3}{3}\right) \frac{1}{(2\pi\hbar)^3}, \tag{1.80}$$

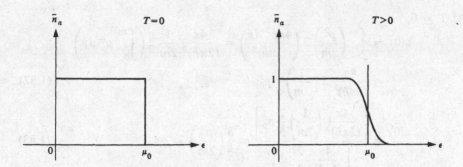

Fig. 1.8 Occupation of states at $T = 0$ and $T > 0$.

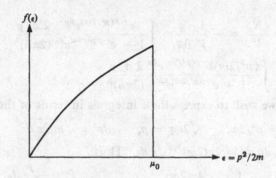

Fig. 1.9 The density of states as a function of energy.

where $\mu_0 = p_0^2/2m$. For most "reasonable" values of μ_0 and T, $\mu_0 \gg kT$ (for example, μ_0/k for copper is $82{,}000°K^*$).

Since

$$2 \frac{4\pi p^2 \, dp}{(2\pi\hbar)^3} = \frac{4\pi(2m)^{3/2}}{(2\pi\hbar)^3} \sqrt{\varepsilon} \, d\varepsilon,$$

the density of states with respect to energy is

$$f(\varepsilon) = \frac{4\pi(2m)^{3/2}}{(2\pi\hbar)^3} \sqrt{\varepsilon}.$$

This function is shown in Fig. 1.9.

$$u \equiv \frac{U}{V} = \frac{1}{V}\left(\frac{\partial \beta g}{\partial \beta} + \mu N\right) = \int \left(\frac{p^2}{2m}\right) \frac{e^{-\beta(p^2/2m - \mu)}}{1 + e^{-\beta(p^2/2m - \mu)}} \frac{2 \, d^3 p}{(2\pi\hbar)^3}. \qquad (1.81)$$

---

* Also, in states of very condensed matter, as in a neutron star, μ_0/k of the neutron, proton, and electron liquids are $\sim 10^{10} - 10^{11}°K$ whereas $T \sim 10^8°K$.

At $T = 0$,

$$u(0) = \int_0^{p_0} \left(\frac{p^2}{2m}\right) 2 \left(\frac{4\pi p^2 \, dp}{(2\pi\hbar)^3}\right) = \frac{4\pi}{(2\pi\hbar)^3} \frac{1}{2m} 2 \left(\int_0^{p_0} p^4 \, dp\right)$$

$$= \frac{4\pi}{(2\pi\hbar)^3} 2 \left(\frac{1}{2m}\right) \frac{1}{5} p_0^5, \tag{1.82}$$

$$\frac{u(0)}{\rho} = \frac{\left[\dfrac{4\pi}{(2\pi\hbar)^3} \left(\dfrac{1}{2m}\right) \dfrac{2}{5} p_0^5\right]}{\left[\dfrac{4\pi}{(2\pi\hbar)^3} \dfrac{2}{3} p_0^3\right]} = \frac{3}{5}\left(\frac{p_0^2}{2m}\right) = \tfrac{3}{5}\mu_0. \tag{1.83}$$

We will now tackle the case of finite T. For $T \neq 0$, what is μ and what is the specific heat, etc.?

$$\rho = \frac{\langle N \rangle}{V} = -\frac{1}{V} \frac{\partial g}{\partial \mu} = \int \frac{e^{-\beta(p^2/2m)} e^{\beta\mu}}{1 + e^{-\beta p^2/2m} e^{\beta\mu}} \frac{2 \, d^3p}{(2\pi\hbar)^3}, \tag{1.78}$$

$$u = \int \frac{(p^2/2m) e^{-\beta(p^2/2m)} e^{\beta\mu}}{1 + e^{-\beta p^2/2m} e^{\beta\mu}} \frac{2 \, d^3p}{(2\pi\hbar)^3}. \tag{1.81}$$

For convenience, we wish to express these integrals in terms of the energy ε:

$$\varepsilon = p^2/2m, \qquad \sqrt{2m\varepsilon} = p, \qquad dp = \sqrt{m/2\varepsilon} \, d\varepsilon.$$

d^3p becomes $4\pi p^2 \, dp \, d\varepsilon = 2\pi(2m)^{3/2}\sqrt{\varepsilon} \, d\varepsilon$. Thus,

$$\rho = \int_0^\infty \frac{1}{e^{\beta(\varepsilon-\mu)} + 1} \frac{4\pi(2m)^{3/2}}{(2\pi\hbar)^3} \sqrt{\varepsilon} \, d\varepsilon = a \int_0^\infty \frac{\sqrt{\varepsilon} \, d\varepsilon}{e^{\beta(\varepsilon-\mu)} + 1}, \tag{1.84}$$

and

$$u = \int_0^\infty \frac{1}{e^{\beta(\varepsilon-\mu)} + 1} \frac{4\pi(2m)^{3/2}}{(2\pi\hbar)^3} \varepsilon^{3/2} \, d\varepsilon = a \int_0^\infty \frac{\varepsilon^{3/2} \, d\varepsilon}{e^{\beta(\varepsilon-\mu)} + 1}, \tag{1.85}$$

where

$$a = \frac{4\pi(2m)^{3/2}}{(2\pi\hbar)^3}.$$

In both cases, the integral that must be evaluated is

$$I = \int_0^\infty \frac{g(\varepsilon) \, d\varepsilon}{e^{\beta(\varepsilon-\mu)} + 1},$$

where $g(\varepsilon) = a\sqrt{\varepsilon}$ for ρ, and $g(\varepsilon) = a\varepsilon^{3/2}$ for μ. Then

$$I = \int_0^\infty \frac{g(\varepsilon) \, d\varepsilon}{e^{\beta(\varepsilon-\mu)} + 1} = \int_\mu^\infty \frac{g(\varepsilon) \, d\varepsilon}{e^{\beta(\varepsilon-\mu)} + 1} + \int_0^\mu g(\varepsilon) \, d\varepsilon - \int_0^\mu \frac{g(\varepsilon) \, d\varepsilon}{e^{-\beta(\varepsilon-\mu)} + 1}.$$

In the first term let $x = \beta(\varepsilon - \mu)$, and in the third term let $x = -\beta(\varepsilon - \mu)$.

$$I = \int_0^\mu g(\varepsilon) \, d\varepsilon + \int_0^\infty \frac{g(\mu + x/\beta)}{e^x + 1} \frac{dx}{\beta} - \int_0^{\beta\mu} \frac{g(\mu - x/\beta)}{e^x + 1} \frac{dx}{\beta}. \tag{1.86}$$

For what we have called "reasonable" values of ρ and T, not much error will arise if we replace $g(\mu \pm x/\beta)$ by $g(\mu) \pm (x/\beta)g'(\mu)$ and $\int_0^{\beta\mu}$ by $\int_0^\infty$. Then

$$I \approx \int_0^\mu g(\varepsilon)\, d\varepsilon + \frac{2}{\beta^2} g'(\mu) \int_0^\infty \frac{x\, dx}{e^x + 1} = \int_0^\mu g(\varepsilon)\, d\varepsilon + \frac{\pi^2 g'(\mu)}{6\beta^2}. \tag{1.87}$$

(The integral

$$\int_0^\infty \frac{x\, dx}{e^x + 1}$$

can be evaluated by expanding $xe^{-x}/(1 + e^{-x})$ in a power series, then integrating and summing; the result is $\pi^2/12$.) So,

$$I = \int_0^\mu g(\varepsilon)\, d\varepsilon + \frac{\pi^2}{6} k^2 T^2 g'(\mu). \tag{1.88}$$

Remember that for $I = \rho$, $g(\varepsilon) = a\sqrt{\varepsilon}$, and for $I = u$, $g(\varepsilon) = a\varepsilon^{3/2}$.

$$\rho = a \int_0^\mu \sqrt{\varepsilon}\, d\varepsilon + \frac{\pi^2}{6} (kT)^2 \frac{a}{2\sqrt{\mu}}$$

$$= \left(\frac{2a}{3}\right) \mu^{3/2} + \frac{\pi^2}{6} (kT)^2 \frac{a}{2\sqrt{\mu}}. \tag{1.89}$$

But we have also $\rho = (2a/3)\mu_0^{3/2}$ from Eq. (1.80). Thus

$$\mu \approx \mu_0 \left(1 - \frac{\pi^2}{12} \frac{k^2 T^2}{\mu_0^2}\right) \tag{1.90}$$

$$u \approx a \int_0^\mu \varepsilon^{3/2}\, d\varepsilon + \frac{\pi^2}{6} (kT)^2 \frac{3a}{2} \sqrt{\mu}$$

$$= \frac{2}{5} a\mu^{5/2} + \frac{\pi^2}{6} (kT)^2 \frac{3a}{2} \sqrt{\mu}$$

$$\approx \frac{2}{5} a\mu_0^{5/2} - a\mu_0^{5/2} \frac{\pi^2}{12} \frac{(kT)^2}{\mu_0^2} + \frac{a\pi^2}{4} (kT)^2 \sqrt{\mu_0}$$

$$= u_0 + \frac{a\pi^2}{6} \sqrt{\mu_0}\, (kT)^2 = u_0 + \gamma T^2; \qquad \gamma = \frac{a\pi^2}{6} \sqrt{\mu_0}\, k^2. \tag{1.91}$$

$$u = u_0 + \gamma T^2; \qquad U = uV = U_0 + \gamma V T^2 = U_0 + \gamma' T^2$$

$$C_V = \frac{\partial U}{\partial T} = 2\gamma' T,$$

$$C_V \text{ of a metal} = \underset{\text{elect.}}{2\gamma' T} + \underset{\substack{\text{lattice} \\ \text{vibrations}}}{\alpha T^3} \quad \text{at low } T$$

At very low T the $\gamma'T$ term dominates, and the contribution of the electron gas to the specific heat is noticeable.

C_v of a metal $= 2\gamma'T + 3R$ at high T. Since the part of the specific heat due to lattice vibration is constant, the $\gamma'T$ term is again detectable.

Note that at very high T, we have to include higher terms in the expansion of $g(\varepsilon)$, and the part of the specific heat due to the electrons is no longer proportional to T.

We can demonstrate more easily than we have done above that the internal energy is proportional to T^2. At any temperature less than the Fermi temperature almost all the states with energy less than the Fermi energy are filled, and almost all states with energy greater than the Fermi energy are empty. As the electron gas is raised to temperature T from $T = 0$, the average energy than *can* be imparted to an electron is about kT.

Also, only those electrons within about kT of the Fermi level will be excited, because electrons with less energy have no place to go—the states are filled. The number of electrons with energies between μ_0 and $\mu_0 - kT$ is

$$N = \int_{\mu_0 - kT}^{\mu_0} a\varepsilon^{1/2} \, d\varepsilon.$$

For $\mu_0 \gg kT$, N is proportional to kT. Thus the internal energy added is proportional to $(kT)kT = k^2T^2$.

CHAPTER 2

DENSITY MATRICES

2.1 INTRODUCTION TO DENSITY MATRICES

When we solve a quantum-mechanical problem, what we really do is divide the universe into two parts—the system in which we are interested and the rest of the universe. We then usually act as if the system in which we are interested comprised the entire universe. To motivate the use of density matrices, let us see what happens when we include the part of the universe outside the system.

Let x describe the coordinates of the system, and let y describe the rest of the universe. Let $\varphi_i(x)$ be a complete set of wave functions. The most general wave function can be written

$$\psi(x, y) = \sum_i C_i(y)\varphi_i(x). \tag{2.1}$$

At this point we will convert to Dirac notation.

Let $|\varphi_i\rangle$ be a complete set of vectors in the vector space describing the system, and let $|\theta_i\rangle$ be a complete set for the rest of the universe.

$$\varphi_i(x) = \langle x|\varphi_i\rangle \quad \text{and} \quad \theta_i(y) = \langle y|\theta_i\rangle.$$

The most general wave function can be written

$$|\psi\rangle = \sum_{ij} C_{ij}|\varphi_i\rangle|\theta_j\rangle \tag{2.2}$$

$$\psi(x, y) = \langle y|\langle x|\psi\rangle = \sum_{ij} C_{ij}\langle x|\varphi_i\rangle\langle y|\theta_j\rangle$$

We can obtain Eq. (2.1) by taking

$$C_i(y) = \sum_j C_{ij}\langle y|\theta_j\rangle.$$

Now let A be an operator that acts only on the system; that is to say, A does not act on the $|\theta_j\rangle$. When A acts on product states (for example, $|\psi\rangle$) we really mean $A|a\rangle|b\rangle \equiv (A|a\rangle)|b\rangle$. In such a case A does not equal

$$\sum_{ii'} A_{ii'}|\varphi_i\rangle\langle\varphi_{i'}|,$$

but equals

$$\sum_{ii'j} A_{ii'} |\varphi_i\rangle |\theta_j\rangle \langle\theta_j| \langle\varphi_{i'}|.$$

Then

$$\langle A \rangle = \langle \psi | A | \psi \rangle = \sum_{\substack{ij \\ i'j'}} C_{ij}^* C_{i'j'} \langle\theta_j| \langle\varphi_i | A | \varphi_{i'}\rangle |\theta_{j'}\rangle$$

$$= \sum_{iji'} C_{ij}^* C_{i'j} \langle\varphi_i | A | \varphi_{i'}\rangle$$

$$= \sum_{ii'} \langle\varphi_i | A | \varphi_{i'}\rangle \rho_{i'i} \tag{2.3}$$

where

$$\rho_{i'i} = \sum_j C_{ij}^* C_{i'j} = \text{density matrix.} \tag{2.4}$$

We define the operator ρ to be such that $\rho_{i'i} = \langle\varphi_{i'}|\rho|\varphi_i\rangle$. ρ operates only on the system described by x.

$$\langle\psi|A|\psi\rangle = \sum_i \langle\varphi_i| A \sum_{i'} |\varphi_{i'}\rangle\langle\varphi_{i'}|\rho|\varphi_i\rangle$$

$$= \sum_i \langle\varphi_i|A\rho|\varphi_i\rangle = \text{Tr } \rho A \tag{2.5}$$

Where we have used the result

$$\sum_{i'} |\varphi_{i'}\rangle\langle\varphi_{i'}| = 1 \quad \text{(by completeness arguments).}$$

From Eq. (2.4), it is obvious that ρ is hermitian. Therefore it can be diagonalized with a complete orthonormal set of eigenvectors $|i\rangle$ and real eigenvalues w_i,

$$\rho = \sum_i w_i |i\rangle\langle i|. \tag{2.6}$$

If we let A be 1, we obtain

$$\sum_i w_i = \text{Tr } \rho = \langle A \rangle = \langle\psi|\psi\rangle = 1. \tag{2.7}$$

If we let A be $|i'\rangle\langle i'|$ we have

$$w_{i'} = \text{Tr } \rho A = \langle A \rangle = \langle\psi|A|\psi\rangle = \sum_j (\langle\psi|i'\rangle|\theta_j\rangle)(\langle\theta_j|\langle i'|\psi\rangle)$$

$$= \sum_j |(\langle i'|\langle\theta_j|)|\psi\rangle|^2. \tag{2.8}$$

Therefore,

$$w_i \geq 0 \quad \text{and} \quad \sum_i w_i = 1. \tag{2.9}$$

We now consider the concept of a density matrix independent of the preceding motivation. First let us reformulate quantum mechanics:

Any system is described by a density matrix ρ, where ρ is of the form $\sum_i w_i |i\rangle\langle i|$ and

a) the set $|i\rangle$ is a complete orthonormal set of vectors.

b) $w_i \geq 0$.

c) $\sum_i w_i = 1$.

d) Given an operator A, the expectation of A is given by

$$\langle A \rangle = \text{Tr } \rho A.$$

Notice that

$$\langle A \rangle = \text{Tr } \rho A = \sum_{i'} \langle i' | \rho A | i' \rangle = \sum_{i'i} w_i \langle i' | i \rangle \langle i | A | i' \rangle$$

$$= \sum_i w_i \langle i | A | i \rangle. \tag{2.10}$$

Since $\langle i|A|i\rangle =$ the expectation value of A in the state $|i\rangle$, it is obvious from (b), (c) and Eq. (2.10) that we can interpret the w_i as the probability that the system is in state i. If all but one of the w_i are zero, we say that the system is in a *pure state*; otherwise it is in a *mixed state*. It is easy to show that a necessary and sufficient condition for a pure state is $\rho = \rho^2$. If a system is in a pure state, $|i_{\text{pure}}\rangle$ we can express the matrix as

$$\rho = |i_{\text{pure}}\rangle\langle i_{\text{pure}}|,$$
$$\rho_{ij} = \langle \varphi_i | \rho | \varphi_j \rangle = \langle \varphi_i | i_{\text{pure}} \rangle \langle i_{\text{pure}} | \varphi_j \rangle$$
$$= \langle \varphi_i | i_{\text{pure}} \rangle (\langle \varphi_j | i_{\text{pure}} \rangle)^* \tag{2.11}$$

More generally,

$$\rho_{ij} = \sum_k w_k \langle \varphi_i | k \rangle \langle \varphi_j | k \rangle^*. \tag{2.11a}$$

If it is possible to discuss the system in the x-representation, we can write

$$\rho(x', x) \stackrel{\text{def.}}{=} \langle x' | \rho | x \rangle = \sum_i w_i \langle x' | i \rangle \langle i | x \rangle$$

$$= \sum_i w_i i(x') i^*(x), \tag{2.12}$$

which, for a pure state $|i\rangle$, becomes

$$\rho(x', x) = i(x') i^*(x) \tag{2.13}$$

In the x-representation we write

$$\langle A \rangle = \text{Tr } \rho A = \int dx \langle x | \rho A | x \rangle.$$

But

$$\langle x|\rho A|x\rangle = \langle x|\rho \left(\int dx'|x'\rangle\langle x'| \right) A|x\rangle$$

$$= \int dx'\langle x|\rho|x'\rangle\langle x'|A|x\rangle = \int dx'\rho(x, x')A(x', x).$$

$$\langle A\rangle = \int dx\, dx'\rho(x, x')A(x', x). \tag{2.14}$$

Considering again the problem of a system plus the rest of the universe, we can easily show

$$\langle A\rangle = \int \psi^*(x', y)A(x', x)\psi(x, y)\, dx\, dx'\, dy.$$

We see that in this case

$$\rho(x, x') = \int \psi(x, y)\psi^*(x', y)\, dy. \tag{2.15}$$

From our example in which we split the universe into two parts, we see that pure states are not general enough to describe a quantum mechanical system that does not include the whole universe. It is unknown whether or not the universe is in a pure state. To reformulate quantum mechanics in terms of the more general density matrices, it would be appropriate next to find the equation of motion of ρ. But first, as a simple example of a density matrix, let us try to describe polarized and unpolarized light.

Consider a beam of light travelling in the z-direction. First define

$$\begin{pmatrix} 1 \\ 0 \end{pmatrix}: \text{Wave function for the } x\text{-polarized state}$$

$$\tag{2.16}$$

$$\begin{pmatrix} 0 \\ 1 \end{pmatrix}: \text{Wave function for the } y\text{-polarized state}$$

Any pure state can be written as a linear combination of the two states in Eq. (2.16):

$$\begin{pmatrix} a \\ b \end{pmatrix} = a \begin{pmatrix} 1 \\ 0 \end{pmatrix} + b \begin{pmatrix} 0 \\ 1 \end{pmatrix}, \tag{2.17}$$

where

$$|a|^2 + |b|^2 = 1.$$

When we use Eq. (2.11), the density matrix for the pure state Eq. (2.17) becomes

$$\rho = \begin{pmatrix} aa^* & ab^* \\ ba^* & bb^* \end{pmatrix}. \tag{2.18}$$

In using Eq. (2.11), we took

$$|\varphi_1\rangle = \begin{pmatrix} 1 \\ 0 \end{pmatrix} \quad \text{and} \quad |\varphi_2\rangle = \begin{pmatrix} 0 \\ 1 \end{pmatrix}.$$

Now to examine four pure states and the density matrices corresponding to them:

The x-polarized state: Let $a = 1$ and $b = 0$ in Eq. (2.17). Equation (2.18) gives

$$\rho_{x\text{-pol.}} = \begin{pmatrix} 1 & 0 \\ 0 & 0 \end{pmatrix}. \tag{2.19}$$

The y-polarized state: Let $a = 0$ and $b = 1$.

$$\rho_{y\text{-pol.}} = \begin{pmatrix} 0 & 0 \\ 0 & 1 \end{pmatrix}. \tag{2.20}$$

The 45°-polarized state: $a = 1/\sqrt{2}$ and $b = 1/\sqrt{2}$ gives

$$\rho_{45°\text{-pol.}} = \begin{pmatrix} \frac{1}{2} & \frac{1}{2} \\ \frac{1}{2} & \frac{1}{2} \end{pmatrix}. \tag{2.21}$$

The 135°-polarized state: $a = -1/\sqrt{2}$ and $b = 1/\sqrt{2}$ gives

$$\rho_{135°\text{-pol.}} = \begin{pmatrix} \frac{1}{2} & -\frac{1}{2} \\ -\frac{1}{2} & \frac{1}{2} \end{pmatrix}. \tag{2.22}$$

The ρ's in the last four equations are for pure states. Now consider the following two mixed states:

1. *Mixture of 50% x-polarized and 50% y-polarized states:* From Eq. (2.11a) the density matrix for this mixture is

$$\rho = \tfrac{1}{2}\rho_{x\text{-pol.}} + \tfrac{1}{2}\rho_{y\text{-pol.}} = \begin{pmatrix} \frac{1}{2} & 0 \\ 0 & \frac{1}{2} \end{pmatrix}. \tag{2.23}$$

2. *Mixture of 50% 45°-polarized and 50% 135°-polarized states:* The density matrix for this mixture is

$$\rho = \tfrac{1}{2}\rho_{45°\text{-pol.}} + \tfrac{1}{2}\rho_{135°\text{-pol.}} = \begin{pmatrix} \frac{1}{2} & 0 \\ 0 & \frac{1}{2} \end{pmatrix}. \tag{2.24}$$

Thus these two mixtures have the same density matrix and show the same physical effect. Note that a given pure state (for example, x-polarization) determines a state vector (wave function) only up to a phase factor, whereas the density matrix is determined uniquely.

2.2 ADDITIONAL PROPERTIES OF THE DENSITY MATRIX

Recall that the density operator can be written

$$\rho = \sum_i w_i |i\rangle\langle i|, \tag{2.25}$$

where the system is in state $|i\rangle$ with probability w_i. As time changes, the possible states of the system also change; so

$$\rho(t) = \sum_i w_i |i(t)\rangle\langle i(t)|. \tag{2.26}$$

It is easy to find how $|i(t)\rangle$ changes in time, for we can expand $|i(0)\rangle$ in eigenstates of the Hamiltonian, H, and we know how those eigenstates change in time. Let $H|E_n\rangle = E_n|E_n\rangle$.

$$|i(t)\rangle = \sum_n |E_n\rangle\langle E_n|i(0)\rangle. \tag{2.27}$$

It follows that

$$|i(t)\rangle = \sum_n |E_n\rangle e^{-iE_n t}\langle E_n|i(0)\rangle.$$

For convenience we use units in which $\hbar = 1$. If we define $f(H)$ for any function f by the equation $f(H)|E_n\rangle = f(E_n)|E_n\rangle$, we can write

$$|i(t)\rangle = \sum_n e^{-iHt}|E_n\rangle\langle E_n|i(0)\rangle = e^{-iHt}|i(0)\rangle. \tag{2.28}$$

Notice that, according to our definition of $f(H)$, if $f(E_n)$ can be expanded in a power series for any eigenvalue of H, then $f(H)$ can be expanded in the same power series.

Substituting Eq. (2.28) into Eq. (2.26), we get

$$\rho(t) = \sum_i w_i e^{-iHt}|i(0)\rangle\langle i(0)|e^{iHt} = e^{-iHt}\rho(0)e^{iHt}. \tag{2.29}$$

Taking the derivative of ρ with respect to time, we get

$$\dot\rho = -i(H\rho - \rho H). \tag{2.30}$$

You may be suspicious about treating H like a number when taking the derivative of terms such as e^{-iHt}, but such a procedure can be justified by expanding e^{-iHt} in a power series. Alternatively, we can obtain Eq. (2.30) from Eq. (2.29) by taking the derivative of $\langle E_n|\rho(t)|E_m\rangle$. In any case, Eq. (2.30) is true, and it plays the same role for density matrices that Schrödinger's equation plays for wave functions. It should be noticed that an observable A in the Heisenberg representation obeys

$$\dot A = +i(HA - AH). \tag{2.31}$$

The signs of Eq. (2.30) and Eq. (2.31) are opposite. We could put Planck's constant back into the equations by replacing every H by $H/\hbar$. From Eq. (2.29) we obtain

$$\operatorname{Tr} \rho^n(t) = \operatorname{Tr} \left[e^{-iHt}(\rho^n(0)e^{iHt}) \right] = \operatorname{Tr} \left[(\rho^n(0)e^{iHt})e^{-iHt} \right]$$
$$= \operatorname{Tr} \rho^n(0) \qquad (2.32)$$

where we use the fact that $\operatorname{Tr} AB = \operatorname{Tr} BA$. It follows that $\operatorname{Tr} \rho^n$ is constant in time, and consequently $\operatorname{Tr} f(\rho)$ for any function f is constant. This result can also be obtained by noticing that

$$\rho^n = \sum_i w_i^n |i\rangle\langle i|,$$

so

$$f(\rho) = \sum_i f(w_i)|i\rangle\langle i|. \qquad (2.33)$$

$\operatorname{Tr} f(\rho) = \sum_i f(w_i)$, and the w_i's are constant. [being the eigenvalues of ρ and thus unchanged by a unitary transformation such as Eq. (2.29)].

It is possible to define "entropy" by the equation

$$\text{"}S\text{"} = - \sum_i w_i \ln w_i. \qquad (2.34)$$

A pure state has "S" = 0, whereas very impure states have large positive "entropy." The quotation marks are used because, by our definition, the "entropy" cannot increase in time but must remain constant.

Next let us try to get a bit more experience in dealing with density matrices. In ordinary quantum mechanics it was easy to find such things as the expectation value of the position, the expectation value of momentum, and the probability that a system is in a given state.

(a) Expectation value of position: According to Eq. (2.14), if A is an operator,

$$\langle A \rangle = \int dx \, dx' \rho(x, x') A(x', x).$$

In this case $A = x$, and

$$A(x', x) = \langle x'|A|x \rangle = x\langle x'|x \rangle = x \, \delta(x' - x), \qquad (2.35)$$

so

$$\langle x \rangle = \int dx \, x\rho(x, x). \qquad (2.36)$$

(b) Expectation value of momentum: Now we must take $A = p$. Since

$$\langle x| p|\psi \rangle = \frac{\hbar}{i} \frac{\partial}{\partial x} \langle x|\psi \rangle,$$

we have

$$A(x', x) = \frac{\hbar}{i} \frac{\partial}{\partial x} \delta(x' - x). \tag{2.37}$$

Integrate by parts; for any function f we have

$$\int f(x) \frac{\partial}{\partial x} \delta(x' - x) \, dx = -\int \delta(x' - x) \frac{\partial}{\partial x} f(x) \, dx = -f'(x')$$

$$\langle p \rangle = -\int dx \, dx' \frac{\hbar}{i} \delta(x' - x) \frac{\partial}{\partial x} \rho(x, x')$$

$$= -\frac{\hbar}{i} \int dx \left[\frac{\partial}{\partial x} \rho(x, x') \right]_{x' = x} \tag{2.38}$$

(c) **Probability that a system is in state $|\chi\rangle$:** Finding the probability that a system is found in the state $|\chi\rangle$ is equivalent to finding the expectation value of the operator $|\chi\rangle\langle\chi|$. To see this for the case of a pure state, notice that if the system is in a state $|\psi\rangle$, the probability of experimentally finding it in state $|\chi\rangle$ is

$$|\langle\chi|\psi\rangle|^2 = \langle\psi|\chi\rangle\langle\chi|\psi\rangle = \langle\psi|(|\chi\rangle\langle\chi|)|\psi\rangle.$$

Therefore, the probability that the system is found in state $|\chi\rangle$ can be written

$$P = \text{probability} = \text{Tr} \, \rho|\chi\rangle\langle\chi| = \text{Tr} \sum_i w_i |i\rangle\langle i|\chi\rangle\langle\chi|$$

$$= \sum_i w_i |\langle i|\chi\rangle|^2 = \langle\chi|\rho|\chi\rangle. \tag{2.39}$$

The equation

$$P = \sum_i w_i |\langle i|\chi\rangle|^2$$

is in accord with our interpretation of the w_i as the probability that the system actually is in state $|i\rangle$.

The probability density at value x_0 for a coordinate is the probability that the system is in state $|x_0\rangle$, which by Eq. (2.39) is $\rho(x_0, x_0)$. For a pure state $|\psi\rangle$, $\rho(x_0, x_0) = |\psi(x_0)|^2$.

2.3 DENSITY MATRIX IN STATISTICAL MECHANICS

When $|\varphi_i\rangle$ is an eigenket (eigenfunction), and E_i is the corresponding eigenvalue of the Hamiltonian H of the system, the probability that the system is in the state $|\varphi_i\rangle$ is

$$(1/Q)e^{-\beta E_i}. \tag{2.40}$$

Thus the density matrix is

$$\rho = \sum_n w_n |\varphi_n\rangle\langle\varphi_n|, \qquad \text{where} \qquad w_n = \frac{1}{Q} e^{-\beta E_n}. \tag{2.41}$$

Alternatively, the coordinate representation is

$$\rho(x, x') = \sum_i \frac{1}{Q} e^{-\beta E_i} \varphi_i(x)\varphi_i^*(x'). \tag{2.42}$$

Because $H|\varphi_n\rangle = E_n|\varphi_n\rangle$, we can write Eq. (2.41) as

$$\rho = \frac{1}{Q} \sum_n e^{-\beta H} |\varphi_n\rangle\langle\varphi_n| = \frac{e^{-\beta H}}{Q} \tag{2.43}$$

where

$$e^{-\beta F} = Q = \sum_n e^{-\beta E_n} = \text{Tr } e^{-\beta H}, \tag{2.44}$$

so

$$\rho = \frac{e^{-\beta H}}{\text{Tr } e^{-\beta H}} . \tag{2.45}$$

The average energy of the system, U, can be written

$$U = \text{Tr } \rho H = \frac{\text{Tr } [He^{-\beta H}]}{\text{Tr } [e^{-\beta H}]} . \tag{2.46}$$

When we know the free energy and the energy of the system, the entropy is derived from

$$F = U - TS. \tag{2.47}$$

It can be shown that, if H_0 is any other Hamiltonian,

$$F \leq \langle H \rangle_0 - TS_0, \tag{2.48}$$

where

$$\langle H \rangle_0 = \frac{\text{Tr } [He^{-\beta H_0}]}{\text{Tr } [e^{-\beta H_0}]}, \tag{2.49}$$

and S_0 is the entropy calculated from H_0. F in Eq. (2.48) is the true free energy corresponding to the true Hamiltonian H, as written in Eq. (2.44). When we write the free energy corresponding to H_0 as

$$F_0 = \langle H_0 \rangle_0 - TS_0 \tag{2.50}$$

and subtract Eq. (2.50) from Eq. (2.48), we may write

$$F \leq F_0 + \frac{\text{Tr } [(H - H_0)e^{-\beta H_0}]}{\text{Tr } [e^{-\beta H_0}]} . \tag{2.51}$$

This theorem will be proved later, in Section 2.11.

Now, we regard the density matrix as a function of β:

$$\rho(\beta) = e^{-\beta H}/\mathrm{Tr}\ e^{-\beta H}. \tag{2.42}$$

The unnormalized ρ is defined by

$$\rho_U(\beta) = e^{-\beta H}. \tag{2.52}$$

In place of ρ_U we will hereafter write ρ. Next we will show that $\rho(\beta)$ obeys the differential equation

$$-\partial\rho/\partial\beta = H\rho. \tag{2.53}$$

The proof is as follows. In the energy representation, Eq. (2.52) can be written as

$$\rho_{ij} = \delta_{ij}e^{-\beta E_i}. \tag{2.54}$$

Differentiating this equation, we have

$$-\partial\rho_{ij}/\partial\beta = \delta_{ij}E_i e^{-\beta E_i} = E_i\rho_{ij}, \tag{2.55}$$

from which we obtain Eq. (2.53).

The initial condition for Eq. (2.53) is

$$\rho(0) = 1. \tag{2.56}$$

We can write Eq. (2.53) in the position representation as follows:

$$-\partial\rho(xx';\beta)/\partial\beta = H_x\rho(xx';\beta). \tag{2.57}$$

Here the subscript x on H_x indicates that H_x operates on x in $\rho(xx';\beta)$. The initial condition is

$$\rho(xx';0) = \delta(x - x'). \tag{2.58}$$

2.4 DENSITY MATRIX FOR A ONE-DIMENSIONAL FREE PARTICLE

We want to solve the differential equation discussed in Section 2.3 for a one-dimensional free particle. The Hamiltonian is

$$H = p^2/2m, \tag{2.59}$$

and the differential equation (from Eq. (3.57)) is

$$\frac{-\partial\rho(x, x', \beta)}{\partial\beta} = \frac{-\hbar^2}{2m}\frac{\partial^2}{\partial x^2}\rho(x, x', \beta). \tag{2.60}$$

This is a diffusion-type equation, and we can write down the solution readily:

$$\rho(x, x'; \beta) = \sqrt{\frac{m}{2\pi\hbar^2\beta}} \exp\left[-\left(\frac{m}{2\hbar^2\beta}\right)(x - x')^2\right]. \tag{2.61}*$$

The numerical factor is chosen such that

$$\rho(xx'; 0) = \delta(x - x'). \tag{2.62}$$

For a linear system of length L, the trace of Eq. (2.61) leads to

$$e^{-\beta F} = \int \rho(x, x)\, dx = L\sqrt{\frac{mkT}{2\pi\hbar^2}}. \tag{2.63}$$

This is the partition function for the system.

Problem: Show that the partition function for a three-dimensional system with N particles is

$$e^{-\beta F} = V^N(\sqrt{mkT/2\pi\hbar^2})^{3N}. \tag{2.64}$$

Another way of deriving Eq. (2.61) is as follows:

$$\rho(x, x'; \beta) = \sum_m e^{-\beta E_m}\psi_m(x)\psi_m^*(x'). \tag{2.65}$$

If the particle is in a large box of volume V, we take

$$\sum_m \to \int \frac{V d^3 p}{(2\pi\hbar)^3} \quad \text{and} \quad \psi_m(x) \to \psi_p(x) = \frac{1}{\sqrt{V}} e^{i\mathbf{p}\cdot\mathbf{x}/\hbar}.$$

Then Eq. (3.65) becomes

$$\rho(x, x'; \beta) = \int \frac{d^3 p}{(2\pi\hbar)^3} e^{-\beta p^2/2m} e^{i\mathbf{p}/\hbar\cdot(x-x')}$$

$$= \left(\frac{m}{2\pi\hbar^2\beta}\right)^{3/2} e^{-m|x-x'|^2/2\hbar^2\beta}, \tag{2.66}$$

as in Eq. (2.61).

2.5 LINEAR HARMONIC OSCILLATOR

The Hamiltonian for a linear harmonic oscillator is

$$H = p^2/2m + mw^2x^2/2, \tag{2.67}$$

---

* Note that for x,x' constant, no expansion in powers of β exists near $\beta = 0$. This is, indeed, an important feature of ρ_{free} and will influence the construction of perturbation series.

so that the differential equation for ρ is

$$\frac{-\partial\rho}{\partial\beta} = \frac{-\hbar^2}{2m}\frac{\partial^2}{\partial x^2}\rho + \frac{m\omega^2}{2}x^2\rho. \tag{2.68}$$

Let us write

$$\xi = \sqrt{\frac{m\omega}{\hbar}}\,x, \qquad f = \frac{\hbar\omega}{2}\beta = \frac{\hbar\omega}{2kT}, \tag{2.69}$$

which simplifies Eq. (2.68) to

$$\frac{-\partial\rho}{df} = \frac{-\partial^2\rho}{\partial\xi^2} + \xi^2\rho. \tag{2.70}$$

The initial condition is

$$\rho = \delta(x - x') \qquad \text{for } f = 0, \tag{2.71}$$

or

$$\rho = \sqrt{\frac{m\omega}{\hbar}}\,\delta(\xi - \xi') \qquad \text{for } f = 0. \tag{2.72}$$

The factor comes from the scale change in the δ-function. In general, $\delta(x - x_0) = |f'(x)|\delta[f(x) - f(x_0)]$ for f monotonic.

For low f (high temperature) the particle should act almost like a free particle, as its probable kinetic energy is so high. Therefore we expect that, for low f, the density matrix for a harmonic oscillator will be given approximately by

$$\rho(\xi, \xi', f) \approx \sqrt{m\omega/(4\pi\hbar f)}\,e^{-(\xi - \xi')^2/4f}. \tag{2.73}*$$

Knowing the Gaussian dependence of ρ_{free} on ξ, we try the following form:

$$\rho = \exp\{-[a(f)\xi^2 + b(f)\xi + c(f)]\}. \tag{2.74}$$

Use of Eq. (2.74) in Eq. (2.70) leads us to the equation

$$a'\xi^2 + b'\xi + c' = (1 - 4a^2)\xi^2 - 4ab\xi + 2a - b^2, \tag{2.75}$$

where a prime ($'$) means a derivative with respect to f. From Eq. (2.75),

$$a' = 1 - 4a^2, \tag{2.76a}$$

$$b' = -4ab, \tag{2.76b}$$

$$c' = 2a - b^2. \tag{2.76c}$$

---

* This is only true for $\xi \neq \xi'$. When they are equal, $\rho(\xi, \xi', f)$ is equal to ρ_{free} only for $f = 0$. See Section 3.10 for details. Obviously, when T is large and the classical motion has large amplitude, the larger the amplitude the less will the oscillator feel a force between fixed x and x'.

Equation (2.76a) is integrated to give

$$a = (\tfrac{1}{2}) \coth 2(f - f_0),$$

where the initial condition Eq. (2.73) requires that the integration constant f_0 vanish:

$$a = (\tfrac{1}{2}) \coth 2f. \tag{2.77}$$

Equations (2.76b and c) are integrated as

$$b = A/\sinh 2f. \tag{2.78}$$

$$c = (\tfrac{1}{2}) \ln (\sinh 2f) + (A^2/2) \coth 2f - \ln B,$$

where A and B are constants. Use Eqs. (2.77) and (2.78) to get

$$\rho = \frac{B}{\sqrt{\sinh 2f}} \exp\left[-\left(\frac{\xi^2}{2} \coth 2f + \frac{A\xi}{\sinh 2f} + \frac{A^2}{2} \coth 2f \right) \right]. \tag{2.79}$$

When we let $f \to 0$, Eq. (2.79) gives

$$\rho \to \frac{B}{\sqrt{2f}} \exp\left[-\frac{\xi^2 + 2A\xi + A^2}{4f} \right]. \tag{2.80}$$

For this result to agree with Eq. (2.73) we see that

$$A = -\xi', \qquad B = \sqrt{\frac{m\omega}{2\pi\hbar}}. \tag{2.81}$$

Using Eq. (2.81) in Eq. (2.79), we obtain the final form:

$$\rho(x, x'; \beta) = \sqrt{\frac{m\omega}{2\pi\hbar \sinh 2f}}$$

$$\times \exp\left\{ \frac{-m\omega}{2\hbar \sinh 2f} \left[(x^2 + x'^2) \cosh 2f - 2xx' \right] \right\} \tag{2.82}$$

or

$$\rho(x, x'; \beta) = \sqrt{\frac{m\omega}{2\pi\hbar \sinh (\hbar\omega/kT)}}$$

$$\times \exp\left\{ \frac{-m\omega}{2\hbar \sinh (\hbar\omega/kT)} \left[(x^2 + x'^2) \cosh \frac{\hbar\omega}{kT} - 2xx' \right] \right\}. \tag{2.83}$$

The special case when $x = x'$ gives

$$\rho(x, x; \beta) = \sqrt{\frac{m\omega}{2\pi\hbar \sinh 2f}} \exp\left(\frac{-m\omega}{\hbar} x^2 \tanh f \right). \tag{2.84}$$

This is the probability for finding the system at x. Notice that it is of Gaussian form.

Equation (2.84) is used to calculate

$$\langle x^2 \rangle = \frac{\int x^2 \rho(x, x)\, dx}{\int \rho(x, x)\, dx} = \frac{\hbar}{2m\omega} \coth f = \frac{\hbar}{2m\omega} \coth \frac{\hbar\omega}{2kT}. \qquad (2.85)$$

The average of the potential energy is thus

$$\langle \text{Potential energy} \rangle = (\tfrac{1}{2})m\omega^2 \langle x^2 \rangle = (\hbar\omega/4) \coth f. \qquad (2.86)$$

We may compare this result with the average of total energy calculated by means of the partition function. We know

$$\langle \text{Energy} \rangle = \frac{\hbar\omega}{2} + \frac{\hbar\omega e^{-\hbar\omega/kT}}{1 - e^{-\hbar\omega/kT}} = \frac{\hbar\omega}{2} \frac{1 + e^{-2f}}{1 - e^{-2f}} = \frac{\hbar\omega}{2} \coth f. \qquad (2.87)$$

Equation (2.86) is exactly half of Eq. (2.87). Therefore we know that

$$\langle \text{Kinetic energy} \rangle = \langle \text{Potential energy} \rangle = (\hbar\omega/4) \coth f. \qquad (2.88)$$

The partition function is derived from Eq. (2.84):

$$e^{-\beta F} = \int \rho(x, x)\, dx = \sqrt{\frac{m\omega}{2\pi\hbar \sinh 2f}} \sqrt{\frac{\pi\hbar}{m\omega \tanh f}} = \frac{1}{2 \sinh f} \qquad (2.89)$$

which leads to

$$F = \frac{1}{\beta} \ln (2 \sinh f) = kT \ln (e^{\hbar\omega/2kT} - e^{-\hbar\omega/2kT})$$

$$= \frac{\hbar\omega}{2} + kT \ln (1 - e^{-\hbar\omega/kT}). \qquad (2.90)$$

This is the free energy, as already derived.

We may examine the limit of high temperature, or small f in Eq. (2.84). In this limit

$$\rho(x, x; \beta) \rightarrow \exp\left(\frac{-m\omega^2 x^2/2}{kT}\right) = \exp\left(\frac{-V(x)}{kT}\right), \qquad (2.91)$$

except for the factor in front of the exponential. This result, Eq. (2.91), agrees with classical mechanics.

The low-temperature limit is found by inspection of Eq. (2.83):

$$\rho(x, x') \rightarrow \sqrt{\frac{m\omega}{\pi\hbar}} \exp\left(\frac{-\hbar\omega}{2kT}\right) \exp\left(\frac{-m\omega x^2}{2\hbar}\right) \exp\left(\frac{-m\omega^2 x'^2}{2\hbar}\right). \qquad (2.92)$$

This limit is calculated as

$$\rho(x, x') = \sum_i e^{-\beta E_i} \varphi_i(x) \varphi_i^*(x') \rightarrow e^{-\beta E_0} \varphi_0(x) \varphi_0^*(x'), \qquad (2.93)$$

because when $\beta \to \infty$, only the ground state has an effective contribution. Equation (2.92) agrees with Eq. (2.93) because we know that for the ground state

$$\varphi_0(x) = \left(\frac{m\omega}{\pi\hbar}\right)^{1/4} e^{(-m\omega/2\hbar)x^2}. \tag{2.94}$$

2.6 ANHARMONIC OSCILLATOR

Consider the potential energy shown by the curve in Fig. 2.1, given by the following function:

$$V(x) = \frac{m\omega^2}{2} x^2 + kx^3. \tag{2.95}$$

We will consider the area near $x = 0$ only, so that the region $V(x) < 0$ does not come into the calculation.

Because of the anharmonicity, when the temperature increases the mean position of the oscillation moves out, as shown in Fig. 2.2. Bearing this in mind,

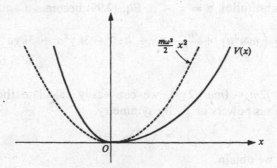

Fig. 2.1 Anharmonic oscillator $(k < 0)$.

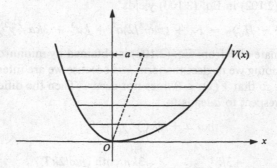

Fig. 2.2 Mean position of the oscillation is displaced.

we will treat the problem using the minimum principle discussed previously. The principle is written as

$$F \leq F_0 + \langle H - H_0 \rangle_0. \tag{2.96}$$

Here F is the true free energy and H is the true Hamiltonian. When H_0 is any Hamiltonian, $\langle \rangle_0$ means the average taken in a system characterized by H_0. F_0 is the free energy of the system of H_0.

In the present problem we take for H_0

$$H_0 = p^2/2m + (m\omega^2/2)(x - a)^2, \tag{2.97}$$

which describes a harmonic oscillator with its center displaced by an amount a. The ω in Eq. (2.97) is the same as that in Eq. (2.95). From Eq. (2.90) we find that

$$F_0 = kT \ln (2 \sinh (\hbar\omega/2kT)). \tag{2.98}$$

Now we have only to find

$$\langle H - H_0 \rangle_0 = \left\langle \frac{m\omega^2}{2} x^2 + kx^3 - \frac{m\omega^2}{2} (x - a)^2 \right\rangle_0. \tag{2.99}$$

If we make the substitution $y = x - a$, Eq. (2.99) becomes

$$\langle H - H_0 \rangle_0 = \left\langle m\omega^2 ay + \frac{m\omega^2}{2} a^2 + ky^3 + 3ky^2a + 3kya^2 + ka^3 \right\rangle_0. \tag{2.100}$$

Because $H_0 = p^2/2m + (m\omega^2/2)y^2$, we can easily calculate the expectation values of the various powers of y. By symmetry,

$$\langle y \rangle_0 = \langle y^3 \rangle_0 = 0. \tag{2.101}$$

From Eq. (2.85), we obtain

$$\langle y^2 \rangle_0 = (\hbar/2m\omega) \coth (\hbar\omega/2kT). \tag{2.102}$$

Substituting Eq. (2.102) in Eq. (2.100) yields

$$F \leq F_0 + \langle H - H_0 \rangle_0 = F_0 + (m\omega^2/2)a^2 + ka^3 + 3ka \langle y^2 \rangle_0. \tag{2.103}$$

The best estimate of F from Eq. (2.103) is obtained by minimizing the right-hand side. In so doing we neglect the term ka^3 because we are interested only in small oscillations, so that $V(x) < 0$ does not occur. Then the differentiation of Eq. (2.103) with respect to a leads to

$$m\omega^2 a + 3k\langle y^2 \rangle_0 = 0, \tag{2.104}$$

that is,

$$a = \frac{-3k}{m\omega^2} \langle y^2 \rangle_0 = -\frac{3k\hbar \coth (\hbar\omega/2kT)}{2m^2\omega^3}. \tag{2.105}$$

In this case, Eq. (2.103) becomes

$$F \le F_0 - (m\omega^2/2)a^2.$$
(2.106)

In Eq. (2.105), the displacement of the mean position, a, has a simple interpretation. It is the point at which the average force on the oscillator vanishes. This result is seen as follows: from Eq. (2.95),

$$\text{Force} = m\omega^2 x + 3kx^2 = m\omega^2(y + a) + 3ky^2 + \cdots$$
(2.107)

$$\langle \text{Force} \rangle = m\omega^2 a + 3k\langle y^2 \rangle + \cdots.$$
(2.108)

Therefore, when we neglect the higher powers of a, we arrive at Eq. (2.104).

Problem: Consider a general bounding potential $V(x)$ like that shown in Fig. 2.3. The Hamiltonian is

$$H = p^2/2m + V(x).$$
(2.109)

In order to apply the variation principle in Eq. (2.96), try the following H_0:

$$H_0 = \frac{p^2}{2m_0} + \frac{m_0\omega_0^2}{2}(x - a)^2,$$
(2.110)

and vary a, m_0, and ω_0. Show that

i) $m_0 = m$;
(2.111)

ii) ω_0 and a are determined from

$$\langle V'(x) \rangle_0 = 0,$$
(2.112)

$$\langle xV'(x) \rangle_0 = \langle p^2/m \rangle_0.$$
(2.113)

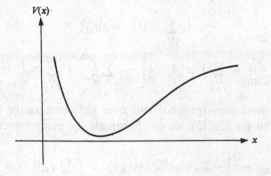

Fig. 2.3 A general bounding potential.

Here $\langle\rangle_0$ is taken for H_0. Equation (2.112) comes from the consideration of the absolute location of the system, and Eq. (2.113) comes from the scale of this variable. The latter corresponds to the virial theorem, which states that

$$\langle xV'(x)\rangle = \langle p^2/m\rangle. \tag{2.114}$$

To show how Eq. (2.114) is related to considerations of the scale of the variable, let us derive the virial theorem in a more general form.

Consider a system inside a cubical box with sides of length L. The Hamiltonian is taken to be

$$H_L = \sum_i \frac{p_i^2}{2m_i} + \sum_{\substack{ij\\i<j}} V(r_{ij}). \tag{2.115}$$

(We assume that the motion of the system can be described by a potential depending only on the distances separating pairs of the particles making up the system.) $r_{ij} = |r_i - r_j|$.

If the system is bound, then for a sufficiently large box the pressure is essentially zero. But in general,

$$P = \text{Pressure} = -\frac{\partial F}{\partial\,(\text{Vol})} = -\frac{1}{3L^2}\frac{\partial F}{\partial L}. \tag{2.116}$$

Let $F_L = F$ for a box of length L. When we later prove the theorem that $F \le F_0 + \langle H - H_0\rangle_0$, we will in fact prove that $\langle H - H_0\rangle_0$ is the first-order correction in the perturbation expansion for F. We get an expression for P by writing

$$F_{L(1+\varepsilon)} \approx F_L + \text{correction term}$$

or

$$F_{L(1+\varepsilon)} \approx F_L + \langle H_{L(1+\varepsilon)} - H_L\rangle_{H_L}. \tag{2.117}$$

Now $H_{L(1+\varepsilon)}$ differs from H_L only in that $(r_i)_x$ can run from 0 to $L(1 + \varepsilon)$ rather than 0 to L, and similarly for $(r_i)_y$ and $(r_i)_z$. Therefore let us make a change of variables, such that $r_i' = r_i/(1 + \varepsilon)$. As $p_i^2 = -\hbar^2\nabla_i^2$, we obtain

$$H_{L(1+\varepsilon)} = \sum_i \frac{p_i'^2}{2m_i(1 + \varepsilon)^2} + \sum_{\substack{ij\\i<j}} V[(1 + \varepsilon)r_{ij}']$$

$$\approx \sum_i \frac{p_i'^2}{2m_i} + \sum_{\substack{ij\\i<j}} V(r_{ij}') + \varepsilon\left[-2\sum_i \frac{p_i'^2}{2m} + \sum_{\substack{ij\\i<j}} r_{ij}'V'(r_{ij})\right] \tag{2.118}$$

Since the new coordinates (primed) run over the same range as the original (unprimed) ones in Eq. (2.115), we can eliminate the primes over the r_{ij} and P_i and write

$$\langle H_{L(1+\varepsilon)} - H_L\rangle_{H_L} = \varepsilon\left[-2\,\langle\text{kinetic energy}\rangle + \left\langle\sum_{\substack{ij\\i<j}} r_{ij}V'(r_{ij})\right\rangle\right]. \tag{2.119}$$

From Eq. (2.116) and Eq. (2.117),

$$3PV = -L\frac{\partial F}{\partial L} = -L\frac{F_{L(1+\varepsilon)} - F_L}{L\varepsilon} = -\frac{1}{\varepsilon}\langle H_{L(1+\varepsilon)} - H_L\rangle. \qquad (2.120)$$

From Eq. (2.119),

$$3PV = 2\langle KE\rangle - \left\langle \sum_{\substack{ij \\ i<j}} r_{ij}V'(r_{ij})\right\rangle. \qquad (2.121)$$

For the case of a harmonic oscillator, P is zero and we get

$$\langle xV'(x)\rangle = \langle p^2/m\rangle. \qquad (2.122)$$

This result should be compared with Eq. (2.113).

For the case of a particle moving in an inverse-square force field, we get the classical result

$$\langle KE\rangle = -(\tfrac{1}{2})\langle \text{potential energy}\rangle. \qquad (2.123)$$

This method can be generalized. Let H be the Hamiltonian for the original system and let H_ε be the Hamiltonian with the box changed in an infinitesimal manner. Suppose $F_H = F_{H_\varepsilon}$ and suppose that $H_\varepsilon - H = \varepsilon S$. Then the above method gives $\langle S\rangle_H = 0$. For example, in Eq. (2.118) we have the case for which

$$S = \left[-2\sum_i \frac{p_i^2}{2m} + \sum_{\substack{ij \\ i<j}} r_{ij}V'(r_{ij})\right].$$

As another example, suppose that the box undergoes an infinitesimal translation, ε, in the x direction. We get, by taking $x' = x - \varepsilon$,

$$H_\varepsilon \approx \sum_i \frac{p_i'^2}{2m_i} + \sum_{\substack{ij \\ i<j}} V(r_{ij}') + \varepsilon\sum_{\substack{ij \\ i<j}} \frac{\partial V}{\partial x}(r_{ij}'), \qquad (2.124)$$

so that

$$S = \sum_{\substack{ij \\ i<j}} \frac{\partial V}{\partial x}(r_{ij}) \qquad (2.125)$$

and

$$\left\langle \sum_{\substack{ij \\ i<j}} \frac{\partial V}{\partial x}(r_{ij})\right\rangle = 0. \qquad (2.126)$$

Compare Eq. (2.126) with Eq. (2.112).

2.7 WIGNER'S FUNCTION

The density matrix may be written in the coordinate representation and in the momentum representation as follows:

$$\rho(x, x') = \sum_i e^{-\beta E_i} \varphi_i(x) \varphi_i^*(x'), \tag{2.127}$$

$$\rho(p, p') = \sum_i e^{-\beta E_i} \varphi_i(p) \varphi_i^*(p')$$

$$= \int \rho(x, x') e^{(-i/\hbar)(p \cdot x - p' \cdot x')} \, dx \, dx'. \tag{2.128}$$

The diagonal element

$$\rho(x, x) \equiv P(x) \tag{2.129}$$

represents the probability for finding the particle at x (neglecting the normalization constant $1/\text{Tr}\,\rho$). Similarly

$$\rho(p, p) \equiv P(p) \tag{2.130}$$

is proportional to the probability for finding the particle at p in momentum space.

The expression Eq. (2.130) is used in calculating the average value of the kinetic energy:

$$\left\langle \frac{p^2}{2m} \right\rangle = \frac{\text{Tr}\,[(p^2/2m)\rho]}{\text{Tr}\,[\rho]} = \frac{\int \rho(p, p)(p^2/2m)(dp/2\pi\hbar)}{\int \rho(p, p)(dp/2\pi\hbar)}. \tag{2.131}$$

In classical mechanics, the density function $f(p, x)$ in phase space has the following property:

$$P(p) = \int f(p, x) \, dx,$$

$$P(x) = \int f(p, x) \frac{dp}{2\pi\hbar}. \tag{2.132}$$

We ask ourselves if there is any function $f(p, x)$ in quantum mechanics that satisfies Eq. (2.132). One answer is Wigner's function $f_W(p, x)$:

$$f_W(p, x) = \int \rho\left(x + \frac{\eta}{2}, x - \frac{\eta}{2}\right) e^{-i(p\eta/\hbar)} \, d\eta. \tag{2.133}$$

This function is derived by regarding $\rho(x, x')$ as a function of $(x + x')/2$ and $(x - x')/2$, and then writing $(x + x')/2$ as x and making a Fourier transform with respect to $(x - x')/2$.

Let us check that Eq. (2.133) satisfies Eq. (2.132). First,

$$\int f_W(p, x) \frac{dp}{2\pi\hbar} = \int \rho\left(x + \frac{\eta}{2}, x - \frac{\eta}{2}\right)\left(\int e^{-i(p\eta/\hbar)} \frac{dp}{2\pi\hbar}\right) d\eta$$

$$= \int \rho\left(x + \frac{\eta}{2}, x - \frac{\eta}{2}\right) \delta(\eta) \, d\eta = \rho(x, x) = P(x). \qquad (2.134)$$

Thus the second part of Eq. (2.132) is satisfied. Next, using Eq. (2.128) and Eq. (2.130), we have

$$P(p) = \rho(p, p) = \int \rho(x, x')e^{-(i/\hbar)p(x - x')} \, dx \, dx'$$

$$= \int \rho\left(y + \frac{\eta}{2}, y - \frac{\eta}{2}\right) e^{-ip\eta/\hbar} \, dy \, d\eta = \int f_W(p, y) \, dy, \qquad (2.135)$$

where we have made the change of variables:

$$x = y + \eta/2$$
$$x' = y - \eta/2. \qquad (2.136)$$

From Eq. (2.132) we see that, if $h(p, x)$ is either a function only of p or a function only of x, then

$$\langle h(p, x) \rangle = \int f_W(p, x)h(p, x) \frac{dp}{2\pi\hbar} \, dx. \qquad (2.137)$$

However, Eq. (2.137) is not true for a general function, $h(p, x)$.

Although Wigner's function $f_W(p, x)$ satisfies Eq. (2.132), it cannot be regarded as the probability for finding the particle at the point x and with the momentum p, because $f_W(p, x)$ can become negative for some values of p and x.

Problem: Calculate Wigner's function for a harmonic oscillator. This is a special case where $f_W(p, x)$ is always positive.

In Eq. (2.133), we derived Wigner's function from the density matrix. We can derive the latter from the former:

$$\int f_W\left(p, \frac{x + x'}{2}\right) e^{i[p/\hbar](x - x')} \frac{dp}{2\pi\hbar} = \rho(x, x'). \qquad (2.138)$$

Thus $f_W(p, x)$ and $\rho(x, x')$ contain the same information.

Problem: Show that the momentum representation of the free-particle density matrix is

$$\rho(p, p') = 2\pi\hbar\delta(p - p')e^{-(p^2/2m)/kT}. \qquad (2.139)$$

2.8 SYMMETRIZED DENSITY MATRIX FOR N PARTICLES

We will now discuss the density matrix for a system of many identical Bose or Fermi particles.

The symbols	will mean
D	Distinguishable
S	Symmetric
A	Antisymmetric

First consider the case of free particles. The Hamiltonian is

$$H = \frac{-\hbar^2}{2m} \sum_k \nabla_k^2. \tag{2.140}$$

The density matrix ρ_D satisfies the equation

$$\partial \rho_D / \partial \beta = -H \rho_D. \tag{2.141}$$

The solution is

$$\rho_D(x_1, x_2, \ldots x_N; x_1', x_2', \ldots x_N'; \beta) = \left(\frac{m}{2\pi\hbar^2\beta}\right)^{3N/2} \exp\left[-\frac{m}{2\hbar^2\beta} \sum_k (x_k - x_k')^2\right]. \tag{2.142}$$

In general when particles are interacting with each other we can write

$$\rho_D(x_1, x_2, \ldots x_N; x_1', x_2', \ldots x_N'; \beta) = \sum_{\text{all states}} e^{-\beta E_i} \psi_i(x_1 \cdots x_N) \psi_i^*(x_1' \cdots x_N'). \tag{2.143}$$

When the particles obey Bose statistics, then

$$\rho_S(x_1 \ldots; x_1' \ldots x_N'; \beta) = \sum_{\substack{\text{Symmetric} \\ \text{states only}}} e^{-\beta E_i} \psi(x_1 \cdots x_N) \psi^*(x_1' \cdots x_N'). \tag{2.144}$$

We introduce the notation P to indicate the permutation of particles. For example, for a permutation

$$P: \begin{array}{ccccc} 1 & 2 & 3 & 4 & 5 \\ \downarrow & \downarrow & \downarrow & \downarrow & \downarrow \\ 5 & 4 & 3 & 1 & 2 \end{array}$$

we have

$$Px_1 = x_5.$$

When the state ψ_i is symmetric,

$$\psi_i(Px_k) = \psi_i(x_k).$$

We can construct ρ_S starting from ρ_D. As an example consider the case of two particles:

$$\rho_D(x_1 x_2; x_1' x_2') = \sum_{\text{all states}} e^{-\beta E_i} \psi_i(x_1, x_2) \psi_i^*(x_1' x_2').$$

Now interchange x_1' and x_2':

$$\rho_D(x_1 x_2; x_2' x_1') = \sum_{\text{all states}} e^{-\beta E_i} \psi_i(x_1 x_2) \psi_i^*(x_2' x_1').$$

Take the average of the two:

$$\tfrac{1}{2}[\rho_D(x_1 x_2; x_1' x_2') + \rho_D(x_1 x_2; x_2' x_1')]$$

$$= \sum_{\text{all states}} e^{-\beta E_i} \psi_i(x_1 x_2) \tfrac{1}{2}[\psi_i^*(x_1' x_2') + \psi_i^*(x_2' x_1')]$$

$$= \sum_{\substack{\text{symmetric} \\ \text{states only}}} e^{-\beta E_i} \psi_i(x_1 x_2) \psi_i^*(x_1' x_2'),$$

because

$$\tfrac{1}{2}[\psi_i^*(x_1' x_2') + \psi_i^*(x_2' x_1')] = \begin{cases} 0 & \text{antisymmetric } \psi_i^* \\ \psi_i^*(x_1' x_2') & \text{symmetric } \psi_i^*. \end{cases}$$

In general, a group-theoretical argument can show that

$$\frac{1}{N!} \sum_P \psi(P x_1 \cdots P x_k \cdots P x_N) = \begin{cases} \psi(x_1 \cdots x_k \cdots x_N) & \text{symmetric} \\ 0 & \text{for any other symmetry.} \end{cases}$$

$$(2.145)$$

Here $\sum_P$ means that we sum over all permutations. When we use Eq. (2.145), the case of N particles can be written as

$$\rho_S(x_1 \cdots x_k \cdots x_N; x_1' \cdots x_k' \cdots x_N')$$

$$= \frac{1}{N!} \sum_P \rho_D(x_1 \cdots x_k \cdots; P x_1' \cdots P x_k' \cdots P x_N'). \qquad (2.146)$$

Thus the partition function becomes

$$e^{-\beta F_S} = \int \rho_S(x_1 \cdots x_k \cdots; x_1 \cdots x_N) \, dx_1 \cdots dx_N$$

$$= \frac{1}{N!} \sum_P \int \rho_D(x_1 \cdots x_k \cdots; P x_1 \cdots P x_k \cdots) \, dx_1 \cdots dx_N. \qquad (2.147)$$

Problem: Derive the partition function for a noninteracting Bose gas. Use Eq. (2.142) in Eq. (2.147) and first write

$$e^{-\beta F_S} = \frac{1}{N!} \left(\frac{m}{2\pi\hbar^2\beta}\right)^{3N/2}$$

$$\times \sum_P \int_V \exp\left\{-\frac{m}{2\hbar^2\beta}\left[(x_1 - Px_1)^2 + \cdots + (x_k - Px_k)^2 + \cdots\right.\right.$$

$$\left.\left.+ (x_N - Px_N)^2\right]\right\} dx_1 \cdots dx_N. \qquad (2.148)$$

In taking $\sum_P$, we start with the identity permutation that gives V^N from the integral and next go to a pairwise interchange of particles, three-particle cyclic permutation, and so on. The combinatorial factors must be taken into account.

Solution: Any permutation can be broken into cycles. For example,

$$P = \begin{pmatrix} 1 & 2 & 3 & 4 & 5 & 6 \\ \downarrow & \downarrow & \downarrow & \downarrow & \downarrow & \downarrow \\ 5 & 3 & 2 & 4 & 6 & 1 \end{pmatrix} = (156)(23)(4) = \begin{pmatrix} 1 & 5 & 6 \\ \downarrow & \downarrow & \downarrow \\ 5 & 6 & 1 \end{pmatrix}\begin{pmatrix} 2 & 3 \\ \downarrow & \downarrow \\ 3 & 2 \end{pmatrix}\begin{pmatrix} 4 \\ \downarrow \\ 4 \end{pmatrix}.$$

Suppose a permutation contains C_v cycles of length v, where $\sum_v vC_v = N$. For $v > 1$, let

$$h_v = \left(\frac{m}{2\pi\hbar^2\beta}\right)^{3v/2} \int dx_1\, dx_2 \cdots dx_v$$

$$\times \exp\left\{-\frac{m}{2\hbar^2\beta}\left[(x_1 - x_2)^2 + (x_2 - x_3)^2 + \cdots\right.\right.$$

$$\left.\left.+ (x_{v-1} - x_v)^2 + (x_v - x_1)^2\right]\right\} \qquad (2.149)$$

$$h_1 = \left(\frac{m}{2\pi\hbar^2\beta}\right)^{3/2} \int dx_1 = V\left(\frac{m}{2\pi\hbar^2\beta}\right)^{3/2}.$$

Then Eq. (2.148) can be written

$$e^{-\beta F_S} = \frac{1}{N!} \sum_P \left(\prod_v (h_v)^{C_v}\right). \qquad (2.150)$$

It is not difficult to show that

$$\int dy \exp\left[-a(x - y)^2\right] \exp\left[-b(y - z)^2\right]$$

$$= \left(\frac{\pi}{a + b}\right)^{3/2} \exp\left[-\frac{ab}{a + b}(x - z)^2\right]. \qquad (2.151)$$

Using Eq. (2.151) we obtain

$$h_v = V \left(\frac{m}{2\pi h^2 \beta v} \right)^{3/2}. \tag{2.152}$$

We must now count the possible permutations. Let $M(C_1, \ldots C_q) =$ the number of permutations that have

$$C_1 \text{ cycles of length } 1$$
$$C_2 \text{ cycles of length } 2$$
$$\vdots$$
$$C_q \text{ cycles of length } q$$

Then

$$e^{-\beta Fs} = \frac{1}{N!} \sum_{C_1, C_2, \ldots, C_q} M(C_1, \ldots, C_q) \prod_v (h_v)^{C_v}. \tag{2.153}$$

As a preliminary to the determination of $M(C_1, \ldots, C_q)$, consider the case of $N = 6$, $C_1 = 3$, $C_3 = 1$, $C_2 = C_4 = C_5 = C_6 = 0$. An example is the permutation $P = (5)(3)(2)(461)$. Notice that the sequence $Q = 5, 3, 2, 4, 6, 1$ corresponds to P. In other words, from the sequence Q, we can uniquely construct the permutation P with $C_1 = 3$, $C_3 = 1$. But from $Q' = 2, 5, 3, 4, 6, 1$, or from $Q'' = 5, 3, 2, 1, 4, 6$, we can also construct in the same way the same P. In fact, it can be easily seen that there are 3! times $3 = 18$ different sequences that can be used in the way described above to construct the permutation $P = (5)(3)(2)(461)$.

Generalizing to any set of C's, we see that there are two ways we can get a new Q that leads to the same P:

a) we can interchange cycles of the same length,

b) we can make cyclic permutation within a given cycle.

There are $\prod_v C_v!$ ways of interchanging cycles of the same length, and $\prod_v v^{C_v}$ cyclic permutations within cycles. Thus for each P there are $\prod_v C_v! \prod_v v^{C_v}$ different Q's that lead to the same P. Because there are $N!$ possible Q's,

$$M(C_1, \ldots, C_q) = \frac{N!}{\prod_v C_v! \, v^{C_v}}.$$

It follows that

$$e^{-\beta Fs} = \sum_{C_1, C_2, \ldots C_q} \prod_v \frac{h_v^{C_v}}{C_v! \, v^{C_v}}, \tag{2.154}$$

where $\sum_v v C_v = N$.

Instead of trying to do the sum in Eq. (2.154), we will find the free energy of the grand canonical ensemble (that is, we will let N vary).

$$e^{-\beta F} = \sum_{N=1}^{\infty} e^{-\beta F_N} e^{+N\mu\beta}. \tag{2.155}$$

As usual, let $\alpha = e^{+\mu\beta}$. Then,

$$e^{-\beta F} = \sum_{C_1, \ldots C_q} \prod_v \frac{h_v^{C_v}}{C_v! \, v^{C_v}} \alpha^{vC_v} = \sum_{C_1, \ldots C_q \ldots} \prod_v \frac{[h_v(\alpha^v/v)]^{C_v}}{C_v!},$$

where now each C_q runs from 0 to ∞. Interchanging $\prod$ and $\sum$, we have

$$e^{-\beta F} = \prod_v \sum_{C_v=0}^{\infty} \frac{[h_v(\alpha^v/v)]^{C_v}}{C_v!} = \prod_v \exp\left(h_v \frac{\alpha^v}{v}\right) = \exp\left(\sum_v h_v \frac{\alpha^v}{v}\right).$$

Then,

$$\beta F = -\sum_{v=1}^{\infty} \frac{h_v \alpha^v}{v} = -\left(\frac{m}{2\pi\hbar^2\beta}\right)^{3/2} V \sum_v \frac{\alpha^v}{v^{5/2}}. \tag{2.156}$$

This result is identical to the one we obtained in a much simpler way earlier.

When the density of the gas is not large, only the identity permutation is important for high temperatures or for the limit of $\hbar \to 0$, since we have such factors as $e^{-[(mkT/2\hbar^2)(x-x')^2]}$. This means that the quantum effect appears only for low temperatures.

For the antisymmetric case,

$$\rho_A(x_1 \cdots x_k \cdots x_N; x_1' \cdots x_k' \cdots x_N')$$

$$= \frac{1}{N!} \sum_P (-1)^P \rho_D(x_1 \cdots x_N; Px_1' \cdots Px_k' \cdots Px_N'),$$

where

$$(-1)_P = \begin{cases} 1 \text{ for even permutation} \\ -1 \text{ for odd permutation}. \end{cases}$$

Problem: Modify the argument leading to Eq. (2.156) so as to obtain the corresponding result for a noninteracting Fermi gas.

2.9 DENSITY SUBMATRIX

For N particles, the density matrix can be written

$$\rho(x_1, x_2, \ldots, x_N; x_1', x_2', \ldots x_N') = \sum_i w_i \varphi_i(x_1, \ldots, x_N) \varphi_i^*(x_1', \ldots, x_N'). \tag{2.157}$$

Notice that we are not now specializing to the case of statistical mechanics. To compute the expectation value of $P_1^2/2m$, we proceed as follows:

$$\left\langle \frac{P_1^2}{2m} \right\rangle = \frac{1}{\text{Tr } \rho} \text{Tr } \rho \, \frac{P_1^2}{2m}$$

$$= -\frac{1}{\text{Tr } \rho} \frac{\hbar^2}{2m} \int \frac{\partial^2}{\partial x_1^2} \rho(x_1, \ldots, x_N; x_1', \ldots, x_N') \Big|_{x_i = x_i'} dx_1 \cdots dx_N$$

$$= -\frac{\hbar^2}{2m} \int dx_1 \frac{\partial^2}{\partial x_1^2} \rho_1(x_1, x_1') \Big|_{x_1 = x_i'} = \text{Tr } \rho_1 \frac{P_1^2}{2m}, \qquad (2.158)$$

where we define

$$\rho_1(x, x') = \frac{\int \rho(x, x_2, \ldots, x_N; x', x_2, \ldots, x_N) \, dx_2 \cdots dx_N}{\text{Tr } [\rho]}. \qquad (2.159)$$

ρ_1 is the one-particle density submatrix. Using this matrix we can calculate, for example,

$$\left\langle \sum_k \frac{p_k^2}{2m} \right\rangle = -N \int \left[\frac{\hbar^2}{2m} \frac{\partial^2}{\partial x'^2} \rho_1(x, x') \right]_{x' = x} dx = N \text{ Tr} \left[\frac{p^2}{2m} \rho_1 \right] \qquad (2.160)$$

(assuming that the particles are essentially identical).

Similarly,

$$\left\langle \sum_k V(x_k) \right\rangle = N \int V(x) \rho_1(x, x) \, dx = N \text{ Tr } [V\rho_1]. \qquad (2.161)$$

The two-particle density submatrix is defined as

$$\rho_2(x_1 x_2; x_1' x_2') = \frac{\int \rho_2(x_1 x_2 x_3 \cdots x_N; x_1' x_2' x_3 \cdots x_N) \, dx_3 \cdots dx_N}{\text{Tr } [\rho]}. \qquad (2.162)$$

From Eq. (2.162), the two-particle potential energy $V(x_1 x_2)$ has expectation

$$\left\langle \sum_{i<j} V(x_i, x_j) \right\rangle = \frac{N(N-1)}{2} \int V(x_1 x_2) \rho_2(x_1 x_2; x_1 x_2) \, dx_1 \, dx_2. \qquad (2.163)$$

Special cases of density submatrices are the distribution functions. The single-particle distribution function is defined as

$$p_1(x) = \rho_1(x; x). \qquad (2.164)$$

The two-particle distribution function is

$$p_2(x, y) = \rho_2(x, y; x, y). \qquad (2.165)$$

In a uniform substance, such as a gas or a liquid, $p_1(x)$ is independent of x, and $p_2(x, y)$ depends only on the difference $(x - y)$.

2.10 PERTURBATION EXPANSION OF THE DENSITY MATRIX

Recall that the density matrix in statistical mechanics satisfies

$$\frac{\partial \rho}{\partial \beta} = -H\rho. \tag{2.166}$$

There are very few Hamiltonians for which we can solve Eq. (2.166) exactly, but it may be that H is close to one such Hamiltonian, H_0.

$$H = H_0 + H_1, \tag{2.167}$$

$$\frac{\partial \rho_0}{\partial \beta} = -H_0\rho_0. \tag{2.168}$$

We would like to use ρ_0 to obtain an approximation to ρ, which we expect to be close to $\rho_0 = e^{-\beta H_0}$. Because ρ is close to $e^{-\beta H_0}$, we expect $e^{H_0\beta}\rho$ to vary slowly with β.

$$\frac{\partial}{\partial \beta}(e^{H_0\beta}\rho) = H_0 e^{H_0\beta}\rho + e^{H_0\beta}\frac{\partial \rho}{\partial \beta} = e^{H_0\beta}H_0\rho - e^{H_0\beta}H\rho = -e^{H_0\beta}H_1\rho. \tag{2.169}$$

Integrating Eq. (2.169) from zero to β, and remembering that if $\beta = 0$, we find

$$e^{H_0\beta}\rho = 1,$$

$$e^{H_0\beta}\rho(\beta) - 1 = -\int_0^\beta e^{H_0\beta'}H_1\rho(\beta')\,d\beta'. \tag{2.170}$$

Therefore,

$$\rho(\beta) = \rho_0(\beta) - \int_0^\beta \rho_0(\beta - \beta')H_1\rho(\beta')\,d\beta'. \tag{2.171}$$

The last term in Eq. (2.171) is small if H_1 is small, and it is a correction term to the approximate equation $\rho \approx \rho_0$. If the correction term is small, we can use an approximate $\rho(\beta')$ to obtain a much more accurate approximation to $\rho(\beta)$. For example, if we make the approximation $\rho(\beta') \approx \rho_0(\beta')$, we have

$$\rho(\beta) \approx \rho_0(\beta) - \int_0^\beta \rho_0(\beta - \beta')H_1\rho_0(\beta')\,d\beta'. \tag{2.172}$$

In using Eq. (2.172) as a new approximation to $\rho(\beta')$, we can find a still better approximation to ρ. Continuing in this manner we get

$$\rho(\beta) = \rho_0(\beta) - \int_0^\beta d\beta' [\rho_0(\beta - \beta')H_1\rho_0(\beta')]$$

$$+ \int_0^\beta d\beta' \int_0^{\beta'} d\beta'' [\rho_0(\beta - \beta')H_1\rho_0(\beta' - \beta'')H_1\rho_0(\beta'')]$$

$$- \int_0^\beta d\beta' \int_0^{\beta'} d\beta'' \int_0^{\beta''} d\beta''' [\quad\quad] + \cdots \quad\quad (2.173)$$

We can easily rewrite Eq. (2.173) in coordinate representation. For example,

$$\rho(x, x'; \beta) = \langle x|\rho(\beta)|x' \rangle$$

$$\approx \rho_0(x, x'; \beta) - \int_0^\beta \langle x|\rho_0(\beta - \beta')\left(\int |x''\rangle\langle x''|dx''\right) H_1\rho_0(\beta')|x' \rangle \, d\beta',$$

$$(2.174)$$

where we use

$$\int_{-\infty}^\infty |x''\rangle\langle x''| \, dx'' = 1.$$

If $H_1 = V(x)$, then

$$\langle x''|H_1\rho_0(\beta')|x' \rangle = V(x'')\rho_0(x'', x'; \beta'),$$

and Eq. (2.174) becomes

$$\rho(x, x'; \beta) = \rho_0(x, x', \beta)$$

$$- \int_{-\infty}^\infty \int_0^\beta \rho_0(x, x''; \beta - \beta')V(x'')\rho_0(x'', x'; \beta') \, d\beta' \, dx'' + \cdots$$

$$(2.175)$$

2.11 PROOF THAT $F \leq F_0 + \langle H - H_0 \rangle_0$

Now that we have a perturbation theory, we are in a position to prove as promised in Section 2.3, that $F \leq F_0 + \langle H - H_0 \rangle_0$. Let $V = H - H_0$. Since F is determined by the equation $e^{-\beta F} = \text{Tr } e^{-\beta(H_0 + V)}$, the most obvious approach is to find an approximation to F by using the perturbation expansion of

$$e^{-\beta(H_0 + V)} = \rho.$$

$$e^{-\beta(H_0 + V)} = e^{-\beta H_0} - \int_0^\beta e^{-(\beta - u)H_0} V e^{-uH_0} \, du$$

$$+ \int_0^\beta \int_0^{u_1} du_1 \, du_2 \, e^{-(\beta - u_1)H_0} V e^{-(u_1 - u_2)H_0} V e^{-u_2 H_0} - \cdots. \quad (2.176)$$

Taking the trace of Eq. (2.176) we obtain

$$
e^{-\beta F} = \text{Tr } e^{-\beta(H_0+V)} = e^{-\beta F_0} - \int_0^\beta \text{Tr } \left[e^{-(\beta-u)H_0} V e^{-uH_0} \right] du
$$

$$
+ \int_0^\beta \int_0^{u_1} du_1 \, du_2 \, \text{Tr } \left[e^{-(\beta-u_1)H_0} V e^{-(u_1-u_2)H_0} V e^{-u_2 H_0} \right] - \cdots .
$$

$$(2.177)$$

Using the fact that $\text{Tr } AB = \text{Tr } BA$, we find that

$$
e^{-\beta F} = e^{-\beta F_0} - \int_0^\beta du \, \text{Tr } e^{-\beta H_0} V
$$

$$
+ \int_0^\beta \int_0^{u_1} du_1 \, du_2 \, \text{Tr } \left[e^{-\beta H_0} e^{(u_1-u_2)H_0} V e^{-(u_1-u_2)H_0} V \right] - \cdots . \quad (2.178)
$$

It is possible to simplify the second-order term. Let $w = u_1 - u_2$ and let $x = u_1$. Then the integral becomes

$$
\int_A \text{Tr } \left[e^{-\beta H_0} e^{wH_0} V e^{-wH_0} V \right], \tag{2.179}
$$

where A is the shaded region in Fig. 2.4. Now let $w' = \beta - w$ and let $x' = \beta - x$. Then the integral becomes

$$
\int_{A'} \text{Tr } \left[e^{-\beta H_0} e^{wH_0} V e^{-wH_0} V \right]. \tag{2.180}
$$

Averaging Eqs. (2.179) and (2.180), we find, from Eq. (2.178),

$$
e^{-\beta F} = e^{-\beta F_0} - \beta \, \text{Tr } \left[e^{-\beta H_0} V \right] + \frac{\beta}{2} \int_0^\beta dw \, \text{Tr } \left[e^{-\beta H_0} e^{wH_0} V e^{-wH_0} V \right] + \cdots .
$$

$$(2.181)$$

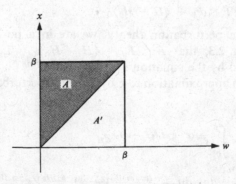

Fig. 2.4 The shaded area A is the region over which Eq. (2.179) is integrated.

Let $|m\rangle$ and $|n\rangle$ be eigenstates of H_0, and let E_m and E_n be their respective eigenvalues. Then

$$\text{Tr}\,[e^{-\beta H_0}V] = \sum_n \langle n|e^{-\beta H_0}V|n\rangle = \sum_n e^{-\beta E_n}V_{nn}. \tag{2.182}$$

$$\text{Tr}\,[e^{-\beta H_0}e^{wH_0}Ve^{-wH_0}V] = \sum_n \langle n|e^{-\beta H_0}e^{wH_0}V \sum_m |m\rangle\langle m|e^{-wH_0}V|n\rangle$$

$$= \sum_{nm} e^{-\beta E_n}e^{w(E_n-E_m)}V_{nm}V_{mn}, \tag{2.183}$$

where we have used the fact that $\sum_m |m\rangle\langle m| = 1$. Equation (2.181) becomes

$$e^{-\beta F} = e^{-\beta F_0} - \sum_n \beta e^{-\beta E_n}V_{nn} + \frac{\beta}{2}\sum_{mn}\frac{e^{-\beta E_m} - e^{-\beta E_n}}{E_n - E_m}|V_{mn}|^2 + \cdots. \tag{2.184}$$

Note that if $m = n$,

$$\frac{e^{-\beta E_m} - e^{-\beta E_n}}{E_n - E_m} = e^{-\beta E_n}\frac{e^{\beta(E_n - E_m)} - 1}{E_n - E_m} = \beta e^{-\beta E_n}.$$

In order to get an expansion for F, we write $F = F_0 + F_1 + F_2 + \cdots$, where F_1 is a first-order perturbation and F_2 is a second-order perturbation. Then

$$e^{-\beta F} = e^{-\beta F_0}e^{-\beta(F_1 + F_2 + \cdots)}$$

$$= e^{-\beta F_0}(1 - \beta F_1 - \beta F_2 + \frac{\beta^2}{2}(F_1 + F_2 + \cdots)^2 + \cdots)$$

$$= e^{-\beta F_0}\left[1 - \beta F_1 + \left(\frac{\beta^2}{2}F_1^2 - \beta F_2\right) + \cdots\right]. \tag{2.185}$$

up to and including the second order. We have a perturbation expansion for F if we set

$$-e^{-\beta F_0}\beta F_1 = -\sum_n \beta e^{-\beta E_n}V_{nn}$$

and
$$\tag{2.186}$$

$$e^{-\beta F_0}\left(\frac{\beta^2}{2}F_1^2 - \beta F_2\right) = \frac{\beta}{2}\sum_{mn}\frac{e^{-\beta E_m} - e^{-\beta E_n}}{E_n - E_m}|V_{mn}|^2.$$

Solving these equations we obtain

$$F_1 = \frac{\text{Tr}\,Ve^{-\beta H_0}}{\text{Tr}\,e^{-\beta H_0}} = \langle V\rangle_{H_0} \tag{2.187}$$

and

$$F_2 = \frac{\beta}{2}\left[\frac{\sum_n V_{nn}e^{-\beta E_n}}{\sum_n e^{-\beta E_n}}\right]^2 - \frac{1}{2}e^{\beta F_0}\sum_{\substack{mn\\m\neq n}}\frac{e^{-\beta E_m} - e^{-\beta E_n}}{E_n - E_m}|V_{mn}|^2$$

$$-\frac{\beta}{2}\left[\frac{\sum_n |V_{nn}|^2 e^{-\beta E_n}}{\sum_n e^{-\beta E_n}}\right]. \tag{2.188}$$

With the Cauchy-Schwarz inequality,

$$\left| \sum_n a_n b_n \right|^2 \le \left[\sum_n |a_n|^2 \right]\left[\sum_n |b_n|^2 \right],$$

it is easy to show that for any set of positive real numbers, w_n,

$$\left[\frac{\sum_n w_n a_n}{\sum_n w_n} \right]^2 \le \frac{\sum_n w_n |a_n|^2}{\sum_n w_n}.$$

Applying this inequality to Eq. (2.188), we have

$$F_2 \le -\frac{1}{2} e^{\beta F_0} \sum_{\substack{mn \\ m \neq n}} \frac{e^{-\beta E_m} - e^{-\beta E_n}}{E_n - E_m} |V_{mn}|^2 \le 0. \qquad (2.189)$$

Now, let

$$H(\alpha) = H_0 + \alpha V,$$

and let $F(\alpha)$ be the free energy calculated from $H(\alpha)$. Then $F(0) = F_0$ and $F(1) = F$. From the preceding perturbation analysis,

$$F(\alpha) = F_0 + \alpha F_1 + \alpha^2 F_2 + 0(\alpha^3).$$

So $F'(0) = F_1$. If we can prove that

$$F''(\alpha) \le 0 \text{ for all } \alpha, \qquad (2.190)$$

then $F(\alpha)$ will be concave downward (see Fig. 2.5) and thus it lies below the line $F(0) + \alpha F'(0) = F_0 + \alpha F_1$. From this we can conclude that

$$F \le F_0 + F_1 = F_0 + \langle V \rangle_{H_0},$$

or

$$F \le F_0 + \langle H - H_0 \rangle_{H_0}, \qquad (2.191)$$

which was to be proved.

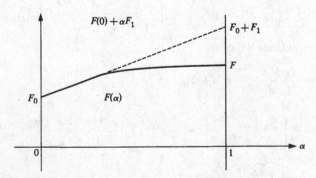

Fig. 2.5 The situation when $F(\alpha)$ is concave downward.

To prove Eq. (2.190) we apply the perturbation analysis to $F(\alpha)$ for each α; that is, we write

$$F(\alpha + \gamma) = F_0(\alpha) + \gamma F_1(\alpha) + \gamma^2 F_2(\alpha) + 0(\gamma^3),$$

where $F_0(\alpha) = F(\alpha)$, and $\gamma F_1(\alpha)$ and $\gamma^2 F_2(\alpha)$ are as in Eqs. (2.187) through (2.189), with H_0 replaced by $H(\alpha)$ and H by $H(\alpha + \gamma) = H(\alpha) + \gamma V$. The result of Eq. (2.189) is then

$$F_2(\alpha) \leq 0,$$

so that

$$F''(\alpha) = \frac{d^2}{d\gamma^2} F(\alpha + \gamma) \bigg|_{\gamma = 0} = 2F_2(\alpha) \leq 0.$$

This proves Eq. (2.190) and thus completes the proof of Eq. (2.191).

Applying this same theorem to F_0, we obtain

$$F_0 \leq F + \langle -V \rangle_H. \tag{2.192}$$

Combining Eqs. (2.191) and (2.192), we see that

$$F_0 + \langle V \rangle_H \leq F \leq F_0 + \langle V \rangle_{H_0}. \tag{2.193}$$

A more accurate approximation F is obtained by including F_2:

$$F \approx F_0 + \langle V \rangle_0 + \frac{\beta}{2} \langle V \rangle_0^2 - \frac{1}{2} \left\langle \int_0^\beta e^{wH_0} V e^{-wH_0} V \, dw \right\rangle_0. \tag{2.194}$$

CHAPTER 3

PATH INTEGRALS

3.1 PATH INTEGRAL FORMULATION OF THE DENSITY MATRIX

Earlier it has been shown that the density matrix satisfies the equation

$$\hbar \frac{\partial \rho(u)}{\partial u} = -H\rho(u). \tag{3.1}$$

and has the formal solution

$$\rho(u) = e^{-Hu/\hbar}, \tag{3.2}$$

where it will be noted that u has been redefined to be $\beta\hbar$, which has the dimensions of time. Furthermore, the "time" u can be broken up into divisions of length $\varepsilon(n\varepsilon = u)$ so that ρ can be developed incrementally as follows:

$$\rho(u) = e^{-H\varepsilon/\hbar}e^{-H\varepsilon/\hbar} \cdots e^{-H\varepsilon/\hbar}$$
$$= \rho_\varepsilon \rho_\varepsilon \cdots \rho_\varepsilon (n \text{ factors}). \tag{3.3}*$$

In the coordinate representation for the density matrix $\rho(x, x'; u)$ the solution is represented as

$$\rho(x, x'; u) = \int \cdots \int \rho(x, x_{n-1}; \varepsilon)\rho(x_{n-1}, x_{n-2}; \varepsilon) \cdots \rho(x_1, x'; \varepsilon)dx_1 \cdots dx_{n-1}. \tag{3.4}$$

The representation in Fig. 3.1 is useful in the interpretation of the above expression. The particle travels from x' to x through a series of intermediate steps, $x_1, x_2 \ldots, x_{n-1}$, which define a "path". The total amplitude $\rho(x, x', u)$ for the particle to begin at x' and end up at x is given by a sum over all possible paths, that is, for all possible values of the intermediate positions x_i. As the time increment ε is allowed to approach zero, the number of integrations on the intermediate variables become infinite and Eq. (3.4) can be written symbolically

$$\rho(x, x'; U) = \iint \Phi[x(u)] \, \mathscr{D}x(u), \tag{3.5}$$

---

* The reason we want to reduce ε as much as possible is, of course, that the asymptotic behavior of ρ at high temperatures can be easily found (see below).

72

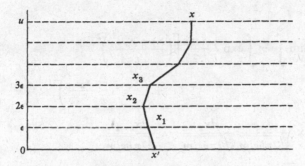

Fig. 3.1 A particle travelling from x' to x along a path x_1, x_2 etc.

where

$$\Phi[x(u)] = \lim_{\substack{\varepsilon \to 0 \\ n\varepsilon = u}} \rho(x, x_{n-1}; \varepsilon)\rho(x_{n-1}, x_{n-2}; \varepsilon)\cdots \rho(x_1, x'; \varepsilon), \qquad (3.6)$$

and

$$\mathscr{D}x(u) = \lim_{n \to \infty} dx_1\, dx_2 \cdots dx_{n-1}.$$

Although all possible paths are to be considered in Eq. (3.5), the main contribu‾tion comes from the limited set of paths for which none of the $\rho(x, x'; \varepsilon)$ is too small. For example, we know that for low ε,

$$\rho(x, x'; \varepsilon) \approx \sqrt{\frac{m}{2\pi\hbar\varepsilon}}\, e^{-(m/2\hbar\varepsilon)(x-x')^2},$$

which is the density matrix for a free particle. The term $\rho(x, x'; \varepsilon)$ is small if $x - x' > \sqrt{2\hbar/m}\sqrt{\varepsilon}$; so the main contribution comes from reasonably smooth paths. Now let us find Φ more explicitly, first for a free particle and then for one moving subject to a constraint.

Free Particle:

The Hamiltonian is

$$H = -\frac{\hbar^2}{2m}\frac{\partial^2}{\partial x^2}. \qquad (3.7)$$

From previous results we know

$$\rho(x, x'; \varepsilon) = \sqrt{\frac{m}{2\pi\hbar\varepsilon}}\, e^{-(m/2\hbar\varepsilon)(x-x')^2} \qquad (3.8)$$

and

$$\rho(x, x'; U) = \lim_{\varepsilon \to 0} \int \cdots \int \exp \left\{ -\frac{m\varepsilon}{2\hbar} \left[\left(\frac{x - x_{n-1}}{\varepsilon} \right)^2 \right. \right.$$
$$\left. \left. + \left(\frac{x_{n-1} - x_{n-2}}{\varepsilon} \right)^2 \cdots \left(\frac{x_1 - x'}{\varepsilon} \right)^2 \right]_x \right\}$$
$$\times \frac{dx_1}{\sqrt{2\pi\hbar\varepsilon/m}} \frac{dx_2}{\sqrt{2\pi\hbar\varepsilon/m}} \cdots \frac{dx_{n-1}}{\sqrt{2\pi\hbar\varepsilon/m}}. \tag{3.9}$$

Thus, for a free particle,

$$\Phi[x(u)] = \lim_{\varepsilon \to 0} \exp \left\{ -\frac{m\varepsilon}{2\hbar} \left[\left(\frac{x - x_{n-1}}{\varepsilon} \right)^2 + \cdots + \left(\frac{x_1 - x'}{\varepsilon} \right)^2 \right] \right\}. \tag{3.10}$$

As the interval ε is made smaller,

$$\frac{x_k - x_{k-1}}{\varepsilon} \to \left. \frac{dx(u)}{du} \right|_{u=k\varepsilon} \equiv \left. \dot{x}(u) \right|_{u=k\varepsilon}$$

and Eq. (3.10) becomes

$$\Phi[x(u)] = \exp \left\{ -\frac{1}{\hbar} \int_0^U \frac{m}{2} \left[\dot{x}(u) \right]^2 du \right\}. \tag{3.11}$$

Particle Moving in One Dimension in a Potential $V(x)$:

$$H = -\frac{\hbar^2}{2m} \frac{\partial^2}{\partial x^2} + V(x).$$

The equation for ρ is explicitly

$$-\hbar \frac{\partial \rho}{\partial u} = -\frac{\hbar^2}{2m} \frac{\partial^2}{\partial x^2} \rho + V(x)\rho. \tag{3.12}$$

In an infinitesimal "time" interval ε the particle may be considered as moving freely with a small correction due to the presence of the potential $V(x)$. To a very good approximation

$$\delta\rho = -\frac{V(x)}{\hbar} \varepsilon \rho_{\text{free}}. \tag{3.13}$$

This result is demonstrated as follows: From perturbation theory we have

$$\delta\rho(x, x'; \varepsilon) \approx -\int_{-\infty}^{\infty} dx'' \int_0^{\varepsilon} \rho_0(x, x''; \varepsilon - u) V(x'') \rho_0(x'', x'; u) \frac{du}{\hbar}, \tag{3.14}$$

where in our case, $\rho_0 = \rho_{\text{free}}$. For low ε, because ρ_{free} is a very localized Gaussian, most of the contribution in the integral over x'' occurs near $x'' = x_0$, where it can be shown that

$$x_0 = \frac{ux + (\varepsilon - u)x'}{\varepsilon}.$$

So we can, for small ε, write Eq. (3.14) as

$$\delta\rho(x, x'; \varepsilon) \approx -\int_0^\varepsilon \frac{du}{\hbar} \, V(x_0) \int_{-\infty}^\infty dx'' \, \rho_0(x, x''; \varepsilon - u)\rho_0(x'', x'; u)$$

$$= -\int_0^\varepsilon \frac{du}{\hbar} \, V(x_0)\rho_{\text{free}}(x, x'; \varepsilon). \tag{3.15}$$

Now, if $x \approx x'$, x_0 is also close to x and $V(x_0)$ is constant over the range of integration. Equation (3.15) goes over to Eq. (3.13):

$$\delta\rho(x, x'; \varepsilon) \approx -\frac{\varepsilon}{\hbar} \, V(x)\rho_{\text{free}}(x, x'; \varepsilon)$$

under the assumption that $|x - x'|$ and ε are small.

Of course, as $|x - x'|$ increases, our conclusion that $\int_0^\varepsilon V(x_0) \, du \approx \varepsilon V(x)$ becomes less accurate, and it might be thought preferable to write something like

$$\varepsilon\left[\frac{V(x) + V(x')}{2}\right] \quad \text{or} \quad \varepsilon V\left(\frac{x + x'}{2}\right)$$

in place of $\varepsilon V(x)$. But for our purposes, Eq. (3.13) is adequate. So

$$\rho(\varepsilon) \approx \rho_{\text{free}}(\varepsilon)\left[1 - \frac{V(x)}{\hbar}\varepsilon\right] \approx \rho_{\text{free}}(\varepsilon)e^{-V(x)\varepsilon/\hbar}. \tag{3.16}$$

Thus

$$\rho(x, x'; \varepsilon) \approx \sqrt{\frac{m}{2\pi\hbar\varepsilon}} \exp\left[-\frac{m(x - x')^2}{2\hbar\varepsilon} - \frac{\varepsilon V(x)}{\hbar}\right]. \tag{3.17}$$

To find $\rho(x, x'; u)$ it is only necessary to apply the correction $e^{-V(x)\varepsilon/\hbar}$ to each element of the "path" of the free particle from x to x'. Thus, as $\varepsilon \to 0$,

$$\rho(x, x'; U) = \iint \exp\left\{-\frac{1}{\hbar}\int_0^U \left[\frac{m\dot{x}(u)^2}{2} + V(x(u))\right] du\right\} \mathcal{D}x(u) \tag{3.18}$$

$$\equiv \iint \Phi[x(u)] \, \mathcal{D}x(u),$$

where $x(0) \equiv x'$ and $x(U) = x$. Thus

$$\Phi[x(u)] = \exp\left\{-\frac{1}{\hbar} \int_0^U \left[\frac{mx'(u)^2}{2} + V(x(u))\right] du\right\}. \qquad (3.18a)$$

The free energy is derived from

$$e^{-\beta F} = \int \rho(x, x) \, dx$$

$$= \int \left\{\int\int \exp\left\{-\frac{1}{\hbar} \int_0^U \left[\frac{m\dot{x}(u)^2}{2} + V(x(u))\right] du\right\} \mathcal{D}x(u)\right\}_{x(0)=x(U)=x} dx$$

$$(3.19)$$

where $U = \beta\hbar$.

Having derived the general expression for Φ the question now is, which "paths" are the most important? If the potential is not very dependent upon position, then the more direct paths contribute the most. This is because longer paths, with higher mean square "velocities," $[\dot{x}(u)]^2$, make the exponent of the weighting functional $\Phi[x(u)]$ more negative. On the other hand, suppose the potential does depend on position. For simplicity, let the potential increase for larger values of x. Then a direct path such as 1 in Fig. 3.2 does not necessarily give the largest contribution, because the integrated value of the potential is higher than over another path such as 2 or 3. Consider a path such as 2 that deviates widely from the direct path. Then $V(x)$ decreases but $\dot{x}$ increases over the path. In this case we would expect that the increased velocity $\dot{x}$ would more than compensate for the decreased potential over the path. Thus, the most important path would be one for which any smaller integrated value of potential energy is more than compensated for by an increase in kinetic energy (symbolically represented by 3 in Fig. 3.2).

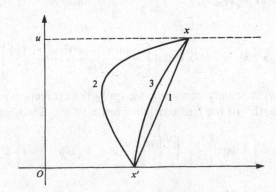

Fig. 3.2 Three paths between x' and x.

In the classical limit, the temperature is high and U is small. We want the classical partition function, and we would like to see how it is related to the quantum-mechanical partition function. From Eq. (3.18)

$$\rho(x, x; U) = \iint_{x(0)=x(U)=x} \exp\left\{-\frac{1}{\hbar}\int_0^U \left[\frac{m\dot{x}(u)^2}{2} + V(x(u))\right] du\right\} \mathscr{D}x(u).$$
(3.20)

If we assume that the most important paths are those for which $V(x(u)) \approx V(x)$, Eq. (3.19) can be approximated by

$$\rho(x, x; U) \approx \exp\left[-\frac{U}{\hbar} V(x)\right] \iint_{x(0)=x(U)=x} \exp\left[-\frac{1}{\hbar}\int_0^U \frac{m\dot{x}(u)^2}{2}\right] \mathscr{D}x(u)$$

$$= \exp\left[-\frac{U}{\hbar} V(x)\right] \rho_{\text{free}}(x, x; U)$$

$$= \sqrt{\frac{m}{2\pi\hbar U}} \exp\left[\frac{-m(x - x')^2}{\hbar U} - \frac{UV(x)}{\hbar}\right]$$
(3.21)

Equation (3.21) is just a specialization of Eq. (3.17). But from the derivation of Eq. (3.21) it is easy to see the direction of the error in the classical limit. We assumed that the most important paths had $x(u) \approx x$, whereas in fact the most important paths are those for which $\int_0^U [m\dot{x}(u)^2/2 + V(x(u))] du$ is a minimum, which occurs when $(x(u))$ goes over paths that tend to let $V(x(u))$ be lower than $V(x)$. It follows, then, that

$$\rho_{\text{classical}}(x, x; U) < \rho(x, x; U).$$
(3.22)

The classical partition function is

$$e^{-\beta F_{\text{cl}}} = \int_{-\infty}^{\infty} \rho_{\text{cl}}(x, x; U) \, dx = \sqrt{\frac{m}{2\pi\hbar U}} \int_{-\infty}^{\infty} e^{-UV(x)/\hbar} dx.$$
(3.23)

From Eq. (3.22) we obtain

$$F_{\text{cl}} \gtrsim F.$$
(3.24)

To find the condition for Eq. (3.21) to be a good approximation to the path integral, note that the paths for which

$$\frac{1}{\hbar}\int_0^U \frac{m\dot{x}^2}{2} \, du \gg 1$$

do not contribute much to the path integral. Using the Cauchy-Schwarz inequality, we find

$$\left|\int_a^b fg\right|^2 \le \int_a^b |f|^2 \int_a^b |g|^2.$$

With $f = \dot{x}$ and $g = 1$, we have for $0 \leq u_1 \leq U$:

$$|x(u_1) - x(0)|^2 = \left|\int_0^{u_1} \dot{x}(u)\right|^2 \leq \left[\int_0^{u_1} \dot{x}^2(u)\right]u_1 \leq \left(\int_0^{u_1} \dot{x}^2\right)U. \qquad (3.25)$$

So for the important paths, $(m/2\hbar U)|x(u_1) - x(0)|^2 < 1$. If d is the maximum value of $|x(u_1) - x(0)|$ for $0 < u_1 < U$, then the paths that contribute significantly satisfy

$$\frac{mkTd^2}{2\hbar^2} \leq 1. \qquad (3.26)$$

For the hydrogen atom at $1°K$ Eq. (3.26) yields $d \leq 9$ Å. For oxygen at $1°K$, $d \leq 2.5$ Å; for oxygen at $100°K$, $d \leq 0.25$ Å.

If the potential varies so slowly that it is approximately unchanged even when the path varies by more than d, then Eq. (3.21) is good. Thus, for heavy particles at normal temperatures, systems may be treated by classical statistical mechanics. However, the classical approximation does not apply to liquid helium, nor does it apply to electrons in a metal, nor even to solids if the temperature is too low.

A few things should be noticed about this formulation of statistical mechanics:

1. We have discussed only the case of one coordinate. However, the extension to N degrees of freedom is obvious; if $x^{(1)}, x^{(2)}, \ldots, x^{(N)}$ are the coordinates, $\mathscr{D}x(u)$ becomes

$$\mathscr{D}x(u) = \lim_{n \to \infty} dx_1^{(1)} dx_1^{(2)} \cdots dx_1^{(N)} dx_2^{(1)} \cdots dx_2^{(N)} \cdots dx_n^{(N)}. \qquad (3.27)$$

2. If "u" is replaced by "iu" in Eq. (3.1), we have Schrödinger's equation. In a similar manner to that of statistical mechanics, quantum mechanics can be formulated in terms of path integrals. For mathematicians, statistical mechanics is easier to deal with because the exponentials in the path integrals are exponentials of real quantities.

3. We have not dealt with relativity (in which particles appear and disappear) and spin (which is a discrete, rather than a continuous, variable) in the path-integral formulation. If such effects exist, we can only get approximations. Actually, spin *can* be put into the formalism, but our interpretation of $\mathscr{D}x(u)$ as a set of paths differing by an infinitesimal amount no longer holds.

3.2 CALCULATION OF PATH INTEGRALS

We will discuss two techniques for the calculation of path integrals:

1. Integrals whose exponents are quadratic in $\dot{x}$ and x will be calculated exactly.

2. Perturbation expansions.

Consider the element

$$f(x_2, x_1, U) = \iint \exp \left\{ -\frac{1}{\hbar} \int_0^U \frac{m[\dot{x}(u)]^2}{2} \, du \right\} \mathscr{D}x(u) \qquad (3.28)$$

where $x(0) = x_1$ and $x(U) = x_2$. This is the density matrix element for a free particle. Let us expand each path about the straight-line path in the xu-plane. Call the straight line path $x(u)$. Then

$$x(u) = x(u) + y(u) \qquad (3.29)$$

and

$$f(x_2, x_1, U) = \iint \exp \left[-\frac{1}{\hbar} \frac{m}{2} \int_0^U (\dot{x}(u) + \dot{y}(u))^2 \, du \right] \mathscr{D}y(u). \qquad (3.30)$$

Now

$$\frac{dx(u)}{du} = \frac{x_2 - x_1}{U} = v = \text{velocity} \qquad (3.31)$$

so that

$$\int_0^U [\dot{x} + \dot{y}]^2 \, du$$

$$= \int_0^U (v^2 + 2v\dot{y} + \dot{y}^2) \, du = v^2 U + 2v[y(U) - y(0)] + \int_0^U \dot{y}^2 \, du.$$

$$(3.32)$$

But $y(U) = y(0) = 0$. Therefore

$$f(x_2, x_1, U) = \exp \left[-\frac{mv^2 U}{2\hbar} \right] \iint \exp \left[-\frac{m}{2\hbar} \int_0^U \dot{y}^2 \, du \right] \mathscr{D}y. \qquad (3.33)$$

Thus we have split the path integral into a factor dependent upon the endpoints and U, and a path integral dependent only upon U. Writing

$$\iint_{y(0) = y(U) = 0} \exp \left[-\frac{m}{2\hbar} \int_0^U \dot{y}^2 \, du \right] \mathscr{D}y = F(U),$$

we get

$$\rho(x_2, x_1, U) = F(U) \exp \left[-m(x_2 - x_1)^2 / 2\hbar U \right]. \qquad (3.34)$$

We can obtain an equation for $F(U)$ by using

$$\rho(x, y; u_1 + u_2) = \int dx' \, \rho(x, x'; u_2) \rho(x', y; u_1). \qquad (3.35)$$

The reader should derive that equation for $F(u)$ and show that its most general continuous solution is

$$F(U) = \sqrt{m/2\pi\hbar U}\; e^{\alpha U}, \tag{3.36}$$

where α is an arbitrary number. Physically, the value of α is immaterial, because a change in the zero of the potential does not change the physics; it does, however, change the value of α. As a matter of fact, since the expectation value of any quantity, A, is given by

$$\langle A \rangle = \frac{\text{Tr } \rho A}{\text{Tr } \rho}, \tag{3.37}$$

$F(U)$ need not be found at all. It cancels out in the division. We know that α is zero, however, from other methods of calculating ρ (see Eq. (3.8)).

Now consider the path integral for the harmonic oscillator:

$$f(x_2, x_1, U) = \iint \exp\left\{-\frac{1}{\hbar}\int_0^U \left[\frac{m\dot{x}^2}{2} + \frac{m\omega^2}{2}\, x^2\right] du\right\} \mathscr{D}x.$$

Let us expand each path about the path that makes the largest contribution to P. For this path,

$$\int_0^U G(u)\, du = \int_0^U \left[\frac{m\dot{x}^2}{2} + \frac{m\omega^2}{2}\, x^2\right] du = \text{minimum},$$

where $x(0) = x_1$, $x(U) = x_2$. Thus, our integrand must satisfy the Euler equation,

$$\frac{\partial}{\partial u}\frac{\partial G(u)}{\partial \dot{x}} - \frac{\partial G}{\partial x} = 0 \quad \text{or} \quad m\ddot{x} - m\omega^2 x = 0. \tag{3.38}$$

$$x = Ae^{+\omega u} + Be^{-\omega u}.$$

For any path, we write

$$x = x + y$$

so $y(0) = 0$, $y(U) = 0$ and the integral over u becomes

$$\int_0^U \left[\frac{m}{2}(\dot{x} + \dot{y})^2 + \frac{m\omega^2}{2}(x + y)^2\right] du$$

$$= \int_0^U \frac{m}{2}\left[\dot{x}^2 + 2\dot{x}\dot{y} + \dot{y}^2 + \omega^2 x^2 + \omega^2 2xy + \omega^2 y^2\right] du$$

$$= \int_0^U \frac{m}{2}(\dot{x}^2 + \omega^2 x^2)\, du + \frac{m}{2}\int_0^U (2\dot{x}\dot{y} + 2\omega^2 xy)\, du$$

$$+ \frac{m}{2}\int_0^U (\dot{y}^2 + \omega^2 y^2)\, du.$$

Now

$$\int_0^U 2\dot{x}\dot{y}\,du = 2y\dot{x}\bigg|_0^U - \int_0^U 2\ddot{x}y\,du.$$

Because $y(0) = 0$ and $y(U) = 0$, we have

$$\int_0^U 2(\dot{x}\dot{y} + \omega^2 xy)\,du = \int_0^U 2y(-\ddot{x} + \omega^2 x)\,du = 0,$$

since x satisfies the Euler equation. Thus, by expanding about the classical path, we have eliminated the integrand linear in the deviation. Thus we may write

$$f(x_2, x_1; U) = \exp\left[-\frac{1}{\hbar}\int_0^U \left(\frac{m}{2}\dot{x}^2 + \frac{m\omega^2}{2}x^2\right)du\right] F(U),$$

where

$$F(U) = \iint \exp\left[-\frac{1}{\hbar}\int_0^U \left(\frac{m}{2}\dot{y}^2 + \frac{m\omega^2}{2}y^2\right)du\right] \mathscr{D}y(u),$$

with $y(0) = 0$ and $y(U) = 0$, is a factor independent of the endpoints x_1 and x_2. Thus expansion about the classical path allows us to separate the path integral into a factor dependent upon U and the endpoints, and another path integral dependent upon U alone. Thus the important dependence on x_1 and x_2 can be found by simply solving the minimizing differential equations (Eq. 3.38) subject to the end conditions $x(0) = x_1$ and $x(U) = x_2$ and calculating the integral

$$\int_0^U \left(\frac{m}{2}\dot{x}^2 + \frac{m\omega^2}{2}x^2\right) du.$$

In fact even this integral can be simplified. We have

$$\int_0^U \frac{m}{2}\dot{x}^2\,du = \frac{m}{2}x\dot{x}\bigg|_0^U - \frac{m}{2}\int_0^U x\ddot{x}\,du.$$

Therefore

$$\int_0^U \frac{m}{2}(\dot{x}^2 + \omega^2 x^2)\,du = \frac{m}{2}x\dot{x}\bigg|_0^U + \int_0^U \frac{mx}{2}(-\ddot{x} + \omega^2 x)\,du = \frac{m}{2}x\dot{x}\bigg|_0^U.\text{*}$$

As a new problem, suppose we have two interacting systems, one of which is a harmonic oscillator. Let x describe the position of the oscillator and q that of the other system. A possible Hamiltonian is

$$H = \frac{m\dot{x}^2}{2} + \frac{m\omega^2}{2}x^2 + \frac{M\dot{q}^2}{2} + V(q) - \gamma xq.$$

---

* You can compare your result with Eq. (2.83).

The partition function is

$$
Q = \iint \exp\left[-\frac{1}{\hbar} \int_0^U \left(\frac{m\dot{x}^2}{2} + \frac{m\omega^2 x^2}{2} - \gamma q(u)x(u) \right) du \right]
$$
$$
\times \exp\left[-\frac{1}{\hbar} \int_0^U \left(\frac{M\dot{q}^2}{2} + V(q) \right) du \right] \mathscr{D}x\, \mathscr{D}q.
$$

We cannot, in general, evaluate this path integral, but we can integrate out the x variable. To do so we must find

$$
F[q(u)] = \iint \exp\left[-\frac{1}{\hbar} \int_0^U \left(\frac{m\dot{x}^2}{2} + \frac{m\omega^2 x^2}{2} - \gamma q x \right) du \right] \mathscr{D}x.
$$

This integral can be evaluated by the same method as that used for a harmonic oscillator. More generally, we can evaluate

$$
F_1[f; x, x'] = \iint_{\substack{x(0)=x \\ x(U)=x'}} \exp\left[-\frac{1}{\hbar} \int_0^U \left(\frac{m\dot{x}^2}{2} + \frac{m\omega^2 x^2}{2} + if(u)x(u) \right) du \right] \mathscr{D}x.
$$

The result is

$$
F_1(f; x, x') = \sqrt{m\omega/2\pi\hbar \sinh \omega U}\; e^{-\Phi/\hbar}, \tag{3.39}
$$

where

$$
\Phi = \frac{1}{4m\omega} \int_0^U \int_0^U e^{-\omega|u-u'|} f(u)f(u')\, du\, du' + \frac{m\omega}{2 \sinh \omega U}
$$
$$
\times \left[(x^2 + x'^2) \cosh \omega U - 2xx' + 2A(xe^{\omega U} - x') + 2B(x'e^{\omega U} - x) \right.
$$
$$
\left. + (A^2 + B^2)e^{\omega U} - 2AB \right], \tag{3.40}
$$

where

$$
A = \frac{i}{2m\omega} \int_0^U e^{-\omega u} f(u)\, du,
$$

$$
B = \frac{i}{2m\omega} \int_0^U e^{-\omega(U-u)} f(u)\, du. \tag{3.41}
$$

Now let

$$
\mathfrak{E}[f] = \left\langle \exp\left[-\frac{i}{\hbar} \int_0^U f(u)x(u)\, du \right] \right\rangle_{\text{harm. osc.}}
$$
$$
= \frac{\int F_1(f; x, x)\, dx}{\int F_1(0; x, x)\, dx}. \tag{3.42}
$$

Eventually one will arrive at the result:

$$\mathfrak{E}[f] = \exp\left[-\frac{1}{4m\omega h}\int_0^U\int_0^U \frac{\cosh \omega(|u - u'| - \frac{1}{2}U)}{\sinh \frac{1}{2}\omega U}f(u)f(u')\,du\,du'\right].$$

(3.43)

If we define f outside the interval $0 \le u \le U$ by letting f be periodic with period U, we may write

$$\mathfrak{E}[f] = \exp\left[-\frac{1}{4m\omega h}\int_{-\infty}^{\infty}\int_0^U e^{-\omega|u-u'|}f(u)f(u')\,du\,du'\right].$$ (3.43a)

An alternative way of finding $F_1(f; x, x)$ is to Fourier analyze $f(u)$ and $x(u)$:

$$x(u) = a_0 + \sum_{n=1}^{\infty}\left(a_n \cos \frac{2\pi n u}{\beta} + b_n \sin \frac{2\pi n u}{\beta}\right),$$

$$f(u) = f_0 + \sum_{n=1}^{\infty}\left(f_n \cos \frac{2\pi n u}{\beta} + g_n \sin \frac{2\pi n u}{\beta}\right).$$

The path integral

$$F_1[f] = \int F_1(f; x, x)\,dx$$

can then be written as

$$F_1[f] = \iiint \cdots \int da_0\,da_1\cdots db_1\,db_2\cdots$$

$$\times \exp\left[-\frac{1}{h}\int_0^U\left(\frac{m\dot{x}^2}{2} + \frac{m\omega^2}{2}x^2 + if(u)x(u)\right)\right]du,$$

and can easily be evaluated.*

---

* The equation

$$\frac{1}{2a^2} + \sum_{n=1}^{\infty}\frac{\cos n\theta}{n^2 + a^2} = \frac{\pi}{2a}\frac{\cosh (\pi - |\theta|)a}{\sinh \pi a}$$

is useful. $F_1[f]$ as found by the above method has an unknown multiplicative constant as a consequence of the Jacobian for the change of variables implicit in $\mathcal{D}x(u) \rightarrow da_1 \cdots db_1\,db_2\cdots$; but Eq. (3.43) can be found unambiguously by the above method.

For a free particle, we can find $F_1[f]$ either by letting $\omega \to 0$ or by direct calculation. The result is

$$\iint \exp\left[-\frac{1}{\hbar}\int_0^U \left[\tfrac{1}{2}m\dot{x}(u)^2 + if(u)x(u)\right] du\right] \mathscr{D}x(u)$$

$$= \sqrt{\frac{2\pi\hbar m}{U}} \,\delta\left[\int_0^U f(u)\,du\right] \exp\left\{\frac{1}{4m\hbar}\left(\int_0^U \int_0^U |u - u'|f(u)f(u')\,du\,du'\right.\right.$$

$$\left.\left. + \frac{2}{U}\left[\int_0^U uf(u)\,du\right]^2\right)\right\}. \qquad (3.44)$$

We will make use of this last result later.

3.3 PATH INTEGRALS BY PERTURBATION EXPANSION

Suppose that we have found the density matrix element $\rho_0(x_2, x_1, U)$ for some system of potential V_0, and we wish to know $\rho(x_2, x_1, U)$ for another system with potential $V_0 + V'$, where $V' \ll V_0$. Then

$$\rho(x_2, x_1, U) = \iint \exp\left\{-\frac{1}{\hbar}\int_0^U \left[\frac{m\dot{x}^2}{2} + V_0(x) + V'(x)\right] du\right\} \mathscr{D}x\right\}$$

$$= \iint \left[\exp\left\{-\frac{1}{\hbar}\int_0^U \left[\frac{m\dot{x}^2}{2} + V_0(x)\right] du\right\}\right.$$

$$\left. \times \exp\left\{-\frac{1}{\hbar}\int_0^U V'(x)\,du\right\}\right] \mathscr{D}_x$$

Expanding the second exponential, we have

$$= \iint \exp\left\{-\frac{1}{\hbar}\int_0^U \left[\frac{m\dot{x}^2}{2} + V_0(x)\right] du\right\}$$

$$\times \left[1 - \frac{1}{\hbar}\int_0^U V'[x(u)]\,du + \frac{1}{2\hbar^2}\left(\int_0^U V'[x(u)]\,du\right)^2 + \cdots\right] \mathscr{D}x$$

$$= \rho_0 - \frac{1}{\hbar}\iint \int_0^U \exp\left\{-\frac{1}{\hbar}\int_0^u \left[\frac{m\dot{x}^2}{2} + V_0(x)\right] du'\right\}$$

$$\times \exp\left\{-\frac{1}{\hbar}\int_u^U \left[\frac{m\dot{x}^2}{2} + V_0(x)\right] du'\right\} V'[x(u)]\,du\,\mathscr{D}x + \cdots$$

$$= \rho_0 - \frac{1}{\hbar}\int_0^U du \iint \exp\left\{-\frac{1}{\hbar}\int_0^u \left[\frac{m\dot{x}^2}{2} + V_0(x)\right] du'\right\}$$

$$\times V'[x(u)] \exp\left\{-\frac{1}{\hbar}\int_u^U \left[\frac{m\dot{x}^2}{2} + V_0(x)\right] du'\right\} + \cdots$$

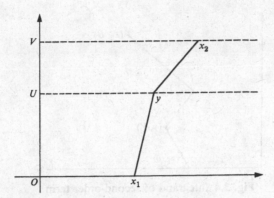

Fig. 3.3 Integrand of first-order term

In order to calculate the path integral, we first integrate over all paths for which $x(u) = y$ (with $x(0) = x_1$, $x(U) = x_2$) and then integrate over y.

$$\rho(x_2, x_1 U) = \rho_0 - \frac{1}{\hbar} \int_0^U \int_{\text{all } y} \rho_0(x_2, y, U - u)V(y)\rho_0(y, x_1, u) \, dy \, du + \cdots$$

Pictorially, the integrand may be represented as shown in Fig. 3.3.

We can think of the zeroth-order term, ρ_0, as being the contribution to the density matrix that results from the particle going from x_1 to x_2 in "time" U without being affected by the potential. Then the first-order term, which has its integrand represented in the Fig. 3.3, may be thought of as the contribution resulting from the particle being essentially free, but being "scattered" once at some "time" u and some position y. Similarly, the second-order term can be considered as the contribution from two "scatterings", and can be written as

$$+\frac{1}{\hbar} \int_0^U du \int_0^u dv \int_{-\infty}^{\infty} dy \int_{-\infty}^{\infty} dz$$
$$\times \rho_0(x_2, z, U - u)V(z)\rho_0(z, y, u - v)V(y)\rho_0(y, x_1, v)$$

Pictorially, this integrand may be represented as shown in Fig. 3.4.

Notice that we have not found any new results. We have already derived this perturbation expansion without the use of path integrals.

Problem: Find to what order the nth-order integral expansion of $\rho(x_2, x_1, U)$ solves

$$-\frac{\partial \rho}{\partial u} = -\frac{\hbar^2}{2m} \frac{\partial^2 \rho}{\partial x^2} + [V_0(x) + V'(x)]\rho,$$

knowing that

$$-\frac{\partial \rho_0}{\partial u} = -\frac{\hbar^2}{2m} \frac{\partial^2 \rho_0}{\partial x^2} + V_0(x)\rho_0.$$

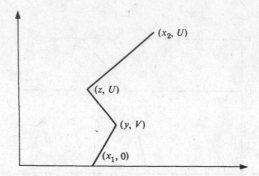

Fig. 3.4 Integrand of second-order term.

3.4 VARIATIONAL PRINCIPLE FOR THE PATH INTEGRAL

For the free energy we know the relation

$$F \le F_0 + \langle H - H_0 \rangle_0, \tag{3.45a}$$

or

$$F \le \langle H \rangle_0 - TS_0. \tag{3.45b}$$

This is written in terms of the Hamiltonians H and H_0. We ask ourselves if we can write a corresponding minimum relation using the path integral.

In the path-integral form, the partition function is written as

$$e^{-\beta F} = \iint_{x(0)=x(U)} e^{-S[x(u)]} \mathscr{D}x(u). \tag{3.46}$$

Suppose we have another "S", S_0, that is easier to work with. (Here S_0 does not mean entropy). Then Eq. (3.46) may be written as

$$e^{-\beta F} = \frac{\iint e^{-(S-S_0)} e^{-S_0} \mathscr{D}x}{\iint e^{-S_0} \mathscr{D}x} e^{-\beta F_0}, \tag{3.47}$$

where

$$e^{-\beta F_0} \equiv \iint e^{-S_0} \mathscr{D}x. \tag{3.48}$$

The first factor of Eq. (3.47) has the form of an average of $\exp\left[-(S - S_0)\right]$ with e^{-S_0} as the weight for a certain path $x(u)$. We may write Eq. (3.47) as

$$e^{-\beta F} = \langle e^{-(S-S_0)} \rangle_{S_0} e^{-\beta F_0}. \tag{3.49}$$

Now we suppose that S and S_0 are real and use the inequality

$$\langle e^{-f} \rangle \ge e^{-\langle f \rangle}. \tag{3.50}$$

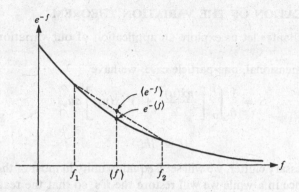

Fig. 3.5 Geometrical interpretation of $\langle e^{-f} \rangle \geq e^{-\langle f \rangle}$.

The geometrical interpretation of this relation is as shown in Fig. 3.5. In Fig. 3.1, $\langle e^{-f} \rangle$ is always above the curve e^{-f}. Note that Eq. (3.50) does not depend on how the f's are distributed. When we apply Eq. (3.50) to Eq. (3.49) we can write

$$e^{-\beta F} \geq e^{-\langle S - S_0 \rangle} \, e^{-\beta F_0}, \tag{3.51}$$

where

$$\langle S - S_0 \rangle \equiv \frac{\iint (S - S_0) e^{-S_0} \, \mathcal{D}x}{\iint e^{-S_0} \, \mathcal{D}x}. \tag{3.52}$$

Usually Eq. (3.52) is simpler to calculate than the first factor in Eq. (3.47). We thus have a theorem:

$$F \leq F_0 + \frac{1}{\beta} \langle S - S_0 \rangle_{S_0}. \tag{3.53}$$

Exercise: When

$$S = \int_0^\beta \left[\frac{m\dot{x}^2(u)}{2} + V(x(u)) \right] du,$$
$$S_0 = \int_0^\beta \left[\frac{m\dot{x}^2(u)}{2} + V_0(x(u)) \right] du, \tag{3.54}$$

show that

$$\frac{1}{\beta} \langle S - S_0 \rangle_0 = \langle V - V_0 \rangle_0. \tag{3.55}$$

Thus Eq. (3.53) contains Eq. (3.45) as a special case. (Hint: from Eq. (3.54)),

$$S - S_0 = \int_0^\beta (V - V_0) \, du. \tag{3.56}$$

The numerator of the right-hand side of Eq. (3.52) becomes an integral over u, the integrand of which is independent of u.

3.5 AN APPLICATION OF THE VARIATION THEOREM

Finally in this chapter let us explore an application of our variation theorem, Eq. (3.53).

In a one-dimensional, one-particle case, we have

$$S = \frac{1}{\hbar} \int_0^U \left[\frac{m\dot{x}(u)^2}{2} + V(x(u)) \right] du, \tag{3.57}$$

where

$$U \equiv \beta\hbar. \tag{3.58}$$

To avoid unnecessary clutter, we will set $\hbar$ equal to unity in most of the following work. Every once in a while we will restore the $\hbar$'s, so that the reader can see where they belong.

The partition function may be written as

$$e^{-\beta F} = \oint \exp\left\{ -\int_0^U \left[\frac{m\dot{x}(u)^2}{2} + V(x(u)) \right] du \right\} \mathscr{D}x(u)\, dx(0). \tag{3.59}$$

Here, we first fix $x(0)$ and $x(U) = x(0)$, and then do the path integral over all paths; then we vary $x(0)$. The zero in $\oint$ indicates that we integrate over $x(0)$.

In the classical limit of high temperature, or small $\hbar$, U is small, so that the path does not deviate much from the initial point $x(0)$. Thus the first approximation in Eq. (3.59) is to replace $V(x(u))$ by the initial value $V(x(0))$ to obtain

$$e^{-\beta F} = \sqrt{\frac{mkT}{2\pi\hbar^2}} \int e^{-V(x)/kT}\, dx. \tag{3.60}$$

This is the well-known classical result.

How can we improve on this approximation to take the quantum effect into account?

First we observe that, because $x(0)$ and $x(U)$ are equal, it appears more natural to use an average

$$x \equiv \frac{1}{U} \int_0^U x(u)\, du \tag{3.61}$$

in place of $x(0)$. Second, the deviation of the path from the classical straight line may be taken into account by some average of $V(x)$ over the path rather than by a constant $V(x(0))$. These considerations lead to the following trial S_0:

$$S_0 = \int_0^U \frac{m\dot{x}(u)^2}{2}\, du + Uw(x). \tag{3.62}$$

Here $w(x)$ is a still undetermined function, which is to be varied later to minimize Eq. (3.53). In the first approximation one would expect to choose $w(x) = V(x)$ but we can do much better with our variational principle.

$$e^{-\beta F_0} = \iint_{\text{all closed paths}} \mathscr{D}x \, \exp\left[-\int_0^U \left[\frac{m\dot{x}^2}{2} + Uw(x)\right] du\right]$$

$$= \int dx \iint_{\substack{\text{all closed paths } x \\ \text{with fixed average } x}} \exp\left[-\int_0^U \left[\frac{m\dot{x}^2}{2} \, du + Uw(x)\right] du\right] \mathscr{D}x.$$

Now let $y = x - x(0)$. Since $\dot{y}(u) = \dot{x}(u)$, we obtain

$$e^{-\beta F_0} = \int dx \, e^{-Uw(x)} \iint_{\substack{\text{all closed paths} \\ y \text{ with } y(0) = 0}} \exp\left[-\int_0^U \frac{m\dot{y}^2}{2}\right] \mathscr{D}y.$$

In the path integral above, there is no restriction on y because $y = x - x(0)$, and we integrate over $x(0)$. The reasons for integrating over the paths chosen can be made clearer by drawing diagrams; the reader should do this if he needs clarification.

$$e^{-\beta F_0} = \int dx \, e^{-Uw(x)} \rho_{\text{free}}(0, 0; U) = \sqrt{\frac{m}{2\pi\hbar U}} \int_{-\infty}^{\infty} e^{-Uw(y)} \, dy. \tag{3.63}$$

It should be noted here that in the classical case Eq. (3.60), $w(y)$ is replaced by $V(y)$.

Next, we calculate $\langle S - S_0 \rangle$. Using Eqs. (3.57) and (3.62), we see by definition

$$\frac{1}{\beta} \langle S - S_0 \rangle$$

$$= \frac{\iint 1/\beta(S - S_0) \exp\left[-\int m\dot{x}^2/2 \, du\right] \exp\left[-Uw(x)\right] \mathscr{D}x}{\iint \exp\left[-\int m\dot{x}^2/2 \, du\right] \exp\left[-Uw(x)\right] \mathscr{D}x}$$

$$= \frac{\iint \left\{1/U \int_0^U V(x(u')) \, du' - w(x)\right\} \exp\left[-\int m\dot{x}^2/2 \, du\right] \exp\left[-Uw(x)\right] \mathscr{D}x}{\iint \exp\left[-\int m\dot{x}^2/2 \, du\right] \exp\left[-Uw(x)\right] \mathscr{D}x}.$$

$$\tag{3.64}$$

Here x is defined in Eq. (3.61) and is functional of $x(u)$. The first term in the numerator is simplified as follows:

$$\iint \frac{1}{U} \int_0^U V(x(u')) \, du' \exp\left[-\int_0^U \frac{m\dot{x}(u)^2}{2} \, du\right] \exp\left[-Uw(x)\right] \mathscr{D}x$$

$$= \int_0^U \frac{du'}{U} \iint V(x(u')) \exp\left[-\int_0^U \frac{m\dot{x}(u)^2}{2} \, du\right] \exp\left[-Uw(x)\right] \mathscr{D}x$$

$$= \int_0^U \frac{du'}{U} \int dx' \, V(x') \iint_{\substack{x(0) = x(U) \\ x(u') = x'}} \exp\left[-\int_{u'}^U \frac{m\dot{x}^2(u)}{2} \, du \int_0^{u'} \frac{m\dot{x}^2(u)}{2} \, du\right]$$

$$\times \exp\left[-Uw(x)\right] \mathscr{D}x(u).$$

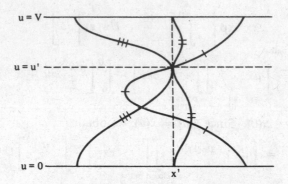

Fig. 3.6a Paths of integration.

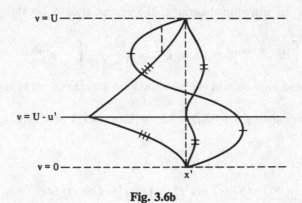

Fig. 3.6b

We are integrating over paths like those in Fig. 3.6a. By making the change of variables

$$V = \begin{cases} u + U - u' & 0 \le u \le u' \\ u - u' & u' \le u \le U \end{cases}$$

we integrate over paths like those in Fig. 3.6b. Because of the condition that $x(u = 0) = x(u = U)$, we do not get a discontinuity in x at $V = U - u'$; therefore

$$\int_0^U \frac{m\dot{x}^2(u)}{2}\, du = \int_0^U \frac{m\dot{x}^2(V)}{2}\, dV.$$

Also,

$$x = \frac{1}{U} \int_0^U x(u)\, du = \frac{1}{U} \int_0^U x(V)\, dV.$$

The first term in the numerator of Eq. (3.64) becomes

$$\int_0^U \frac{du'}{U} \int dx' V(x') \iint\limits_{x(0)=x(U)=x'} \exp\left[-\int_0^U \frac{m\dot{x}^2(V)}{2} dV\right] \exp\left[-Uw(x)\right]$$

$$= \iint\limits_{x(0)=x(U)} V(x(0)) \exp\left[-\int_0^U \frac{m\dot{x}^2(u)}{2} du\right] \exp\left[-Uw(x)\right] \mathscr{D}x(u). \quad (3.65)$$

Thus Eq. (3.64) may be written as

$$\frac{1}{\beta} \langle S - S_0 \rangle_{S_0} = \langle V(x(0)) \rangle - \langle w(x) \rangle. \quad (3.66)$$

To transform further we use the following result (proved previously):

$$\iint \exp\left[-\int_0^U \frac{m\dot{x}(u)^2}{2} du\right] \exp\left[-i\int_0^U f(u)x(u) du\right] \mathscr{D}x(u)$$

$$= \sqrt{\frac{2\pi m}{U}} \delta\left(\int_0^U f(u) du\right) \exp\left\{\frac{1}{4m}\left[\int_0^U \int_0^U |u - u'| f(u)f(u') du du'\right.\right.$$

$$\left.\left. + \frac{2}{U}\left(\int_0^U uf(u) du\right)^2\right]\right\} \quad (3.67)$$

In evaluating Eq. (3.65) we first do the path integral keeping x fixed; then Eq. (3.65) is written

$$\iint V(x(0)) \exp\left[-\int \frac{m\dot{x}(u)^2}{2} du\right] \exp\left[-Uw(x)\right] \mathscr{D}x$$

$$= \int K(y) \exp\left[-Uw(y)\right] dy \quad (3.68)$$

where

$$K(y) = \iint V(x(0)) \exp\left[-\int_0^U \frac{m\dot{x}(u)^2}{2} du\right] \mathscr{D}x \quad (3.69)$$

such that $x = y$. Equation (3.69) may be written

$$K(y) = \iint V(x(0)) \exp\left[-\int_0^U \frac{m\dot{x}(u)^2}{2} du\right] \delta(x - y) \mathscr{D}x(u). \quad (3.70)$$

For this we use the Fourier transform:

$$V(x) \equiv \int v(q) e^{iqx} dq \quad (3.71)$$

to write Eq. (3.70) as

$$K(y) = \int v(q)\,dq \int dk \iint \exp\left[iqx(0)\right] \exp\left[ik(x-y)\right]$$

$$\times \exp\left[-\int_0^U \frac{m\dot{x}^2}{2}\,du\right] \mathcal{D}x. \tag{3.72}$$

Here $\delta(x - y)$ of Eq. (3.70) was also Fourier transformed, and we are ignoring factors such as 2π. Use Eq. (3.61) in Eq. (3.72) to write

$$K(y) = \int v(q)\,dq \int dk \exp\left[-iky\right] \iint \exp\left[iqx(0)\right] \exp\left[i\frac{k}{U}\int_0^U x(u)\,du\right]$$

$$\times \exp\left[-\int_0^U \frac{m\dot{x}^2}{2}\,du\right] \mathcal{D}x. \tag{3.73}$$

Now we see that the path-integral part of this expression can be brought into the form of Eq. (3.67) if we define

$$f(u) \equiv q\delta(u - 0) + \frac{k}{U} \tag{3.74}$$

because

$$\int_0^U f(u)x(u)\,du = qx(0) + \frac{k}{U}\int_0^U x(u)\,du. \tag{3.75}$$

When we use Eq. (3.74) in Eq. (3.67) we see that

$$\int_0^U f(u)\,du = q + k, \tag{3.76}$$

$$\int_0^U uf(u)\,du = \frac{kU}{2}, \tag{3.77}$$

$$\int_0^U \int_0^U |u - u'|f(u)f(u')\,du\,du' = 2\int_0^U du \int_0^u (u - u')f(u')\,du'f(u)$$

$$= 2\int_0^U du \left[\int_0^u (u - u')\frac{k}{U}\,du' + uq\right]f(u)$$

$$= 2\int_0^U du \left[\frac{ku^2}{2U} + uq\right]f(u) = \frac{k^2U}{3} + kqU. \tag{3.78}$$

Use Eqs. (3.76), (3.77), and (3.78) in Eq. (3.67) and also in the path-integral part of Eq. (3.73) to write the latter as

$$K(y) = \int dq v(q) \int dk \exp\left[-iky\right]\delta(q + k) \exp\left[\frac{U}{4m}\left(\frac{k^2}{3} + kq + \frac{k^2}{2}\right)\right]$$

$$= \int dq v(q) \int dk \exp\left[-iky\right]\delta(q + k) \exp\left[-\frac{k^2 U}{24m}\right]$$

$$= \int dk v(-k) \exp\left[-iky\right] \exp\left[-\frac{k^2 U}{24m}\right]. \qquad (3.79)$$

Using the inverse Fourier transform, we find that Eq. (3.79) becomes

$$K(y) = \int dk \exp\left[-iky\right] \int V(z) \exp\left[ikz\right] dz \exp\left[-\frac{k^2 U}{24m}\right]$$

$$= \int dz V(z) \exp\left[-\frac{6m}{U}(y - z)^2\right]. \qquad (3.80)$$

When the correct factors are taken into account, the final form is

$$K(y) = \sqrt{\frac{6mkT}{\pi\hbar^2}} \int dz V(z) \exp\left[-\frac{6mkT(y - z)^2}{\hbar^2}\right]. \qquad (3.81)$$

It should be noticed that $K(y)$ is $V(z)$ averaged over a Gaussian. The root mean square of the Gaussian spread is

$$\frac{\hbar}{\sqrt{12\, mkT}}. \qquad (3.82)$$

This spread is about 1 Å for a helium atom at 2°K, and is a very narrow spread at room temperatures. At the limit of infinite temperature, the Gaussian becomes a delta function, and $K(y) \to V(y)$.

Summarizing the results obtained so far, we have

$$F \leq F_0 + \frac{1}{\beta}\langle S - S_0\rangle_{S_0}, \qquad (3.83a)$$

$$e^{-\beta F_0} = \int \exp\left[-\beta w(y)\right] dy \sqrt{\frac{m}{2\pi\beta\hbar^2}}, \qquad (3.83b)$$

$$\frac{1}{\beta}\langle S - S_0\rangle_{S_0} = \frac{\int (K(y) - w(y))e^{-\beta w(y)}\, dy}{\int e^{-\beta w(y)}\, dy}. \qquad (3.83c)$$

Equation (3.83b) is Eq. (3.63). Equation (3.83c) follows from Eqs. (3.64), (3.65), and (3.68). Further, $K(y)$ in Eq. (3.83c) is given in Eq. (3.81).

Now we ask what is the best choice of $w(y)$? Changing

$$w(y) \rightarrow w(y) + \eta(y) \tag{3.84}$$

where $\eta(y)$ is small, we see from Eqs. (3.83b) and (c)

$$\delta F_0 = \frac{\int \eta e^{-\beta w} \, dy}{\int e^{-\beta w} \, dy} \tag{3.85a}$$

$$\delta \frac{1}{\beta} \langle S - S_0 \rangle = \frac{\int e^{-\beta w}[-\beta \eta(K - w) - \eta \, dy]}{\int e^{-\beta w} \, dy} + \frac{\int e^{-\beta w}(K - w) \, dy \int \beta \eta e^{-\beta w} \, dy}{(\int e^{-\beta w} \, dy)^2}. \tag{3.85b}$$

Thus

$$\delta \left(F_0 + \frac{1}{\beta} \langle S - S_0 \rangle_{S_0} \right) = 0 \tag{3.86}$$

leads to

$$w(y) = K(y). \tag{3.87}$$

This is the best choice of $w(y)$.

In this case Eq. (3.83c) yields

$$\langle S - S_0 \rangle_{S_0} = 0, \tag{3.88}$$

so that, from Eq. (3.83a),

$$F \leq F_{cl\,K} \tag{3.89}$$

where $F_{cl\,K}$ means the classical free energy with the potential $V(y)$ replaced by $K(y)$. Alternatively,

$$e^{-\beta F_{cl\,K}} = \sqrt{\frac{m}{2\pi \hbar U}} \int e^{-\beta K(y)} \, dy \tag{3.90}$$

compare Eq. (3.60). $K(y)$ is defined in Eq. (3.81). $F_{cl\,K}$ is a better approximation than the ordinary classical form F_{cl} in Eq. (3.60).

Example: To find out how good $F_{cl\,K}$ is, let us consider the harmonic oscillator. In this case the potential is

$$V(x) = m \frac{\omega^2 x^2}{2}. \tag{3.91}$$

From Eq. (3.81)

$$K(y) = \sqrt{\frac{6mkT}{\pi \hbar^2}} \frac{m\omega^2}{2} \int z^2 \exp \left[-\frac{6mkT}{\hbar^2} (y - z)^2 \right] dz = \frac{m\omega^2}{2} \left(y^2 + \frac{\hbar^2 \beta}{12m} \right). \tag{3.92}$$

Using this result in Eq. (3.90), we obtain

$$e^{-\beta F_{cl\,K}} = \sqrt{\frac{m}{2\pi\hbar^2\beta}} \int \exp\left[-\beta\frac{m\omega^2}{2}\left(y^2 + \frac{\hbar^2\beta}{12m}\right)\right] dy$$

$$= (1/\hbar\beta\omega)e^{-\hbar^2\beta^2\omega^2/24}. \tag{3.93}$$

Thus, we have the following results:

$$F_{\text{true}} = \frac{1}{\beta}\ln\left[2\sinh\left(\frac{\hbar\omega\beta}{2}\right)\right], \tag{3.94a}$$

$$F_{cl\,K} = \frac{1}{\beta}\left[\ln(\hbar\omega\beta) + \frac{\hbar^2\beta^2\omega^2}{24}\right], \tag{3.94b}$$

$$F_{cl} = \frac{1}{\beta}\ln(\hbar\omega\beta). \tag{3.94c}$$

When we write

$$G \equiv \frac{2F}{\hbar\omega} \quad \text{and} \quad f \equiv \frac{\hbar\omega\beta}{2}. \tag{3.95}$$

Eq. (3.94) becomes

$$G_{\text{true}} = \frac{1}{f}\ln(2\sinh f), \tag{3.95a}$$

$$G_{cl\,K} = \frac{1}{f}\ln(2f) + \frac{f^2}{6}, \tag{3.95b}$$

$$G_{cl} = \frac{1}{f}\ln(2f). \tag{3.95c}$$

Table 3.1 is a numerical comparison of G_{true}, $G_{cl\,K}$ and G_{cl}. This table shows the remarkable improvement $G_{Cl\,K}$ has over G_{cl}. Remember that $f = \frac{1}{2}$ corresponds to the temperature $kT = \hbar\omega$ where the quantum effect is large.

Table 3.1

	$f = 1/2$	$f = 1$	$f = 2$
G_{true}	0.08263	0.8544	0.9908
$G_{cl\,K}$	0.08333	0.8598	1.0264
G_{cl}	0	0.6931	0.6931

The expression $F_{cl\ K}$ in Eq. (3.90) combined with Eq. (3.81) is better than the series expansion of $K(y)$ in terms of $V(y)$, d^2V/dy^2, ...

Nevertheless, $F_{cl\ K}$ is not as useful as Table 3.1 might indicate. First, it cannot be used when quantum-mechanical exchange effects exist. Second, it fails in its present form when the potential $V(y)$ has a very large derivative as in the case of hard-sphere interatomic potential.

CHAPTER 4

CLASSICAL SYSTEM OF N PARTICLES

4.1 INTRODUCTION

The classical partition function of N interacting particles is written as

$$e^{-\beta F} = \frac{1}{N!} \left(\frac{m}{2\pi\hbar^2\beta} \right)^{3N/2} Z_N, \tag{4.1}$$

where

$$Z_N = \int e^{-\beta V(R_1 R_2 \cdots R_N)} \, d^3R_1 \, d^3R_2 \cdots d^3R_N. \tag{4.2}$$

These equations can be derived from the quantum-mechanical results we have obtained previously. From Chapter 2, Section 8, we have

$$e^{-\beta F_s} = \frac{1}{N!} \sum_P \int \rho_D(X_1, \ldots, X_N; PX_1, \ldots, PX_N) \, dX_1 \cdots dX_N$$

for Bose statistics. For Fermi-Dirac statistics,

$$e^{-\beta F_A} = \frac{1}{N!} \sum_P (-1)^P \rho_D(X_1, \ldots, X_N; PX_1, \ldots, PX_N) \, dX_1 \cdots dX_N.$$

In either case, only the identity permutation is important, because at sufficiently high temperatures, factors such as $e^{-mkT/2\hbar^2(X-PX)^2}$ kill the other terms. It follows that:

$$e^{-\beta F} \approx \frac{1}{N!} \int \rho_D(X_1, \ldots, X_N; X_1, \ldots, X_N) \, dX_1 \cdots dX_N.$$

We have already estimated this integral for the case $N = 1$ by the use of path integrals (see Chapter 3, Eq. (3.60)). Generalizing the method used to the case of N particles in three dimensions, we arrive at Eqs. (4.1) and (4.2).

When a gas particle is a polyatomic molecule, the internal motion of a molecule and the motion of the center of gravity can be separated; then the specific heat is a sum of the two contributions. In Eq. (4.2), R_i is regarded as the center of gravity of the ith particle; the partition function for the internal motion is not included here. The latter can be calculated using information about the energy levels obtained from the infrared spectra, or calculated using quantum mechanics.

For a dense system such as a liquid the internal motion of polyatomic molecules and the motion of center of gravity are entangled with each other, and the two are hard to separate.

The following discussion excludes the internal motion of a particle, and the system is regarded as a gas or a liquid of inert gas particles.

The free energy is written from Eq. (4.1) as

$$F = \frac{3N}{2\beta} \ln \beta + \frac{3N}{2\beta} \ln \frac{2\pi h^2}{m} - \frac{1}{\beta} \ln Z_N + \frac{N}{\beta} \ln \frac{N}{e}, \tag{4.3}$$

The e in the last term of Eq. (4.3) is the base of natural logarithms, rather than the electronic charge. This e appears because we used the Stirling formula:

$$N! \approx \left(\frac{N}{e}\right)^N \sqrt{2\pi N} \Rightarrow \ln N! \approx N \left(\ln \frac{N}{e}\right). \tag{4.4}$$

From Eq. (4.3), we have for the internal energy U

$$U = -T^2 \frac{\partial (F/T)}{\partial T} = \tfrac{3}{2}RT + \frac{RT^2}{NZ_N} \frac{\partial Z_N}{\partial T}, \tag{4.5}$$

where

$$R = Nk. \tag{4.6}$$

The equation of state is found to be

$$\frac{PV}{RT} = \frac{V}{NZ_N} \cdot \frac{\partial Z_N}{\partial V}, \tag{4.7}$$

from Eq. (1.6).

Example: For an ideal gas, $V(R_1, \ldots, R_N) = 0$; therefore $Z_N = V^N$ and Eq. (4.7) becomes

$$PV = RT. \tag{4.8}$$

When we assume that the potential energy can be written as a sum of pairwise potentials:

$$V(R_1 R_2 \cdots R_N) = \sum_{\substack{ij \\ \text{pair}}} V(r_{ij}), \tag{4.9}$$

Eq. (4.2) can be written

$$Z_N = \int \exp\left[-\beta \sum_{\text{pair}} V(r_{ij})\right] d^3R_1 \, d^3R_2 \cdots d^3R_N. \tag{4.10}$$

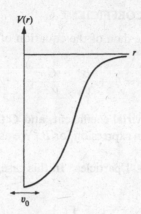

Fig. 4.1 An unrealistic potential; v_0 is the volume of a cluster of N particles.

Note that this expression contains all the information about the phase transitions of the system. In other words, the nature of the phase transitions can, in principle, be discussed by examining the behavior of the integral in Eq. (4.10) purely mathematically, without physical knowledge.

Equation (4.10) suggests that when the number of particles N is finite there is no discontinuity in physical qualities. Only when $N \to \infty$ do we expect discontinuities, and this is the case we are interested in.

The assumption of a central force potential $V(r_{ij})$ in Eq. (4.9) is not exactly justified. The necessity of a noncentral force is illustrated by the following observation: In the solid state, if we assume the nearest-neighbor interaction only, the face-centered cubic and the hexagonal close-packed structures are not distinguishable. When the second nearest neighbors are taken into account the central-force assumption leads to the conclusion that hexagonal close packing has the lower energy, whereas face-centered cubic is actually more stable for solid argon for example, which is inert in its gaseous phase.

The free energy F of Eq. (4.3) is not necessarily a function of V/N only. If we take an unrealistic potential like the one in Fig. 4.1, the energy of the system is roughly of the form CN^2, where C is a constant and N is the number of particles forming a cluster of volume v_0. Thus, we can think of two extreme forms of Z_N:

$$Z_N^{(1)} = e^{-\beta N^2 C} v_0^N,$$
$$Z_N^{(2)} = V^N. \tag{4.11}$$

When the temperature is low, for large N, the contribution of $Z_N^{(1)}$ can be larger than $Z_N^{(2)}$; then F is not a function of V/N only.

4.2 THE SECOND VIRIAL COEFFICIENT

When the gas is not dense, the data of the equation of state are summarized in practice as

$$\frac{PV}{RT} = 1 + \frac{B}{V} + \frac{C}{V^2} + \cdots \tag{4.12}$$

where B is called the second virial coefficient, and C the third virial coefficient. In this section we will derive an expression for B. We assume a pairwise potential as in Section 4.1.

Consider a system of $N + 1$ particles. In this case, Eq. (4.10) of Section 4.1 becomes

$$Z_{N+1} = \int \exp\left[-\beta \sum_i V(X - R_i)\right] \exp\left[-\beta \sum_{\text{pair}} V(r_{ij})\right] d^3X \, d^{3N}R. \tag{4.13}$$

When we integrate over X first, all the R_i's are fixed. If the gas is not dense, the volume for which $V(X - R_i)$ is appreciably different from zero is small; so we may approximate

$$Z_{N+1} \cong VZ_N. \tag{4.14}$$

Equation (4.14) will lead to the ideal-gas equation of state. To obtain a better estimate of Z_{N+1}, we write Eq. (4.13) as

$$Z_{N+1} = \frac{\int \exp\left[-\beta \sum_i V(X - R_i)\right] \exp\left[-\beta \sum_{\text{pair}} V(r_{ij})\right] d^3X \, d^{3N}R}{\int \exp\left[-\beta \sum_{\text{pair}} V(r_{ij})\right] d^{3N}R}$$

$$\times \int \exp\left[-\beta \sum_{\text{pair}} V(r_{ij})\right] d^{3N}R. \tag{4.15}$$

The ratio part of this equation has the form of the weighted average of

$$\int \exp\left[-\beta \sum V(X - R_i)\right] d^3X.$$

We will evaluate this average by neglecting the three-body and higher-order collisions. When X is close to R_i, this pair is assumed to be far away from any of the rest of the R_j. This assumption is equivalent to replacing the weight factor

$$\exp\left[-\beta \sum_{\text{pair}} V(r_{ij})\right]$$

by unity (see Fig. 4.2). When X approaches R_i in Fig. 4.2(b), we have a three-body collision among X, R_i, and R_j. Therefore, we neglect the configuration (b).

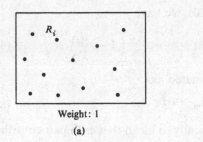

Weight: 1

(a)

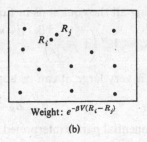

Weight: $e^{-\beta V(R_i - R_j)}$

(b)

Fig. 4.2 Weighting factors in the derivation of the ideal gas equation of state.

Thus, when we neglect the three-body and higher-order collisions, we can approximate Eq. (4.15) as

$$Z_{N+1} \cong \frac{\int \exp\left[-\beta \sum_i V(X - R_i)\right] d^3 X \, d^{3N} R}{\int d^{3N} R} Z_N \qquad (4.16a)$$

$$= \frac{\int d^3 X \left[\int \exp\left[-\beta V(X - R)\right] d^3 R\right]^N}{V^N} Z_N. \qquad (4.16b)$$

Now, we make the transformation:

$$\int e^{-\beta V(X-R)} d^3 R = V - \int \left[1 - e^{-\beta V(X-R)}\right] d^3 R. \qquad (4.17)$$

When we write the second term as a,

$$a \equiv \int_0^\infty \left[1 - e^{-\beta V(r)}\right] 4\pi r^2 \, dr. \qquad (4.18)$$

Using Eqs. (4.17) and (4.18), we write Eq. (4.16b) as

$$Z_{N+1} = \frac{V(V - a)^N}{V^N} Z_N. \qquad (4.19)$$

This is a recurrence relation for Z_N. We can solve for Z_N from Eq. (4.19) as follows:

$$Z_N = V\left(1 - \frac{a}{V}\right)^{N-1} Z_{N-1},$$

$$Z_{N-1} = V\left(1 - \frac{a}{V}\right)^{N-2} Z_{N-2}, \qquad (4.20)$$

. . . .

$$Z_1 = V.$$

Multiplying all the equations in Eq. (4.20), we have

$$Z_N = V^N \left(1 - \frac{a}{V}\right)^{(N-1)+(N-2)+\cdots+1} = V^N \left(1 - \frac{a}{V}\right)^{N(N-1)/2}. \quad (4.21)$$

When N is very large, it can be approximated as

$$Z_N \cong V^N e^{-N^2 a/2V}. \quad (4.22)$$

This exponential part is interpreted physically to mean that each pair contributes a factor $e^{-(a/V)}$.

The equation of state is derived from Eqs. (4.22), and (4.7):

$$\frac{PV}{RT} = \frac{V}{N} \frac{\partial \ln Z_N}{\partial V} = \frac{V}{N} \frac{\partial}{\partial V} \left(N \ln V - \frac{N^2 a}{2V}\right), \quad (4.23)$$

or

$$PV/RT = 1 + Na/2V \quad (4.24)$$

Comparing this with Eq. (4.12), we obtain the expression for the second virial coefficient:

$$B = Na/2. \quad (4.25)$$

This is an exact result, in the sense that the three-body and higher-order collisions that we have neglected have effect only for the third virial coefficient and higher-order terms of Eq. (4.12).

Example: Let us consider a potential as shown in Fig. 4.3. Then Eq. (4.18) is evaluated as

$$a = \int_0^b (1 - 0)4\pi r^2 \, dr + \int_b^\infty [1 - e^{\beta\varphi(r)}]4\pi r^2 \, dr. \quad (4.27)$$

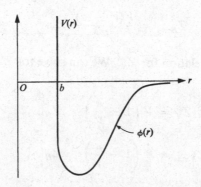

Fig. 4.3 A hypothetical potential.

When the temperature is high we may approximate Eq. (4.27) as

$$a = \frac{4\pi b^3}{3} - \frac{1}{kT} \int_b^\infty \varphi(r) 4\pi r^2 \, dr, \tag{4.28}$$

which we may write in the abbreviated form:

$$a = v_b - \frac{D}{T}. \tag{4.29}$$

When we use this in Eq. (4.24),

$$\frac{PV}{RT} = 1 + \frac{Nv_b}{2V} - \frac{ND}{2VT}. \tag{4.30}$$

We can find out if this expression fits experimental data well by comparing it with the Van der Waals equation of state, which is a good fit to the data. The Van der Waals equation:

$$(P + a_w/V^2)(V - b_w) = RT, \tag{4.31}$$

can be expanded as

$$\frac{PV}{RT} = 1 + \frac{b_w - a_w/RT}{V} + \frac{b_w^2}{V^2} + \cdots . \tag{4.32}$$

If we set $NV_b/2 = b_w$ and $ND/2 = a_w/R$, we arrive at Eq. (4.30) up to the second virial coefficient. If the Van der Waals equation is valid, it gives a second virial coefficient of

$$B_w = b_w - a_w/RT. \tag{4.33}$$

A special feature of this expression is that, because a_w is positive, B_w becomes negative for low temperatures. Experimental data for helium are drawn in Fig. 4.4. The slight decrease for high temperatures does not follow

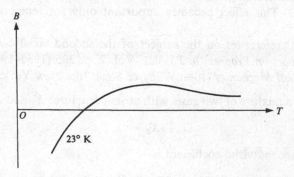

Fig. 4.4 Experimental data for helium gas.

from Eqs. (4.33) or (4.30). It must be remembered that Eq. (4.30) is based on the hard-core assumption of Fig. 4.3. In actuality, the potential $V(r)$ in Eq. (4.18), which is accurate, is slightly soft inside $r = b$, so that for high temperatures, when particles are hitting each other hard, they can come slightly coser than b. Thus there is a decrease in the "hard-core" radius, and hence a decrease in B.

Numerical results: When we use the Slater potential

$$V(r) = e^{-c_i r} - c_2/r^6, \tag{4.34}$$

or the Lennard–Jones potential

$$V(r) = A'/r^n - B'/r^6, \tag{4.35}$$

in Eq. (4.18) we can calculate a. Incidentally, in Eqs. (4.34) and (4.35), the r^{-6} term comes from the Van der Waals attraction.

Table 4.1 lists the numerical results for helium gas (in units of cc/mol).

Table 4.1

$T°K$	*B_{Calc} (CL.)	B_{obs}	†B_{Calc} (Q.M.)
350	10.80	11.60	10.82
250	11.34	11.95	11.16
100	10.75	10.95	
35		4.80	4.44
20	—6.95	−4.00	−5.14
15	−20.0	−14.0	−15.1

* The Slater potential is used. J. G. Kirkwood, *Phys. Rev.* **37**, 832 (1931).

† J. G. Kirkwood, *Phys. Zs.* **33**, 39 (1932).

The quantum-exchange effect on the second virial coefficient was neglected in this section. This effect becomes important only for temperatures below about 1°K.

The general references on the subject of the second virial coefficient are: J. de Boer, *Report on Progress in Physics*, Vol. 2, p. 305 (1948–1949). Mayer-Mayer, *Statistical Mechanics* (John Wiley & Sons, Inc. New York, 1940).

Problem: For a mixture of two gases with atomic fractions X_1 and X_2, such that

$$X_1 + X_2 = 1,$$

show that the second virial coefficient is

$$B = B_{11}X_1^2 + 2B_{12}X_1X_2 + B_{22}X_2^2.$$

4.3 MAYER CLUSTER EXPANSION

From Eq. (4.2), we know that the classical partition function for an N-particle interacting gas is

$$e^{-\beta F} = \frac{1}{N!} \left(\frac{mkT}{2\pi\hbar^2}\right)^{3N/2} Z_N \tag{4.36}$$

$$Z_N = \int e^{-\beta V(R_1, R_2, \ldots, R_N)} \, d^3R_1 \, d^3R_2 \cdots d^3R_N. \tag{4.37}$$

The equation of state can be written, as in Eq. (4.12), as

$$\frac{PV}{RT} = 1 + \frac{B}{V} + \frac{C}{V^2} + \cdots. \tag{4.38}$$

This is the virial expansion; as before, B is the second virial coefficient, C the third, and so on.

This formal expansion can, in principle, give all the virial coefficients. It was originally hoped that this expansion would yield the triple point, and related phenomena. This was not to be. For reasons to be discussed later, the formal program, which we now review, has not proved practical.

In the following work, it is implicitly recognized that the lth virial coefficient is due to l-fold clusters.

$V(R_1, R_2, \ldots, R_N)$ can be written (we assume) as $\sum V(r_{ij})$.

$$Z_N = \int \exp\left[-\beta \sum V(r_{ij})\right] d^N R. \tag{4.39}$$

Let

$$W_N(R_1, R_2, \ldots R_N) = \exp\left[-\beta \sum V(r_{ij})\right]. \tag{4.40}$$

If there is one particle, then

$$W_1(R_1) = e^{-\beta(0)} = 1. \tag{4.41}$$

For two particles,

$$W_2(R_1, R_2) = e^{-\beta V(r_{12})}$$

$$W_3(R_1, R_2, R_3) = \exp\left\{-\beta[V(r_{12}) + V(r_{23}) + V(r_{13})]\right\}, \text{ etc.} \tag{4.42}$$

Let

$$W_1(R_1) = U_1(R_1). \tag{4.43}$$

If there are N points characterized by $W_N(R_1, \ldots, R_N)$ and if points 1 to M are far from points $M + 1$ to N, $W_N(R_1, \ldots, R_N) \approx W_M(R_1, \ldots, R_M) \times W_{N-M}(R_{M+1}, \ldots, R_N)$. For example, if $N = 2$ and R_1 is far from R_2,

$W_2(R_1, R_2) \approx U_1(R_1)U_1(R_2)$. If the points are close to each other, this approximate equation must be corrected by the addition of a quantity, called $U_2(R_1, R_2)$, that is large only when R_1 and R_2 are close.

Thus $W_2(R_1, R_2)$ can be written as

$$W_2(R_1, R_2) = U_1(R_1)U_1(R_2) + U_2(R_1, R_2). \qquad (4.44)$$

Similarly, $W_3(R_1, R_2, R_3)$ can be written as

$$\begin{aligned}
W_3(R_1, R_2, R_3) = {} & U(R_1)U(R_2)U(R_3) \\
& + [U_2(R_1, R_2)U_1(R_3) + U_2(R_2, R_3)U_1(R_1) \\
& \qquad\qquad + U_2(R_1, R_3)U_1(R_2)] \\
& + U_3(R_1, R_2, R_3). \qquad (4.45)
\end{aligned}$$

$U_3(R_1, R_2, R_3)$ is the contribution due to all three particles close to one another; $U_3(R_1, R_2, R_3) \approx 0$ unless the particles are "clustered."

$$W_N(R_1, R_2, \ldots, R_N) = U(R_1) \ldots U(R_N) + \cdots + U_N(R_1, R_2, \ldots, R_N). \quad (4.46)$$

There is another way of expressing the above equations. If we set $U_2(R_i, R_j) = f_{ij}$, then $e^{-\beta V(r_{ij})} = 1 + f_{ij}$ and, from Eq. (4.40),

$$W_N(R_1, R_2, \ldots, R_N) = \prod_{i<j} (1 + f_{ij}). \qquad (4.47)$$

For example, if $N = 3$,

$$\begin{aligned}
W_3(R_1, R_2, R_3) = {} & 1 + f_{12} + f_{13} + f_{23} + f_{12}f_{23} + f_{13}f_{23} \\
& + f_{12}f_{13} + f_{12}f_{23}f_{13}.
\end{aligned}$$

We can express each individual term in this sum as a graph consisting of points and lines. If $N = 3$, then

$$1 \leftrightarrow \underset{1 \quad 2}{\overset{3}{\bullet \quad \bullet}} \qquad f_{12} \leftrightarrow \underset{1 \quad 2}{\overset{3}{\bullet\!\!-\!\!\bullet}} \qquad f_{12}f_{23} \leftrightarrow \underset{1 \quad 2}{\overset{3}{\diagup\!\!\diagdown}} \qquad f_{12}f_{23}f_{13} \leftrightarrow \underset{1 \quad 2}{\triangle}.$$

If G represents a graph, and $W(G)$ is the integral over all space of its corresponding function, then

$$Z_N = \sum_G W(G), \qquad (4.48)$$

where the sum is taken over all possible graphs with N points. For disconnected graphs, such as

$$\underset{1 \quad 2}{\overset{\overset{\textstyle 4}{\diagup\diagdown}}{5\!\!-\!\!\bullet\!\!-\!\!\bullet3}}$$

$W(G)$ is a product of its disconnected parts. A cluster of N points is represented by a sum of all possible graphs in which those points are connected. For example:

$$U_3(R_1, R_2, R_3) = \overset{3}{\underset{1\quad 2}{\diagdown}} + \overset{3}{\underset{1\quad 2}{\wedge}} + \overset{3}{\underset{1\quad 2}{\diagup}} + \overset{3}{\underset{1\quad 2}{\triangle}}.$$

The U_N can also be expressed in terms of the W's. From Eq. (4.43) $U_1(R_1) = W_1(R_1)$. From Eq. (4.43) and (4.44),

$$U_2(R_1, R_2) = W_2(R_1, R_2) - W_1(R_1)W_1(R_2). \tag{4.49}$$

From Eqs. (4.45) and (4.49),

$$U_3(R_1, R_2, R_3) = W_3(R_1, R_2, R_3) - \sum W_2(R_1, R_2)W_1(R_3)$$
$$+ 2W_1(R_1)W_2(R_2)W_3(R_3), \tag{4.50}$$

where

$$\sum W_2(R_1, R_2)W_1(R_3) \equiv W_2(R_1, R_2)W_1(R_3) + W_2(R_2, R_3)W_1(R_1)$$
$$+ W_2(R_1, R_3)W_1(R_2).$$

In general,

$$U_l(R^l) = U_l(R_1, \ldots, R_l)W_l(R_1, \ldots, R_l) - \sum W_2(R_i, R_j)W_{l-2}(R_k, \ldots, R_l) + \cdots$$
$$+ (-1)^{l-1}(l-1)! \, W_1(R_1) \cdots W_1(R_l), \tag{4.51}$$

Here the coefficient for a term with k groups is $(-1)^{k-1}(k-1)!$.

Thus, from Eq. (4.40), all the W_N are known. From Eq. (4.51), $U(R^l)$ in terms of the W_i is given; thus the $U_l(R^l)$ can be found.   (4.52)

In any particular partition of the N particles, there will be m_l clusters of size l; and of course $\sum_l lm_l = N$.

For example (as is evident from Eq. (4.45)), a three-particle gas can be partitioned into

1. 1 cluster of 3, $m_3 = 1, m_2 = m_1 = 0$.
2. 3 clusters of 1, $m_1 = 3, m_2 = m_3 = 0$.
3. 1 cluster of 2 and 1 cluster of 1 (this can happen 3 ways), $m_1 = m_2 = 1$, $m_3 = 0$.

From the above discussion, with Eqs. (4.45) and (4.46),

$$W_N(R^N) = S_{(m_l)} \prod U_l(R^l) \qquad \sum lm_l = N. \tag{4.53}$$

$S_{(m_l)}$ means the sum over all possible partitions of N, that is, all divisions of particles into groups.

From Eqs. (4.37), (4.39), and (4.40),

$$X_N \equiv \frac{Z_N}{N!} = \int \frac{W_N}{N!} \, d^N R. \tag{4.54}$$

Now construct the expression

$$b_l = \frac{1}{V l!} \int U_l(R^l) \, dR_1 \cdots dR_l. \tag{4.55}$$

This equation is known as *Mayer's cluster integral*. From Eqs. (4.51), (4.53), and the statement labeled (4.52),

$$X_N = \frac{1}{N!} S_{(m_l)} \prod_l \int U_l(R^l)$$

$$= \frac{1}{N!} S_{(m_l)} \prod_l (V l! \, b_l)^{m_l}.$$

But if $N = 5$, for example, $U(R_1 R_2) U(R_3, R_4) U(R_5)$ gives the same contribution as $U(R_1 R_3) U(R_2 R_5) U(R_4)$, which is a different partition with the same set of m_l. For each possible set of m_l there are several terms, all of which are equal. To see how many such terms there are, note that there are $N!$ permutations of the N coordinates, and each such permutation corresponds to a term in $S_{(m_l)}$ with the same set of m_l. But we are counting each term more than once. We should not count different permutations of the coordinates within a given U, for example, $U(R_1, R_2) \equiv U(R_2, R_1)$. We also do not count changes of order of the U's as separate contributions. For example, $U(R_1, R_2) U(R_3, R_4) U(R_5) = U(R_3, R_4) U(R_1, R_2) U(R_5)$ are not counted separately. Thus for a given set of m_l, there are $N!/\prod_l (l!)^{m_l} m_l!$ terms. So we can write

$$X_N = \frac{Z_N}{N!} = \frac{1}{N!} \sum_{\substack{\text{all possible sets} \\ \text{of } m_l: \, \Sigma l m_l = N}} \prod_l \frac{(V l! \, b_l)^{m_l}}{(l!)^{m_l} m_l!} N!$$

$$= \sum_{\{m_l\}} \prod_l \frac{(V b_l)^{m_l}}{m_l!}. \tag{4.56}$$

It is important to note that (for large volume), b_l is independent of volume. That this is so can be seen as follows: $U_l(R^l)$ vanishes unless the particles are clustered. Holding dR_1 fixed and varying $dR_2, \ldots, dR_l$ then gives some number completely independent of volume, because unless $R_2, \ldots, R_l$ are near R_1, $U_l(R) = 0$.

The integral over dR_1 then gives a quantity proportional to the volume, V, and as b_l was defined with a V, in the denominator, b_l is independent of V.

We must now evaluate $X_N = S_{(m_l)} \prod_l (V b_l)^{m_l}/m_l!$ subject to $\sum m_l l = N$.

Recalling a similar summation encountered earlier (for the quantum statistics of a many-particle system), we note that if the restriction $\sum m_l l = N$ could be removed, X_N can be easily evaluated. Using the same methods as before we define

$$e^{-\beta F_N} = \frac{1}{N!} \left(\frac{m}{2\pi\hbar^2\beta} \right)^{3N/2} Z_N$$

and

$$e^{-\beta g} = \sum_N e^{-\beta(F_N - \mu N)} = \sum_N \alpha^N X_N \tag{4.57}$$

where

$$\alpha = \left(\frac{m}{2\pi\hbar^2\beta} \right)^{3/2} e^{\mu\beta}.$$

But first taking the sum $\sum_{(ml)}$ with $\sum m_l l = N$ and then summing over all N is clearly equivalent to summing over all m_l with no restriction.

$$e^{-\beta g} = \sum_m \prod_l \frac{(Vb_l)^{m_l}\alpha^{lm_l}}{m_l!} \quad \text{(no restrictions)}$$

$$= \sum_{m_1, m_2, \cdots} \frac{(Vb_1)^{m_1}\alpha^{m_1}}{m_1!} \frac{(Vb_2)^{m_2}\alpha^{2m_2}}{m_2!} \cdots$$

$$= \exp\left[V \sum_l \alpha^l b_l \right]$$

$$g = -kTV \sum_l \alpha^l b_l. \tag{4.58}$$

This is known as *Mayer's cluster expansion*.
From Eq. (1.51)

$$\langle N \rangle = -\frac{\partial g}{\partial \mu}$$

$$= +kTV \sum_l l\alpha^{l-1} \frac{\partial \alpha}{\partial \mu} b_l = V \sum_l l\alpha^l b_l.$$

Therefore the equation

$$\rho = \langle N \rangle / V = \sum_l l\alpha^l b_l \tag{4.59}$$

determines α if we know the density.
From Eq. (1.53) the pressure is given by

$$p = -\frac{\partial g}{\partial V}\bigg|_\mu = kT \sum_l \alpha^l b_l.$$

Therefore we get the equation of state

$$\frac{PV}{RT} = \frac{PV}{NkT} = \frac{\sum_l b_l \alpha^l}{\sum_l l b_l \alpha^l}.$$ (4.60)

For small density, α is small—as can be seen from Eq. (4.59). Then $\alpha^{l+1} \ll \alpha^l$; so that expansion (Eq. (4.60)) need only be carried out a few terms. For small l, b_l can be calculated easily. Inverting Eq. (4.59) to get α as a power series in $1/V$, and writing

$$\frac{PV}{RT} = 1 + \frac{B}{V} + \frac{C}{V^2} + \cdots,$$

we have

$$B = -\frac{b_2}{b_1^2} N = \frac{N}{2} \int [1 - e^{-\beta B(r_{12})}] \, dR_2$$ (4.61)

$$C = \frac{4b_2^2 - 2b_1 b_3}{b_1^4} = 4B^2 - \frac{N^2}{3} \iint \left(2 - e^{-\beta V(r_{12})} - e^{-\beta V(r_{23})} - e^{-\beta V(r_{13})} \right.$$

$$\left. + e^{-\beta[V(r_{12}) + V(r_{23}) + V(r_{13})]} \right) dR_2 \, dR_3.$$ (4.62)

When the density becomes large, the Mayer cluster expansion is of little use. This is because at very high density (liquid), the most important term in the expansion becomes the Nth term. That is, almost every particle becomes a member of a very large cluster— and in a liquid all the particles are in an N-fold cluster.

Hard Sphere Gas

The question may be asked: What is the partition function of a hard-sphere gas—that is, a gas composed of small inpenetrable spheres that exert no forces on one another when not touching? Unfortunately, this question is still un-answered. The question may be stated in a different way.

$$Z_N = \oint_{\substack{\text{no two closer than } 2a. \\ a = \text{radius of sphere}}} 1 \, d^N R.$$

But

$$V^N = \int_{\text{no restriction}} d^N R.$$

Thus $\oint d^N R / \int d^N R$ is essentially the probability that if N points are "dropped" into a volume V, no two particles will be closer than $2a$.

4.4 RADIAL DISTRIBUTION FUNCTION

Consider a system of N particles interacting with pair-wise forces. The partition function Z_N is

$$Z_N = \int \exp\left[-\beta \sum_{\text{pair}} V\right] d^{3N}R. \tag{4.63}$$

The probability density for particle 1 being at R_1, particle 2 at R_2, etc., is

$$\left(\frac{1}{Z_N}\right) \exp\left[-\beta V(R_1, \ldots, R_N)\right],$$

where the factor of $1/Z_N$ provides normalization. The probability that a given particle, say particle 1, is at R_1 and the rest of the particles are anywhere is

$$\left(\frac{1}{Z_N}\right) \int \exp\left[-\beta V(R_1, \ldots, R_N)\right] d^3R_2 \, d^3R_3, \ldots, d^3R_N.$$

Because there are N particles, the probability density for one of these particles being at R is

$$n_1(R) = \frac{N}{Z_N} \int \exp\left[-\beta V(R_1, R_2, \ldots, R_N)\right] d^3R_2 \cdots d^3R_N$$

$$= \frac{N}{Z_N} \int \exp\left[-\beta \sum_{\text{pair}} V(r_{ij})\right] d^3R_2 \cdots d^3R_N; \tag{4.64}$$

n_1 is the one-particle density, or *distribution function*.

The two-particle distribution function $n_2(R_1, R_2)$ is defined as

$$n_2(R_1, R_2) = \frac{N(N-1)}{Z_N} \int \exp\left[-\beta \sum_{\text{pair}} V\right] d^3R_3 \cdots d^3R_N; \tag{4.65}$$

n_2 is the probability density for finding a particle at R_1 and another particle at R_2. For a liquid or a gas, $n_2(R_1, R_2)$ is a function of the distance between the two particles.

$$n_2(R_1, R_2) = n_2(r_{12}).$$

The general shape of $n_2(r)$ is as shown in Fig. 4.5. When $r \to \infty$, $n_2(r)$ approaches n_1^2 where n_1 is the one-particle density; this is because when the distance becomes large, the existence of one atom does not influence the distribution of the other atom.

The radial distribution function $n_2(r)$ is related to x-ray or neutron diffraction as shown in Fig. 4.6. The x-ray is coming in along the z direction and is scattered by atoms. The scattered x-ray is observed at P somewhere far from the

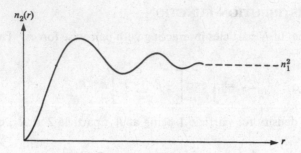

Fig. 4.5 Two-particle distribution function.

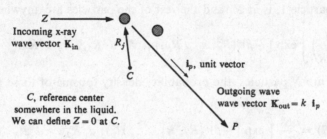

Fig. 4.6 X-ray diffractions.

liquid. The amplitude observed at P is the sum of amplitudes scattered by each atom, the latter being of the form

$$a(\theta)(e^{ikR}/R).$$

Here R is the distance from the scattering atom and $a(\theta)$ indicates the possible angular dependence of scattering, which is not important here. Taking into account the phase of the incoming wave, $e^{ik \cdot Z}$, we write

$$\text{Amplitude} = a(\theta) \sum_j (e^{ikR_{jP}}e^{ikZ_j}/R_{jP}). \qquad (4.66)$$

P is the point of observation. Then, approximately,

$$R_{jP} = R_{CP} - R_j \cdot i_P$$

and we may write

$$kZ_j = K_{in} \cdot R_j.$$

$$ki_P = K_{out}.$$

So Eq. (4.66) becomes

$$\text{Amplitude} = a(\theta) \frac{e^{ikR_{CP}}}{R_{CP}} \sum_j e^{-iK_{out} \cdot R_j} e^{iK_{in} \cdot R_j}$$

or

$$\text{Amplitude} = a(\theta) \frac{e^{ikR_{CP}}}{R_{CP}} \cdot G, \tag{4.67}$$

with

$$G = \sum_j \exp \left[i(K_{in} \cdot R_j - K_{out} \cdot R_j) \right] = \sum_j \exp (iq \cdot R_j),$$

where

$$q \equiv K_{in} - K_{out}.$$

The intensity of the x rays is

$$\text{Intensity} = |\text{Amplitude}|^2 = (|a(\theta)|^2/R_{CP}^2)|G|^2. \tag{4.68}$$

To compute the intensity, we need to know the expectation value of $|G|^2$. Now

$$|G|^2 = \sum_{i,j=1}^{N} e^{iq \cdot (R_i - R_j)}.$$

For a liquid, neglecting the $i = j$ terms in the double sum, we have

$$\langle |G|^2 \rangle = \frac{N(N-1)}{2} \int e^{iq \cdot (R_1 - R_2)} \left[\frac{\exp \left[-\beta \sum_{pair} V(R_{ij}) \right]}{Z_N} \right] d^{3N}R \tag{4.69}$$

because the term in brackets, [ ], is the probability for the configuration R_1, $R_2, \ldots, R_N$. Now, using Eq. (4.65), we write

$$\langle |G|^2 \rangle = \frac{1}{2} \int e^{iq \cdot (R_1 - R_2)} n_2(R_1, R_2) \, d^3R_1 \, d^3R_2. \tag{4.70}$$

When $n_2(R_1, R_2)$ depends on $|R_1 - R_2|$ only, $V = $ the volume of the liquid, and $q = |q|$,

$$\langle |G|^2 \rangle = \frac{1}{2}V \int e^{iq \cdot X} n_2(|X|) \, d^3X = \frac{2\pi V}{q} \int r n_2(r) \sin qr \, dr. \tag{4.71}$$

4.5 THERMODYNAMIC FUNCTIONS

In the classical case, if we know $n_2(r)$, we can derive all the thermodynamic functions.

The internal energy is

$$U = \tfrac{3}{2}RT - \frac{1}{Z_N} \frac{\partial Z_N}{\partial \beta}, \tag{4.72}$$

where Z_N is the partition function defined in Eq. (4.2), now

$$\frac{1}{Z_N} \frac{\partial Z_N}{\partial \beta} = -\int \frac{V(r_{12}) \exp\left[-\beta \sum V\right]}{Z_N} d^{3N}R$$

$$-\int \frac{V(r_{23}) \exp\left[-\beta \sum V\right]}{Z_N} d^{3N}R \cdots$$

$$= -\frac{N(N-1)}{2} \int V(r_{12}) \frac{\exp\left[-\beta \sum V\right]}{Z_N} d^{3N}R.$$

Here, we use Eq. (4.65) to arrive at

$$\frac{1}{Z_N} \frac{\partial Z_N}{\partial \beta} = -\frac{1}{2} \int V(r_{12}) n_2(r_{12})\, d^3R_1\, d^3R_2$$

or

$$U = \frac{3}{2} RT + \frac{\text{Vol}}{2} \int V(r) n_2(r) 4\pi r^2\, dr. \tag{4.73}$$

This equation could have been expected. The first term of Eq. (4.73) is the kinetic energy, and the second term is the potential energy. The latter comes from the pair potential $V(r)$, and the factor 1/2 appears because each pair is counted twice.

The pressure is expressed as follows:

$$pV = RT - \frac{\text{Vol}}{6} \int R \cdot \nabla V(R) n_2(R)\, d^3R \tag{4.74a}$$

or

$$pV = RT - \frac{\text{Vol}}{6} \int r V'(r) n_2(r) 4\pi r^2\, dr. \tag{4.74b}$$

These are classical formulae, which do not hold for liquid helium. Equation (4.74b) is derived as follows. From Eq. (2.121)

$$3pV = 2\langle KE \rangle - \left\langle \sum_{\substack{ij \\ i<j}} r_{ij} V'(r_{ij}) \right\rangle. \tag{4.75}$$

As in Eq. (4.73), $\langle KE \rangle = (3/2)RT$. The second term of Eq. (4.75) can be written

$$\left\langle \sum_{i<j} r_{ij} V'(r_{ij}) \right\rangle = \frac{N(N-1)}{2} \langle r_{12} V'(r_{12}) \rangle$$

$$= \frac{N(N-1)}{2Z_N} \int r_{12} V'(r_{12}) e^{-\beta(V(r_{12})+V(r_{13})+\cdots)}$$

$$= \frac{V}{2} \int r_{12} V'(r_{12}) n_2(r_{12})\, d^3(R_1 - R_2).$$

Equation (4.74b) follows. Equation (4.74a) is nothing but the obvious generalization to the case of potentials that are not spherically symmetric. The product $\bar{R} \cdot \nabla\, V(\bar{R})$ in the integrand of Eq. (4.74a) is called the *virial*.

When the force is homogenuous of degree $(n + 1)$,

$$V(r) = -cr^n,$$

so that

$$rV'(r) = nCr^n = nV(r)$$

we can use these in Eqs. (4.73) and (4.74b) to see that

$$3pV + nU = (3 + \tfrac{3}{2}n)RT. \tag{4.76}$$

This relation is called the *virial theorem*.

Problem: Show that at low densities

$$n_2(r) = 1 - e^{-\beta V(r)}.$$

Using this result in Eq. (4.74), derive the virial coefficients in the virial expansion.

4.6 THE BORN–GREEN EQUATION FOR n_2

In Section 4.5 we pointed out that one can find all thermodynamic functions given $n_2(r)$. In Section 4.1 of this chapter we showed how we could find (in principle) $e^{-\beta F}$ given the potential between pairs of particles, and from F we can also get all thermodynamic functions. But to get $e^{-\beta F}$ we must integrate over something like 10^{23} variables. Computers cannot do such an integration; so we have got to find some way of getting the answer approximately. The Mayer cluster expansion was one way, and it was satisfactory for a gas. So far, for a liquid, we have replaced F by $n_2(r)$, which requires six fewer than the 10^{23} integrations for F. Clearly, we still have quite a way to go. One approach toward approximately evaluating our monster integral is by means of the Born–Green equation.

$n_2(R_1, R_2)$, defined in Eq. (4.65) can be transformed as follows:

$$n_2(R_1, R_2) = \frac{N(N-1)}{Z_N} \int \exp\left[-\beta \sum_{\text{pair}} V\right] d^3R_3 \cdots d^3R_N$$

$$= \frac{N(N-1)}{Z_N} \int \exp\left[-\beta \sum_{i \neq 1} V(R_{1i}) - \beta \sum_{\substack{i \neq 1 \\ j \neq 1}} V(R_{ij})\right]$$

$$\times\, d^3R_3 \cdots d^3R_N.'$$

Differentiate the above equation to obtain

$$\nabla_1 n_2(R_1, R_2) = -\frac{\beta N(N-1)}{Z_N} \int \left[\nabla_1 V(R_{12}) + \sum_{\substack{i \neq 1 \\ i \neq 2}} \nabla_1 V(R_{1i}) \right]$$

$$\times \exp\left(-\beta \sum_{\text{pair}} V\right) d^3R_3 \cdots d^3R_N. \tag{4.77}$$

When we define

$$n_3(R_1, R_2, R_3) = \frac{N!}{(N-3)! \, Z_N} \int \exp\left[-\beta \sum_{\text{pair}} V\right] d^3R_4 \cdots d^3R_N, \tag{4.78}$$

we can write Eq. (4.77) as

$$\nabla_1 n_2(R_1, R_2) = -\beta \nabla_1 V(R_{12}) n_2(R_1, R_2)$$

$$-\beta \int \nabla_1 V(R_{13}) n_3(R_1, R_2, R_3) \, d^3R_3. \tag{4.79}$$

This is the _Born–Green equation._*

The special feature of this equation is that both $n_2(R_1, R_2)$ and $n_3(R_1, R_2, R_3)$ are involved. When we go further and write the equation for n_3 we need n_4. The equation for n_4 includes n_5, and so on. Thus, we have a hierarchy of equations. This kind of hierarchy appears in other branches of physics—for example, in meson field theory—but no general, accurate way of treating the set of equations has been established. When we make a guess for n_5, say, and cut off the hierarchy there and calculate back n_2, we might get a good estimate of n_2; but that an error in the guess of n_5 becomes unimportant as we approach n_2 has not been proved.

An ingenious way of closing the hierarchy is due to Kirkwood. When R_3 is very far from R_1 and R_2,

$$n_3(R_1, R_2, R_3) \to n_2(R_1, R_2)n_1.$$

Considerations such as this led Kirkwood to the approximation

$$n_3(R_1, R_2, R_3) = \frac{n_2(R_1 R_2)n_2(R_2 R_3)n_2(R_3 R_1)}{n_1^3}, \tag{4.80}$$

which is called the _Kirkwood approximation_,† or the _superposition approximation._ When we use Eq. (4.80) in Eq. (4.79) we can solve for n_2.

---

* Born M. and Green H. S., _Proc. Roy. Soc._ (_London_) **A188**, 10 (1946). See also J. Yvon, _Actualities Scientifique et Industrielles_ (Herman and Cie, Paris 1935), No. 203.
† Kirkwood J. G. and Boggs E. M., _J. Chem. Phys._, **10**, 394 (1942).

4.7 ONE-DIMENSIONAL GAS

Our attempts to evaluate the integrals involved in the thermodynamics of liquids and gases have not met with complete success. The Kirkwood approximation might not be bad, but we do not know exactly how to generalize it, or whether a generalization would give better results. What we want is a method that would give progressively more accurate results as we increased the computational labor. No such method is available today.

We are faced with a problem that is too hard for us. We may be able to gain some insight by picking a soluble problem that has some of the features of the insoluble one.

For example, consider gas particles constrained to move on a line (Fig. 4.7). First we assume only nearest-neighbor interaction. The partition function is

$$Z_N(L) = \int\limits_{0<x_1<x_2<\,\cdots\,<x_N<L} e^{-\beta[V(x_1-x_2)+V(x_2-x_3)+\,\cdots\,]}dx_1\,dx_2\,\cdots\,dx_N, \quad (4.81)$$

The length L is fixed.

We would be making a good start toward solving our problem if we could get a recursion relation with Z_{N+1} expressed in terms of Z_N. Since we cannot even do this, we will modify the problem in such a way that the macroscopic properties of the system are not changed but a recursion relation for the partition function can be found. We put one extra atom before x_1 and fix its position to be at y (see Fig. 4.8.) The resulting partition function we call $P_N(y, L)$.

$$P_N(y, L) = \int\limits_{y<x_1<x_2<\,\cdots\,<L} e^{-\beta[V(y-x_1)+V(x_1-x_2)+\,\cdots\,]}dx_1\,dx_2\,\cdots\,dx_N. \quad (4.82)$$

In Eq. (4.82) y is fixed. Now, consider the case shown in Fig. 4.9. From Eq. (4.82) we can write

$$P_{N+1}(z, L) = \int\limits_{z<y<x_1<\,\cdots\,<L} e^{-\beta[V(z-y)+V(y-x_1)+\,\cdots\,]}dy\,dx_1\,\cdots\,dx_N. \quad (4.83)$$

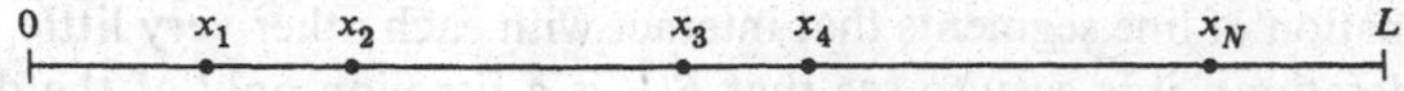

Fig. 4.7 Gas particles constrained to move in one dimension only.

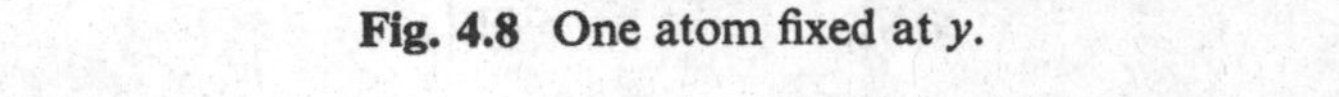

Fig. 4.8 One atom fixed at y.

Fig. 4.9 One atom fixed at z.

In the integral, z is fixed, but y varies. From Eq. (4.83)

$$
P_{N+1}(z, L) = \int\limits_{z<y} e^{-\beta V(z-y)} \, dy
$$

$$
\times \int\limits_{y<x_1<\cdots<L} e^{-\beta[V(y-x_1)+V(x_1-x_2)+\cdots]} \, dx_1 \cdots dx_N
$$

$$
= \int\limits_{z<y<L} P_N(y, L) e^{-\beta V(z-y)} \, dy. \tag{4.84}
$$

Using $P_N(y, L)$, we can construct the grand partition function

$$
e^{-\beta g} = \sum_N e^{\beta \mu N} e^{-\beta F_N} = \sum_N e^{\beta \mu N} \frac{1}{N!} \left(\frac{m}{2\pi \hbar^2 \beta} \right)^{N/2} Z_N.
$$

In our case, $Z_N/N! = P_N(y, L)$ (the factor $1/N!$ comes from the fact that we have in the integral for P_N restricted the range to $y < x_1 < \cdots < L$ instead of allowing each particle to be anywhere in $[0, L]$). We can absorb the factor $(m/2\pi \hbar^2 \beta)^{N/2}$ in the $e^{\beta \mu N}$ by adjusting μ. Then

$$
e^{-\beta g} = \sum_N e^{\beta \mu N} P_N(y, L). \tag{4.85a}
$$

As in Section 1.1, g for a combination of several independent subsystems, each with the same μ, is the sum of the g_i for each subsystem. In Section 1.6 we defined $F = g + \mu N$, where N is the expected number of particles in the system. Clearly, F for a combination of independent subsystems is also a sum of the F's for each system.

If there is a sufficiently large number of particles, we can consider the line to be a combination of line segments that interact with each other very little. With such considerations, it is easy to see that F/V is a function only of the density (at a fixed temperature), and F can be written $F = Vf(V/N)$. It is not hard to show that F of the above form is equivalent to the equation $g = -pV$. The proof is as follows: From Section 1.6:

$$
\mu = \left. \frac{\partial F}{\partial N} \right|_V ; \qquad p = - \left. \frac{\partial F}{\partial V} \right|_N .
$$

Using $F = Vf(V/N)$ we then have

$$
\begin{aligned}
g + pV &= F - \mu N + pV \\
&= F - N\left.\frac{\partial F}{\partial N}\right|_V - V\left.\frac{\partial F}{\partial V}\right|_N \\
&= F - N\left[-\frac{V^2}{N^2}f'\left(\frac{V}{N}\right)\right] - V\left[f\left(\frac{V}{N}\right) + \frac{V}{N}f'\left(\frac{V}{N}\right)\right] \\
&= F - Vf\left(\frac{V}{N}\right) \\
&= 0.
\end{aligned}
$$

Because the "volume" of the system is $L - y$, we can now write Eq. (4.85a) as

$$
e^{\beta p(L-y)} = \sum_N e^{\beta \mu N} P_N(y, L). \tag{4.85b}
$$

Multiply Eq. (4.84) by $e^{(N+1)\mu\beta}$ and sum over N to obtain

$$
e^{p(L-z)\beta} = e^{\mu\beta}\int_z^L e^{p(L-y)\beta}e^{-\beta V(z-y)}\,dy, \tag{4.86}
$$

which leads to

$$
e^{-\beta\mu} = \int_z^L e^{-p(y-z)\beta}e^{-\beta V(z-y)}\,dy,
$$

or

$$
e^{-\beta\mu} = \int_0^\infty e^{-p\beta x}e^{-\beta V(x)}\,dx. \tag{4.87}
$$

The integral is insensitive to the upper limit, so we have replaced that limit with ∞. This is the desired result from which we now can find all the thermodynamic quantities. For example

$$
\langle N \rangle = -\frac{\partial g}{\partial \mu} = \frac{\partial p(\mu, \beta)\,\mathrm{Vol}}{\partial \mu} = \mathrm{Vol}\,\frac{\partial p}{\partial \mu} = L\,\frac{\partial p}{\partial \mu}.
$$

Equation (4.87) can be written as

$$
0 = \beta\mu + \log\int_0^\infty e^{-p\beta x}e^{-\beta V(x)}\,dx = f[\mu, \beta, p(\mu, \beta)].
$$

From $0 = df/d\mu$, it follows that

$$
\frac{\partial p}{\partial \mu} = -\frac{\partial f/\partial \mu}{\partial f/\partial p} = \frac{\int_0^\infty e^{-p\beta x}e^{-\beta V(x)}}{\int_0^\infty x e^{-p\beta x}e^{-\beta V(x)}}.
$$

Problem: When the nearest- and next-nearest-neighbor interactions are included, derive the chemical potential. (Hold two atoms fixed near the left end of the line and derive the recursion formula. An integral equation results.)

We see now that we can solve the problem of particles fixed on a line if the potential acts only between the nearest few neighbors. But what if all the atoms in the system interact? For a certain special form of the potential we can still solve the problem. Section 4.8 discusses the case of an $e^{-|x|}$ potential.

4.8 ONE-DIMENSIONAL GAS WITH POTENTIAL OF THE FORM $e^{-|x|}$

We will now discuss the problem of a one-dimensional gas, all particles of which interact repulsively with all others with a potential $V(x) = e^{-|x|}$.

When there is attraction and no repulsion, the preferred configuration occurs when the particles all sit on top of one another. The potential V is proportional to the number of pairs, $N(N - 1)/2$. The more pairs there are, the lower the potential can get, and ultimately all particles collapse to a point. This result does not tell us much, so we assume a repulsive force.

Assume N particles are distributed in a line of length L (Fig. 4.10). We wish to evaluate

$$
e^{-\beta F_N} = \frac{1}{N!} \left(\frac{m}{2\pi\hbar^2\beta} \right)^{N/2} \int\int\int_0^L\int\int \exp\left(-\beta \sum_{i<j} e^{-|x_i - x_j|} \right) dx_1 \cdots dx_N
$$

$$
= \left(\frac{m}{2\pi\hbar^2\beta} \right)^{N/2} \int_{0<x_1<\cdots<x_N<L} \exp\left[-\beta \sum_{i<j} e^{(x_i - x_j)} \right] dx_1 \cdots dx_N
$$

$$
= \left(\frac{m}{2\pi\hbar^2\beta} \right)^{N/2} \int_{0<x_1<\cdots<L} \exp\left\{ -\beta \left[e^{x_1}(e^{-x_2} + \cdots + e^{-x_N}) \right.\right.
$$

$$
\left.\left. + \sum_{j>i\geq 2} e^{(x_i - x_j)} \right] \right\} dx_1 \cdots dx_N.
$$

(4.87)

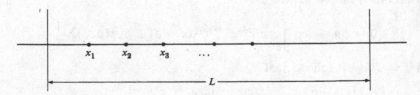

Fig. 4.10 One-dimensional gas.

Suppose, as before, we modify the problem slightly by placing an extra atom before x_1, and fix the position of that atom at y. Let $\alpha = e^y$. Later we will set y equal to zero. The partition function for this system is

$$I_N(\alpha, L) = \int\limits_{0<y<x_1<\cdots<L} \exp\left\{-\beta\left[\alpha(e^{-x_1} + e^{-x_2} + \cdots + e^{-x_N})\right.\right.$$
$$\left.\left. + \sum_{j>i\geq 1} e^{(x_i-x_j)}\right]\right\} dx_1 \cdots dx_2. \qquad (4.88)$$

Then

$$I_{N+1}(\alpha, L) = \int\limits_{0<y<x_0<x_1<\cdots<L} \exp\left\{-\beta\left[\alpha(e^{-x_0} + e^{-x_1} + \cdots + e^{-x_N})\right.\right.$$
$$\left.\left. + \sum_{j>i\geq 0} e^{(x_i-x_j)}\right]\right\} dx_0 \cdots dx_N$$

$$= \int\limits_{0<y<x_0<\cdots<L} \exp\left(-\beta\alpha e^{-x_0}\right)$$
$$\times \exp\left\{-\beta\left[(\alpha + e^{-x_0})(e^{-x_1} + \cdots + e^{-x_N})\right.\right.$$
$$\left.\left. + \sum_{j>i\geq 1} e^{(x_i-x_j)}\right]\right\} dx_0 \cdots dx_N$$

$$= \int_y^L \exp\left(-\beta\alpha e^{-x_0}\right)I_N(\alpha + e^{x_0}, L - y) \, dx_0. \qquad (4.89)$$

As in Section 4.7, we can again modify the problem in a manner that does not affect the macroscopic properties. We allow N to vary, put in a chemical potential, and define

$$e^{-\beta g} = \sum_N I_N(\alpha, L)e^{\beta\mu N}.$$

Because $I_N(\alpha, L)$ is the partition function for a system with one particle fixed at $y = \log \alpha$, its length is $L - \log \alpha$, and $g = -p(L - \log \alpha)$.

Multiply Eq. (4.89) by $e^{(N+1)\mu\beta}$, sum, set $y = 0$ and replace L by ∞. The result after a little algebra is

$$e^{-\mu\beta} = \int_0^\infty \exp\left\{-\beta[e^{-x_0} + p \log(1 + e^{x_0})]\right\} dx_0. \qquad (4.90)$$

From Eq. (4.90) we can obtain all the thermodynamic properties, and we can reasonably assume that the atom fixed at zero does not change those properties. But let us see what we can do without fixing an atom at zero. The force on the first atom, the one located at x_1, is due to the rest of the atoms in the gas, and the effect of the other atoms can be described with a potential involving one extra parameter. Thus the first atom moves in a potential $ce^{x_1} - 1$, where $c = \sum_i e^{-x_i}$. Instead of specifying the position of one of the atoms as an extra parameter in the problem, we will specify c, which determines the potential felt by the atom at x_1. We now calculate $G_N(c)$, which is the partition integral:

$$\int \exp\left[-\beta \sum_{i<j} e^{-(x_j - x_i)}\right] dx_1 \cdots dx_N$$

with the range of integration subject to the restrictions that

$$\sum_i e^{-x_i} = c$$

and $0 < x_1 \cdots < x_N < L$. Mathematically,

$$G_N(c) = \int_{0<x_1<\cdots<x_N<L} \exp\left[-\beta \sum_{i<j} e^{-(x_j - x_i)}\right] \delta\left(c - \sum_i e^{-x_i}\right) dx_1 \cdots dx_N.$$

(4.91)

Note that $G_N(c) = 0$ if $c < 0$, and

$$e^{-\beta F_N} = \left(\frac{m}{2\pi \hbar^2 \beta}\right)^{N/2} \int_0^\infty G_N(c)\, dc.$$

To get a recursion relation we will hold x_1 constant until the rest of the integrations are done.

Equation (4.91) can be rewritten

$$G_N(c) = \int \exp\left[-\beta \sum_i' e^{-(x_i - x_1)} - \beta \sum_{i<j}' e^{-(x_j - x_i)}\right]$$
$$\times \delta\left(c - e^{-x_1} - \sum_i' e^{-x_i}\right) dx_1 \cdots dx_N,$$

(4.92a)

where

$$\sum_i' \equiv \sum_{i \neq 1}.$$

Because of the action of the δ-function, the $\sum_i' e^{-x_i}$ in the exponent of the integrand can be replaced by $c - e^{-x_1}$. To make the argument clearer substitute $x_i' = x_i - x_1$. Thus

$$G_N(c) = \int_0^L dx_1 \int_{x_1}^L \delta\left(c - e^{-x_1} - e^{-x_1} \sum_i' e^{-x_i'}\right)$$
$$\times \exp\left[-\beta \sum_i' e^{-x_i'} - \beta \sum_{i<j}' e^{-(x_j' - x_i')}\right] dx_2 \cdots dx_N.$$

(4.92b)

We know that

$$\delta\left(c - e^{-x_1} - e^{-x_1}\sum_i{}' e^{-x_i}\right) = e^{x_1}\delta\left(c\,e^{x_1} - 1 - \sum_i{}' e^{-x_i}\right)$$

because $\delta(ax) = (1/a)\,\delta(x)$. Equation (4.92b) can be rewritten

$$G_N(c) = \int_0^L dx_1 \exp(x_1) \exp\left[-\beta(ce^{x_1} - 1)\right] \int_0^{L-x_1} \delta\left[ce^{x_1} - 1 - \sum_i{}' e^{-x_i}\right]$$

$$\times \exp\left[-\beta\sum_{i<j}{}' e^{-(x'_j - x)_i}\right] dx'_2 \cdots dx'_N$$

$$\left(= \int_0^L dx_1 \exp(x_1) \exp\left[-\beta(ce^{x_1} - 1)\right] G_{N-1}(ce^{x_1} - 1).\right. \tag{4.93}$$

In general, the solution comes from Eq. (4.93), but it is very difficult to find.
For subsequent use, let us define (for N particles)

$$\sum_N \alpha^N G_N(c) = e^{\beta pL}g(c) \tag{4.94}$$

$$\alpha = e^{\beta\mu}$$

and

$$p = \text{pressure}.$$

Combining Eqs. (4.93) and (4.94), we find that

$$g(c) = \alpha \int_0^\infty dx \exp(x) \exp\left[-\beta(ce^x - 1)\right]e^{-\beta px}g(ce^x - 1), \tag{4.95}$$

where the substitutions $x = x_1$ and extension of the limit $L \to \infty$ have been
made. In obtaining the eigenvalues of this integral equation, a relationship will
be found between α and p; $g(c)$ itself is not of interest. The boundary value on
$g(c)$ is $g(c) = 0$ if $c < 0$. To put Eq. (4.95) in terms of differentials, substitute

$$h(c) = e^{-\beta c}g(c)$$

$$g(c) = \alpha \int_0^\infty dx\, e^x e^{-\beta px} h(ce^x - 1)$$

$$g'(c) = \alpha \int_0^\infty dx\, e^{2x} e^{-\beta px} h'(ce^x - 1)$$

$$= \frac{\alpha}{c} \int_0^\infty e^x e^{-\beta px} \frac{d}{dx}\left[h(ce^x - 1)\right].$$

Integrate by parts:

$$g'(c) = \frac{\alpha}{c} h(ce^x - 1)e^x e^{-\beta px}|_0^\infty - \frac{\alpha}{c} \int_0^\infty h(ce^x - 1)(1 - \beta p)e^{x - \beta px} \, dx$$

$$= \frac{\alpha(\beta p - 1)}{c} g(c) - \frac{\alpha}{c} h(c - 1)$$

$$= \frac{\alpha(\beta p - 1)}{c} g(c) - \frac{\alpha}{c} e^{-\beta(c-1)}g(c - 1). \tag{4.96}$$

Equation (4.96) appears at first glance to be a differential equation, but is not, because $g(c)$ and $g(c - 1)$ enter into it.

For another form, substitute

$$g(c) = e^{-\beta c^2/2}f(c)$$
$$g'(c) = -\beta c e^{-\beta c^2/2}f(c) + e^{-\beta c^2/2}f'(c).$$

Then Eq. (4.96) becomes

$$cf'(c) - \beta c^2 f(c) - \alpha(\beta p - 1)f(c) = -\alpha e^{\beta/2}f(c - 1).$$

We can use the Laplace transformation

$$\varphi(q) = \int_0^\infty e^{-qc}f(c) \, dc$$

to obtain an equation in $\varphi(q)$. Thus:

$$\int_0^\infty e^{-qc}cf'(c) \, dc = -\frac{d}{dq} \int_0^\infty e^{-qc}f'(c) \, dc$$

$$= -\frac{d}{dq} [q\varphi(q)]$$

and

$$\int_0^\infty e^{-qc}c^2 f(c) \, dc = \frac{d^2}{dq^2} \varphi(q).$$

Therefore

$$\frac{d}{dq} [q\varphi(q)] + \beta \frac{d^2}{dq^2} \varphi(q) + \alpha(\beta p - 1)\varphi(q) = \alpha e^{\beta/2}e^{-q}\varphi(q).$$

Substituting

$$\varphi(q) = e^{-q^2/4\beta}\psi(q),$$

we eliminate the first derivative term:

$$-\beta\psi''(q) = \left[-\frac{q^2}{4\beta} - \alpha e^{\beta/2}e^{-q} + \alpha(\beta p - 1) + \tfrac{1}{2} \right]\psi. \tag{4.97}$$

If instead we use Fourier transforms

$$\varphi(\omega) = \int e^{i\omega c} f(c) \, dc$$

and substitute

$$\psi(\omega) = e^{-\omega^2/4\beta} \varphi(\omega),$$

we get

$$\beta \psi''(\omega) = \left(\frac{\omega^2}{4\beta} + \alpha e^{\beta/2} e^{i\omega} + \alpha(\beta p - 1) + \tfrac{1}{2} \right) \psi(\omega). \qquad (4.98)$$

The boundary conditions are not known definitely in this transform notation. However, it may be that the only restriction necessary is that the function $\psi(q)$ or $\varphi(\omega)$ remain finite everywhere. We can now, in principle, get a functional relationship between α and p.

4.9 BRIEF DISCUSSION OF CONDENSATION

For a one-dimensional gas such as the one we have just considered with repulsive forces there is no condensation. Let us consider instead the case where there are both repulsion and attraction, such as the one depicted by the potential of Fig. 4.11. At $T = 0°$ all atoms are adjacent to each other in positions corresponding to V_{min} in Fig. 4.11:

$$OOOOOOOO \quad \cdots \cdot$$

As the temperature is increased there is a continuous transition from the "solid" to an aggregate of groups of atoms: $OOOO \quad OOOOO \quad OOOO$. The energy required for breaking up one group of atoms into two groups is roughly the energy required to break two atoms apart and we call it ε. The probability of such a break is $e^{-\varepsilon/kT}$, and when the temperature is raised it happens more or

less uniformly. In three dimensions one might have a group of N atoms.

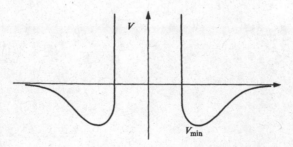

Fig. 4.11 A potential with attraction and repulsion.

The energy required to cut across a group is $\varepsilon = N^{2/3}\Delta$, where Δ is roughly the value of the energy required for taking one atom away from the group. Thus, the probability for splitting up the group is $e^{-N^{2/3}\Delta/KT}$. The larger N is, the higher T is required to be for a finite probability of breaking up a large group of atoms. Thus, in the one-dimensional case, a finite energy is required for breaking up a group of atoms, a finite number of groups exist, and the *fraction* of atoms (essentially infinite in number) in the condensed state (that is, in the largest group) is zero. In the three-dimensional case, because of the probability factor $e^{-N^{2/3}\Delta/KT}$, a finite fraction of the total gas can exist in the condensed state at a finite temperature.

For solids the difference is more subtle. In one dimension, as the temperature is raised, let the mean square error in the position of one atom relative to the ones on each side of it be δ^2. Then, over a group of N atoms, the mean square error in positions of the first and last atoms is $N\delta^2$ from the perfect lattice configuration. The error "propagates" (in a sense), and when temperature is increased, the solid expands monotonically. There is only one path along which information can be transmitted. In three dimensions, the position of an atom in one place influences that of a distant atom much more strongly than in one dimension—there are many paths connecting the two along which forces are being transmitted. Thus to melt a solid T must become high enough that the thermal energy can overcome these "multipath" forces.

CHAPTER 5

ORDER-DISORDER THEORY

5.1 INTRODUCTION

We will now consider a new class of statistical-mechanical problems, problems that do not involve motions of atoms but instead involve their positions in a lattice. More precisely, we will consider alloys made up of two types of atoms. For instance, in the cubic lattice shown in Fig. 5.1, there will be some preferred arrangement for the manner in which the lattice sites are occupied by type A and type B atoms. Let us take the case of zero temperature. Essentially, two types of lattice arrangements are possible, one of which is depicted in the figure, and they can be understood from a physical point of view as follows. Let V_{AB} be the interaction potential between unlike atoms and V_{AA}, V_{BB} the potential between the like atoms. If $V_{AB} > (V_{AA} + V_{BB})/2$, then A atoms prefer to be near A atoms and likewise for the B atoms. The solid would tend to exist in two distinct parts—one of A atoms only, and one of B atoms only. If $V_{AB} < (V_{AA} + V_{BB})/2$, the A and B atoms would tend to alternate in position throughout the solid. If the solid is heated, then gradually A and B atoms will exchange positions in a random way until, at the Curie temperature, the order "melts"; at still higher temperatures, disorder sets in. It is of interest to compute the partition function for such a system:

$$e^{-\beta F} = \sum_{\substack{\text{arrangements} \\ \text{of atoms}}} \exp\left(-\sum_{(i,j,\text{bonds})} \beta V_{ij}\right).$$

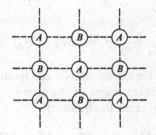

Fig. 5.1 Cubic lattice with two types of atoms, A and B.

Define:

$$\varepsilon_i = +1 \text{ for } A \text{ atom}$$
$$= -1 \text{ for } B \text{ atom}$$

$$N_A, N_B = \text{number of } A, B \text{ atoms, respectively.}$$

Thus,

$$\varepsilon_i \varepsilon_j = +1 \text{ if } AA \text{ pair}$$
$$= +1 \text{ if } BB \text{ pair}$$
$$= -1 \text{ if } AB \text{ pair.}$$

Considering nearest-neighbor interactions only and the total $N_A + N_B$ as constant, we will write the interaction energy in the form

$$V_{ij} = a\varepsilon_i \varepsilon_j + b\left(\frac{\varepsilon_i + \varepsilon_j}{2}\right) + c.$$

Then,

$$V_{AA} = a + b + c$$
$$V_{BB} = a - b + c$$
$$V_{AB} = -a + c$$

and

$$2a = \frac{V_{AA} + V_{BB}}{2} - V_{AB}.$$

Thus, the exponential in the partition-function equation above is

$$\exp\left(-\beta a \sum_{i,j} \varepsilon_i \varepsilon_j\right) \exp\left(-\beta b \sum_{i,j} \frac{\varepsilon_i + \varepsilon_j}{2}\right) \exp\left(-\beta c\right).$$

The last exponential, $e^{-\beta c}$, is a constant and does not enter into calculations.

In the sum $\sum_{\text{bonds}} (\varepsilon_i + \varepsilon_j)/2$, each atom contributes to the sum a certain number of times. The number of times an atom contributes is equal to the number of bonds it makes, that is, the number of nearest neighbors it has. For example, in the case of a two-dimensional square lattice, each atom has four nearest neighbors; so

$$\sum_{\text{bonds}} \frac{\varepsilon_i + \varepsilon_j}{2} = 2(N_A - N_B).$$

In general, if N_A and N_B are kept constant, we can ignore $\sum (\varepsilon_i + \varepsilon_j)/2$ because it does not depend on the arrangement of the atoms. The problem then is to calculate for the partition function the quantity

$$\sum \exp\left(-\beta a \sum_{i,j} \varepsilon_i \varepsilon_j\right) = \sum \exp\left(-H \sum_{\text{bonds}} \varepsilon_i \varepsilon_j\right).$$

But when we sum quantities over all possible arrangements of the atoms, the restriction that $N_A - N_B$ be constant may cause difficulties. To relax this restriction, we put in the usual "chemical potential." In other words, we change the problem (but not the results of the solution of the problem) by saying that $N_A - N_B$ can vary. Associated with variation of $N_A - N_B$ is a proportionate amount of chemical potential, which has the effect of making a certain ratio N_A/N_B the most probable. In such a case, we cannot ignore

$$\sum_{\text{bonds}} \frac{\varepsilon_i + \varepsilon_j}{2} = \sum_{\text{atoms}} \varepsilon_i.$$

Our problem reduces to the calculation of

$$\sum \exp\left(-H \sum_{\text{bonds}} \varepsilon_i \varepsilon_j - J \sum_{\text{atoms}} \varepsilon_i\right),$$

where, in summing over possible arrangements, we still take $N_A + N_B = N$ fixed but no longer fix N_A and N_B, separately. If $J = 0$, we are obviously solving the problem of $\langle N_A \rangle = \langle N_B \rangle$. For practice, we will evaluate the above sum for the one-dimensional case. Then we will consider an approximate method that works for two and three dimensions, and finally we will find the sum exactly for two dimensions. The exact solution for three dimensions has not yet been found.

It should be noted that the above model can be taken as a model of ferromagnetism. We interpret $\varepsilon_1 = +1$ as spin up and $\varepsilon_1 = -1$ as spin down, where the atoms are taken to have half-integer spin. A better model for ferromagnetism is to take the

$$\text{Hamiltonian} = H \sum_{\text{bonds}} S_i \cdot S_j$$

where S_i is the spin operator for the ith atom. This model is not equivalent to the model with $\varepsilon_i \varepsilon_j$ terms. For example, a two-atom system has

$$S_1 \cdot S_2 = \tfrac{1}{2}[(S_1 + S_2)^2 - \tfrac{3}{2}\hbar^2],$$

which has the value $\tfrac{1}{4}\hbar^2$ with *a priori* probability three times as large as that for which the value of $S_1 \cdot S_2$ has value $-\tfrac{3}{4}\hbar^2$. That is, there are three ways in which two atoms of spin $\tfrac{1}{2}$ can combine to give a system with total spin 1, but only one way in which they can combine to give spin 0. On the other hand, $\varepsilon_i \varepsilon_j$, takes on its two possible values with equal *a priori* probability.

Although the two models are different, the qualitative results might be similar. For the rest of this chapter we will discuss the model in which the energy has terms $\varepsilon_i \varepsilon_j$. This is the Ising model.

5.2 ORDER-DISORDER IN ONE-DIMENSION

For practice, let us evaluate

$$e^{-\beta F} = \sum_{\substack{\varepsilon_1 = \pm 1 \\ \varepsilon_2 = \pm 1 \\ \cdots}} \exp\left(-H \sum \varepsilon_i \varepsilon_{i+1} - J \sum \varepsilon_i\right)$$

in one dimension.

First, suppose $J = 0$. Then an easy way of doing the problem is to define a new variable:

$$\eta_i = \begin{cases} 1 & \text{if} \quad \varepsilon_i = \varepsilon_{i+1} \\ -1 & \text{if} \quad \varepsilon_i = -\varepsilon_{i+1} \end{cases}.$$

Then

$$e^{-\beta F} = \sum_{\substack{\eta_1 = \pm 1 \\ \eta_2 = \pm 1 \\ \cdots}} \exp\left[-H \sum_i \eta_i\right] = \left(\sum_{\eta_1 = \pm 1} e^{-H\eta_1}\right)\left(\sum_{\eta_2 = \pm 1} e^{-H\eta_2}\right)\cdots$$

$$= (2 \cosh H)^N.$$

So $F = -N/\beta \ln(2 \cosh H)$, and we have completed the calculation.

If $J \neq 0$, the above method does not work. We have to use what is, by now, a familiar trick. That is, we make up a new problem that, for large N, is almost the same as the old one, but for which a recursion relation can be found. We will try to evaluate the partition function for a system in which the particle on the end has a fixed ε.

Let

$$X_N(\varepsilon_{N+1}) = \sum_{\varepsilon_1, \varepsilon_2 \cdots \varepsilon_N} \exp\left[-H(\varepsilon_1 \varepsilon_2 + \cdots + \varepsilon_N \varepsilon_{N+1})\right.$$

$$\left. -J(\varepsilon_1 + \varepsilon_2 + \cdots + \varepsilon_{N+1})\right].$$

Then

$$X_{N+1}(\varepsilon_{N+1}) = \sum_{\varepsilon_{N+1} = \pm 1} X_N(\varepsilon_{N+1}) \exp\left[-H\varepsilon_{N+1}\varepsilon_{N+2} - J\varepsilon_{N+2}\right].$$

The free energy for a large number of atoms should be proportional to the number of atoms; therefore

$$X_{N+1}(\varepsilon_{N+2}) = e^{\beta\mu} X_N(\varepsilon_{N+2}),$$

where μ is independent of N for large N. The free energy is then $F = -\mu N$. The recursion relation then becomes two simultaneous linear equations

$$e^{\beta\mu} X_N(y) = \sum_{\chi = \pm 1} X_N(\chi) e^{-H\chi y - Jy} \qquad \text{where} \qquad y = \pm 1.$$

If we think of $(X_N(1), X_N(-1))$ as an eigenvector and $e^{\beta\mu}$ as an eigenvalue of the above equation, then it is easy to see what to do next. To find $e^{\beta\mu}$ it is necessary to solve

$$0 = \det \begin{pmatrix} e^{-H-J} - e^{\beta\mu} & e^{H+J} \\ e^{H-J} & e^{-H+J} - e^{\beta\mu} \end{pmatrix}.$$

The result is

$$e^{\beta\mu} = e^{-H} \cosh J \pm e^{H}\sqrt{1 + e^{-4H}\sinh^2 J}.$$

There are two solutions here; which one do we want? If we look back at the case of $J = 0$, it is obvious that $e^{\beta\mu}$ becomes $2\cosh H = e^{-H} + e^{H}$, so the general result is

$$e^{\beta\mu} = e^{-H} \cosh J + e^{H}\sqrt{1 + e^{-4H}\sinh^2 J}.$$

Problem: Try to get the free energy as a function of the ratio between the number of A's and the number of B's.

5.3 APPROXIMATE METHODS FOR TWO DIMENSIONS

You may have noticed that in Section 5.2, none of the equations were considered important enough to number. This is because we were not especially interested in the answers, but instead wanted to examine some possible ways of looking at problems. Perhaps you will be able to improve upon and use some of the methods described here, or maybe you will find these methods useful for solving a completely different problem. In this section we will nibble at the edges of another problem: that of order–disorder for a two-dimensional square lattice. For simplicity, let us take $J = 0$.

One approach is to find a recursion relation. To do this, find the partition function for the case where all the ε's in the last column are fixed. Suppose there are M columns of N particles each, and let $X_M(\varepsilon_1, \varepsilon_2, \ldots, \varepsilon_N)$ be the partition function. As in the one-dimensional case, the free energy is proportional to the number of atoms. It follows that

$$
\begin{aligned}
e^{\beta\mu N} X_M(\varepsilon_1, \ldots, \varepsilon_N) &= X_{M+1}(\varepsilon_1, \ldots, \varepsilon_N) \\
&= \sum_{\theta_1, \ldots, \theta_N} \exp\{-H[\varepsilon_1\theta_1 + \varepsilon_2\theta_2 + \cdots + \varepsilon_N\theta_N]\} \\
&\quad \times \exp\{-H[\varepsilon_1\varepsilon_2 + \cdots + \varepsilon_{N-1}\varepsilon_N]\} X_M(\theta_1, \ldots, \theta_N).
\end{aligned}
$$

$$(5.1)$$

The analogous equation in Section 5.2 required for its solution the diagonalization of a two-by-two matrix. The above equation requires the diagonalization of a 2^N-by-2^N matrix.

Given the matrix equation $MA = \lambda A$ with M hermitian, a well-known way for finding the solution with the minimum or maximum λ is to try to find the extremum of $A \cdot MA / A \cdot A$. This method might be applicable to Eq. (5.1). Let

$$X_N(\varepsilon_1, \ldots, \varepsilon_N) = \exp\left\{-\tfrac{1}{2}H[\varepsilon_1\varepsilon_2 + \cdots + \varepsilon_{N-1}\varepsilon_N]\right\}Z(\varepsilon_1, \ldots, \varepsilon_N).$$

Then Eq. (5.1) becomes

$$e^{\beta\mu N}Z(\varepsilon_1, \ldots, \varepsilon_N) = \sum_{\theta_1, \ldots, \theta_N} \exp\left\{-\tfrac{1}{2}H[\varepsilon_1\varepsilon_2 + \cdots + \varepsilon_{N-1}\varepsilon_N]\right\}$$
$$\times \exp\left\{-\tfrac{1}{2}H[\theta_1\theta_2 + \cdots + \theta_{N-1}\theta_N]\right\}$$
$$\times \exp\left\{H[\varepsilon_1\theta_1 + \cdots + \varepsilon_N\theta_N]\right\}Z(\theta_1, \ldots, \theta_N).$$

We made this substitution in order to get an eigenvalue equation with a symmetric real matrix, that is, a hermitian matrix. Then, to solve this equation, we must find the extremum of I/D, where

$$I = \sum_{\theta_i, \varepsilon_i} Z(\theta_1, \ldots, \theta_N)Z(\varepsilon_1, \ldots, \varepsilon_N) \exp\left\{-\tfrac{1}{2}H[\varepsilon_1\varepsilon_2 + \cdots + \varepsilon_{N-1}\varepsilon_N]\right\}$$
$$\times \exp\left\{-\tfrac{1}{2}H[\theta_1\theta_2 + \cdots + \theta_{N-1}\theta_N]\right\}\exp\left\{-H[\varepsilon_1\theta_1 + \cdots + \varepsilon_N\theta_N]\right\}.$$
$$D = \sum_{\varepsilon_i} |Z(\varepsilon_1, \ldots, \varepsilon_N)|^2.$$

Of course, it is not likely that you will be able to get an exact extremum of I/D. But it should be possible to get an approximate one by cleverly choosing an appropriate function $Z(\varepsilon_1, \ldots, \varepsilon_N)$. To be more systematic, one might choose a class of functions $Z_\alpha(\varepsilon_1, \ldots, \varepsilon_N)$ and maximize I/D with respect to α. $e^{\beta\mu N}$ should be equal to the maximum possible value of I/D.

Problem: Find as large a value of I/D as you can. In evaluating certain sums, try setting $\theta_N = \theta_1$ and $\varepsilon_N = \varepsilon_1$; it might simplify matters if the two-dimensional lattice were wrapped around a cylinder.

We have been considering interactions such that when the temperature is absolute zero, atoms of type A and of type B alternate in the lattice. For the case of two or three dimensions, if the temperature is low enough, the crystal has long-range order. If we know what type of atoms are found at each lattice point in one region of the crystal, we know what type of atoms are likely to be found fairly far off. Note that in one dimension one wrong atom at a lattice point reverses the expected order of atoms on the other side of the defect. Thus, in two or more dimensions, we can speak of α sites and β sites, where α sites are those lattice points at which an A is more likely to be found. In one dimension we cannot do this. For the case of two dimensions, Bethe used the concept of α and β sites to describe order-disorder phenomena approximately.*

---

* Bethe, H. "Statistical Theory of Superlattices," *Proc. Roy. Soc.*, **150**, 552 (1935).

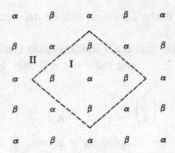

Fig. 5.2 Two-dimensional lattice with α sites and β sites.

Following Bethe, we will work with only a few atoms of the lattice and will try to describe the effect of the rest of the atoms by one parameter. We will look closely at only the atoms pictured within the dotted line (region I) of Fig. 5.2.

The relative probability of a given distribution of atoms in the entire crystal is $\exp(-H \sum_{\text{bonds}} \varepsilon_i \varepsilon_j)$ (we are assuming there are just as many A atoms as B atoms). If we consider only the five atoms in region I we have to allow for the fact that the four outer β sites are affected by region II. We do this approximately by saying that the relative probability of a given distribution of the five atoms is $\varepsilon^n \exp(-H \sum \varepsilon_i \varepsilon_j)$ where n is the number of β sites in region I that have the wrong type of atom (A) on them. The sum is only over the four bonds in region I. ε can be thought of as

$$\varepsilon = \frac{a \ priori \ \text{probability that a } \beta \text{ site in region I has an } A \text{ atom}}{a \ priori \ \text{probability that a } \beta \text{ site in region I has a } B \text{ atom}}.$$

By "*a priori* probability" I mean the probability that would be assumed before the thermodynamic effects of the interaction are taken into account. That is, even if there were no interactions between the atoms of region I, the interaction of the four external atoms of region I with region II would still tend to cause the β sites to be occupied by B atoms. Notice that n is not the total number of sites of region I that have the wrong atom. If the α site in the center has a B atom, we do not multiply the relative probability by another ε because region II does not interact directly with that atom. Note also that the atoms inside region I can affect each other by way of region II, so ε is only an approximate way of describing the effect of region II.

For a given value of ε we can compute w, the relative probability that the α site in region I is occupied by a B atom. We can also compute w', the relative probability that a given β site of region I is occupied by an A atom. We require that $w = w'$, because the probability of a wrong atom occupying a given site is independent of the site. The equation $w = w'$ will allow us to fix ε.

Some of the results of Bethe's method will be derived below.

$$w = \sum_{n=0}^{4} \text{(relative probability that the } \alpha \text{ site in region I is occupied by a } B \text{ atom}$$
$$\text{and there are } n \ \beta \text{ sites occupied by } A \text{ atoms).}$$

There are

$$\binom{4}{n} = \frac{4!}{n! \, (4 - n)!}$$

ways in which $n\beta$ sites may be occupied by A atoms. It is easy to see that

$$\sum_{\text{bonds}} \varepsilon_i \varepsilon_j = 4 - 2n.$$

Then

$$w = \sum_{n=0}^{4} \binom{4}{n} \varepsilon^n e^{-H(4-2n)} = e^{-4H} \sum_{n=0}^{4} \binom{4}{n} (\varepsilon e^{2H})^n = e^{-4H}(1 + \varepsilon e^{2H})^4$$

$$= (e^{-H} + \varepsilon e^{H})^4. \tag{5.2}$$

w is a relative, not an actual, probability. The actual probability that the α site has a B atom is w multiplied by the appropriate normalization factor.

Let r be the relative probability that the α site has an A atom. The equation for r is the same as that for w, except $H \leftrightarrow -H$.

$$r = (e^{H} + \varepsilon e^{-H})^4. \tag{5.3}$$

The normalization factor is $1/(w + r)$; so the actual probability that the site has a B atom is

$$\frac{(e^{-H} + \varepsilon e^{H})^4}{(e^{-H} + \varepsilon e^{H})^4 + (e^{H} + \varepsilon e^{-H})^4}.$$

$w' \propto \frac{1}{4}$ (expected number of β sites of region I with A atoms)

$$w' = \frac{1}{4} \sum_{n=0}^{4} n[\text{(relative probability that } n \ \beta \text{ sites are occupied by } A \text{ atoms and}$$
$$\text{the } \alpha \text{ site has a } B \text{ atom}) + (\text{relative probability that } n \ \beta \text{ sites are}$$
$$\text{occupied by } A \text{ atoms and the } \alpha \text{ site has an } A \text{ atom})]$$

$$= \frac{1}{4} \sum_{n=0}^{4} n \binom{4}{n} \varepsilon^n [e^{-H(4-2n)} + e^{H(4-2n)}]$$

$$= \sum_{n=1}^{4} \binom{3}{n-1} \varepsilon^n [e^{-H(4-2n)} + e^{H(4-2n)}]$$

$$= \varepsilon e^{H}(e^{-H} + \varepsilon e^{H})^3 + \varepsilon e^{-H}(e^{H} + \varepsilon e^{-H})^3. \tag{5.4}$$

Equating w and w' from Eqs. (5.2) and (5.4), we find

$$(e^{-H} + \varepsilon e^{H})^4 = \varepsilon e^{H}(e^{-H} + \varepsilon e^{H})^3 + \varepsilon e^{-H}(e^{H} + \varepsilon e^{-H})^3.$$

So

$$(e^{-H} + \varepsilon e^{H})^{3}(e^{-H} + \varepsilon e^{H} - \varepsilon e^{H}) = \varepsilon e^{-H}(e^{H} + \varepsilon e^{-H})^{3}.$$

Simplifying our equation for ε, we get

$$\varepsilon^{1/3} = \frac{e^{-H} + \varepsilon e^{H}}{e^{H} + \varepsilon e^{-H}}. \qquad (5.5)$$

Notice that if ε satisfies Eq. (5.5), so does $1/\varepsilon$. This fact holds because nowhere in the mathematics of our derivation did we specify that α sites, rather than β sites, are the locations for which A atoms are more probable.

Solving for e^{2H} in terms of ε, we find two solutions: either

$$e^{2H} = \frac{e^{2/3} - \varepsilon^{-2/3}}{\varepsilon^{1/3} - \varepsilon^{-1/3}} = \varepsilon^{1/3} + \varepsilon^{-1/3} \qquad (5.6)$$

or else $\varepsilon = 1$ and H is arbitrary (the latter solution is most easily seen from Eq. (5.5)).

The minimum value of $\varepsilon^{1/3} + \varepsilon^{-1/3}$ is 2; so if H is too small (high temperature) we must take $\varepsilon = 1$. From the definition of ε, $\varepsilon = 1$ corresponds to complete disorder. We can expect discontinuities in certain thermodynamic quantities or their derivatives at the temperature for which complete disorder sets in.

As an example of a thermodynamic calculation using the above model for order-disorder, let us calculate the energy.

Suppose a bond is an A–B bond with probability p. Then with probability p a given bond contributes energy $- H/\beta$, and with probability $(1 - p)$ the bond contributes $+ H/\beta$. The expected value of the energy is

$$E = (\text{number of bonds})\left[p\left(\frac{-H}{\beta}\right) + (1 - p)\left(\frac{+H}{\beta}\right)\right] = \frac{2NH}{\beta}(1 - 2p) \quad (5.7)$$

where N is the number of atoms.

$$\frac{1}{w + r}\left[\varepsilon e^{H}(e^{-H} + \varepsilon e^{H})^{3}\right]$$

is the probability that a given β site of region I is occupied by an A atom given that the α site has a B atom. Except for the normalizing factor $1(w + r)$, this is one of the terms in w'.

$$1 - \frac{1}{w + r}\left[\varepsilon e^{-H}(e^{H} + \varepsilon e^{-H})^{3}\right]$$

is the probability that a given β site of region I is occupied by a B atom, given that the α site has an A atom. The probability that a given bond between an α and a β site is an A–B bond is therefore

$$p = \frac{\varepsilon e^H (e^{-H} + \varepsilon e^H)^3}{w + r} + \left(1 - \frac{\varepsilon e^{-H}(e^H + \varepsilon e^{-H})^3}{w + r}\right). \tag{5.8}$$

H/β is a constant independent of temperature. Using Eqs. (5.7) and (5.8) we can get $C_V = \partial E/\partial T$. We get the entropy from the equation

$$S(T) = S(T) - S(0) = \int_0^T C_V \frac{dT'}{T'}. \tag{5.9}$$

At $T = \infty$, we can compute the entropy exactly from its definition. Recall that $S = -k \sum_n P_n \ln P_n$, where P_n is the probability of state n. There are 2^N states of our N-atom system, and each has equal probability at infinite temperature. So

$$S = -k(2^N)\left(\frac{1}{2^N} \ln \frac{1}{2^N}\right) = Nk \ln 2. \tag{5.10}$$

Our approximate calculation, using Eq. (5.9) gives $S = Nk(0.697)$ instead of $Nk(0.693)$.

5.4 THE ONSAGER PROBLEM

Consider a two-dimensional square lattice of N atoms. Each atom is designated by $+1$ or -1. That is, the ith atom carries a number ε_i, which is $+1$ or -1. The energy of interaction of nearest neighbors is $+ H\varepsilon_i\varepsilon_j$ in units of kT. A state of the system is denoted by giving the ε_i for each atom. The problem is to calculate the free energy of the system.

The partition function

$$Q = \sum_{\text{all states}} e^{-\beta E_K}$$

$$Q = \sum_{\text{all states}} e^{-\Sigma^* H\varepsilon_i\varepsilon_j} = \sum_{\text{all states}} \Pi^* e^{-H\varepsilon_i\varepsilon_j},$$

where Σ^* and Π^* mean the sum and product on pairs of points i and j that are adjacent on the lattice. It is easy to see that if the sign of H is changed, Q remains the same. For convenience we will write

$$Q = \sum_{\text{all states}} \Pi^* e^{H\varepsilon_i\varepsilon_j} \tag{5.11}$$

and take H positive.

Since $\varepsilon_i\varepsilon_j = \pm 1$,

$$\frac{e^H + e^{-H}}{2} + \varepsilon_i\varepsilon_j \frac{e^H - e^{-H}}{2} = e^{+\varepsilon_i\varepsilon_j H}$$

is obviously true. So

$$e^{H\varepsilon_i\varepsilon_j} = \cosh H + \varepsilon_i\varepsilon_j \sinh H$$
$$= \cosh H(1 + \varepsilon_i\varepsilon_j T), \tag{5.12}$$

where $T = \tanh H$. (Note that T does *not* represent temperature in this section.)
From Eqs. (5.11) and (5.12) we find

$$Q = \sum \prod{}^* [\cosh H(1 + \varepsilon_i\varepsilon_j T)]. \tag{5.13}$$

Remembering that for an N atom lattice there are $2N$ bonds, we have

$$Q = \cosh^{2N} H \sum_{\text{all states}} \prod{}^*(1 + \varepsilon_i\varepsilon_j T) \equiv 2^N Q' \cosh^{2N} H. \tag{5.14}$$

We must now find the modified partition function

$$Q' = \frac{1}{2^N} \sum_{\text{all states}} \prod{}^*(1 + \varepsilon_i\varepsilon_j T). \tag{5.15}$$

Expanding the product we use the notation $\sum_l^*$ to indicate a sum over all possible
sets of l different bonds. Then

$$2^N Q' = \sum_{\varepsilon_1 = \pm 1} \cdots \sum_{\varepsilon_N = \pm 1} \{1 + T\Sigma_1^* \varepsilon_i\varepsilon_j + T^2\Sigma_2^*(\varepsilon_i\varepsilon_j)(\varepsilon_{i'}\varepsilon_{j'}) + \cdots\}. \tag{5.16}$$

Note that in each term, the ε's occur in pairs corresponding to nearest
neighbors, and no such pair occurs twice in the same product.

Associate with each pair $\varepsilon_i\varepsilon_j$ a bond connecting the ith and jth atoms (i and j
are nearest neighbors). With each term or product of $2l$ ε's, we associate a graph
or set of l bonds. Because no pair $\varepsilon_i\varepsilon_j$ appears twice in a given term, no bond
appears twice in a given graph. For example, Fig. 5.3 gives the graph associated
with the term

$$(\varepsilon_1\varepsilon_2)(\varepsilon_3\varepsilon_4)(\varepsilon_5\varepsilon_6)(\varepsilon_1\varepsilon_5)(\varepsilon_2\varepsilon_6)(\varepsilon_3\varepsilon_7)(\varepsilon_4\varepsilon_8)(\varepsilon_7\varepsilon_8)(\varepsilon_{10}\varepsilon_{11})(\varepsilon_{10}\varepsilon_{14}).$$

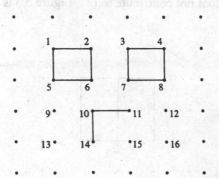

Fig. 5.3 A graph or set of l bonds.

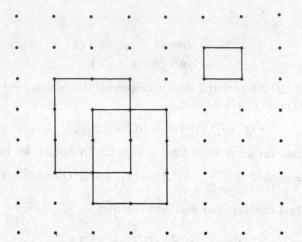

Fig. 5.4 Example of a closed graph

Now,

$$\sum_{\varepsilon=\pm 1} \varepsilon = 0 \qquad \text{and} \qquad \sum_{\varepsilon=\pm 1} \varepsilon^2 = 2. \tag{5.17}$$

From Eqs. (5.16) and (5.17), it is clear that only those terms where each ε_i appears an even number of times will contribute to Q'. This is equivalent to saying that the only graphs that count (or contribute to Q') are those in which each atom or lattice point has an even number (0, 2, or 4) of bonds emanating from it. In other words, the contributing graphs must be superpositions of simple closed polygons having no common sides. Such graphs may be called "closed graphs."

Figure 5.4 gives an example of a closed graph. Note that the graph of Fig. 5.3 is not closed and does not contribute to Q'. Figure 5.5 is another example of

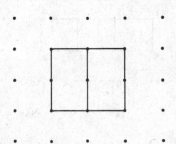

Fig. 5.5 Example of an "Unclosed" Graph.

a nonclosed graph, which contributes zero. Each term in Eq. (5.16) corresponding to a closed graph has each ε_i occuring to an even power (0, 2, or 4), and therefore equals simply T^L, where L is the number of bonds. Summing over all ε_i gives a factor 2^N, which is cancelled by the 2^N on the left side of Eq. (5.16). Each closed graph of length L thus contributes a term T^L to Q'; so

$$Q' = \sum_L g(L)T^L, \tag{5.18}$$

where $g(L)$ is the number of closed graphs of length L that can be drawn on the lattice. Notice that $g(L) = 0$ unless L is even, so the sign of T (and therefore the sign of H) is immaterial, as we expected.

Method of Finding Transition Point

We now digress for a moment to show a method (due to Kramers and Wannier) by which the transition point of the Onsager problem can be found. This method is not powerful enough to yield an analytic expression for the partition function.

Consider a closed graph on the lattice. Label the midpoints of the lattice squares $\mu_i = +1$ if inside the graph and $\mu_i = -1$ if outside. (See Fig. 5.6.)

Consider a bond (---) between midpoints of unlike μ values. Each of these bonds corresponds to a bond of the closed graph (because each -- bond cuts a — bond).

We now have two lattices. The new lattice is formed by putting "atoms" described by μ_i at the midpoint of the old lattice. For the new lattice we can discuss Q at a temperature T' and such that the interaction energy is h in units at kT'. Then

$$Q(h) = \sum_{\text{all states}} \Pi^* e^{h\mu_i\mu_j}.$$

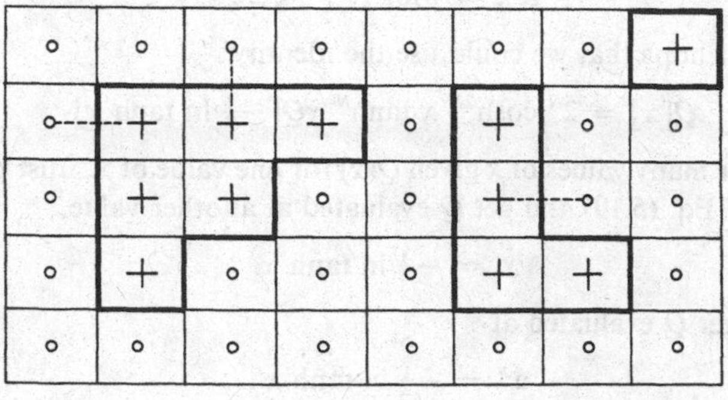

Fig. 5.6 Midpoints of lattice squares labelled + inside the closed graph and − outside it.

If $\mu_i = +\mu_j$, then $\mu_i\mu_j = +1$, and there is a contribution e^{+h}. If $\mu_i = -\mu_j$, then the contribution is e^{-h}.

Now, if in a given state there are L unlike bonds and $2N - L$ like bonds, then its contribution to $Q(h)$ is

$$e^{-Lh}e^{(2N-L)h}$$

so that

$$Q(h) = e^{2Nh} \sum_L m(L)e^{-2Lh}$$

where $m(L)$ is the number of ways by which unlike pairs can be arranged on the lattice.

But from the preceding argument (concerning --- and — bonds), $m(L) = g(L)$, where $g(L)$ is defined by Eq. (5.18).

Thus

$$Q(h) = e^{2Nh} \sum_L g(L)e^{-2Lh},$$

$$Q(H) = 2^N \cosh^{2N} H \sum_L g(L)T^L.$$

Let us now define h by $e^{-2h} = T = tghH$. Then:

$$Q(h) = \tanh^{-N} H \sum_l g(l)T^l$$

and we get the identity

$$Q(H) = 2^N \cosh^{2N} H \tanh^N H\, Q(h).$$

Now, when H is large, h is small; if $Q(H)$ has a singular point, so does $Q(h)$. That is, if there is a transition at H_{cr}, there is also a transition at h_{cr}. If we assume that there is only one transition or Curie point, $H_{cr} = h_{cr}$ and $H_{cr} = -\frac{1}{2}\ln(\tanh H_{cr})$. This can be solved to give

$$H_{cr} = \tfrac{1}{2}\log(1 - \sqrt{2}).$$

You might hope that we could use the identity

$$Q[x] = 2^N \cosh^{2N} x \tanh^N x\, Q[-\tfrac{1}{2}\ln\tanh x] \tag{5.19}$$

to find $Q(x)$ for many values of x given $Q(x)$ for one value of x. Just plug in x on the left side of Eq. (5.19) and get Q evaluated at another value,

$$x' = -\tfrac{1}{2}\ln\tanh x.$$

We can then get Q evaluated at

$$x'' = -\tfrac{1}{2}\ln\tanh x'.$$

Unfortunately, $x'' = x$, so this method is not useful.

Continuation of Onsager Problem

Call $h(l)$ the number of ways that, starting from the atom (origin), we can proceed in l steps (l bonds) through the lattice and return to the origin without using the same bond twice. In $h(l)$ we do not count separately the two or more different paths we can use to form the same polygon.

For carrying out the calculation we shall need the sum over all different l polygons of the crystal. Our $h(l)$ polygons for atom A in Fig. 5.7, say, will be counted again in the $h(l)$ polygons of other atoms, say B in Fig. 5.7. Each polygon will occur l times. Therefore, the number of polygons per atom is $h(l)/l$. Let

$$q = \sum_{l=1}^{\infty} \frac{h(l)}{l} T^l. \tag{5.20}$$

We have

$$Q' = \sum_L g(L)T^L = \sum_{\substack{\text{closed} \\ \text{diagrams}}} T^L,$$

but need an expression for Q' that we can hope to evaluate. We will work up to such an expression by finding an approximation to Q', which we will then correct by means of a trick.

A term from a closed diagram can be thought of as a product of terms made up of graphs formed by polygons. The contribution to Q' from no polygons is simply 1. The contribution from graphs consisting of one polygon is Nq, for the number of graphs consisting of one polygon of length l is $Nh(l)/l$. The contribution from pairs of polygons, one of length l and the other of length l' is $(Nh(l)/l)$ $(Nh(l')/l')T^{l+l'}$ so the contribution from pairs of polygons of any length is about

$$\frac{1}{2} \sum_{l,l'} \left(\frac{Nh(l)}{l}\right)\left(\frac{Nh(l')}{l'}\right) T^{l+l'},$$

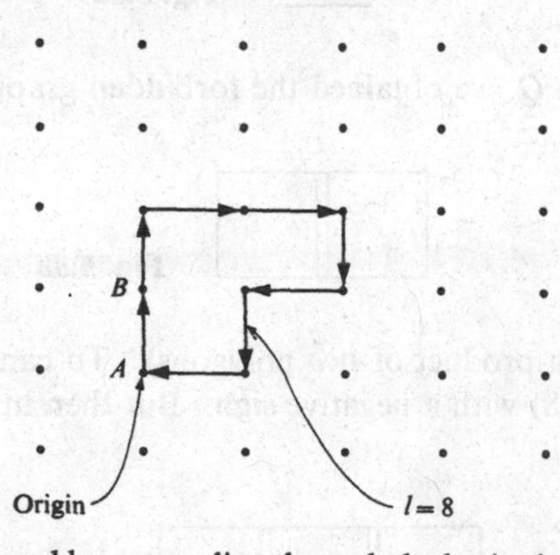

Fig. 5.7 A polygon formed by proceeding through the lattice back to the starting point, without using the same bond twice.

where the factor of 1/2 is to account for the fact that we count each pair twice. So the contribution from pairs is about $\frac{1}{2}N^2q^2$. I say "about" because we have counted such graphs as

even though that pair of squares is not "closed," since one bond appears twice. Furthermore

is a pair of squares and is also a single polygon of length 8, so we have counted it twice. If we continue to ignore such errors, we get

$$Q' \approx 1 + Nq + \tfrac{1}{2}(Nq)^2 + \tfrac{1}{3}!(Nq)^3 + \cdots = e^{Nq}. \tag{5.21}$$

Our next step logically should be to correct q in order to get Q' right. We can correct q by correcting $h(l)$. Let us write down a few of the correction terms.

Because of the presence of polygon 5.8a

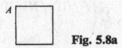

Fig. 5.8a

in the first-order term in Q, we obtained the forbidden graph of 5.8b

Fig. 5.8b

in *second* order (i.e., as a product of *two* polygons). To cancel graph 5.8b, we insert graph 5.8b into $h(8)$ with a negative sign. But then in Q' we seem to get terms such as

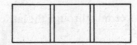

in both second and third order. In second order, there are two terms that form a diagram of type 5.8c, and both have a negative sign:

and

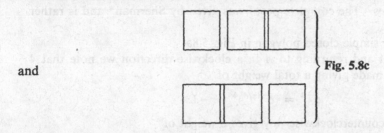

Fig. 5.8c

that is, there are two ways of writing 5.8c as a product of 5.8a and a 5.8b type term. In third order, one of the two second-order terms is canceled by the positive product of three type (a) terms. To cancel the other second order term of type (c), we must insert (c) with a positive sign into $h(12)$.

In Eq. (5.21), we counted polygon 5.8d twice; therefore when we get a rule for what terms belong in our corrected $h(l)$ and what their signs should be, we will have to have an excuse for not counting (d) in the corrected $h(l)$.

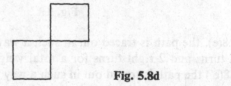

Fig. 5.8d

Now, let us look at where we stand. In evaluating $h(l)$ we expected trouble because of the restriction to certain allowed polygons. In correcting $h(l)$ we found that there must be other polygons with not necessarily positive contribution, and some of the allowed polygons should not be counted. The natural thing to do is assign each polygon an appropriate weight, and not ignore any polygon in the sum. Recall that formerly we did not count separately the two or more different paths we can follow when starting our from an atom and forming a polygon. Now we will count all the different paths, but we will choose the weight of each path so that our corrected "q" will lead to the right value of Q' when we use Eq. (5.21).

A Topological Theorem

Give a weight $\alpha = e^{i\pi/4}$ for a left turn and $\alpha^{-1} = e^{-i\pi/4}$ for a right turn, and keep track of the left and right turns as a closed path (returning to origin) is traversed in the lattice. The topological argument of Kac and Ward is that the closed graphs (or ones that we want to count) will be counted, and the forbidden graphs (e.g., Fig. 5.5) will cancel out when different ways of traversing them are considered. That this theorem is true for simple cases will now be demonstrated by a few examples. The complete proof was given by Sherman* and is rather involved.

Consider the simple closed polygon in Fig. 5.8a.

Starting at A and returning to A in a clockwise direction we note that 4 right turns were made giving a total weight of

$$\alpha^{-4} = (e^{-i\pi/4})^4 = -1.$$

Proceeding in a counterclockwise way gives a weight of

$$(\alpha)^4 = (e^{i\pi/4})^4 = -1.$$

The total weight is thus -2, and the simple polygon is indeed counted. To get the correct weight of 1, we multiply everything by $-1/2$.

Now, consider the polygons in Fig. 5.8e and e':

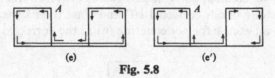

(e)   (e')

Fig. 5.8

In Fig. 5.8(e), the path is traced out in such a way (starting from A) that there are 6 left turns and 2 right turns for a total weight of $\alpha^6(\alpha^{-2}) = (\alpha^4) = -1$. In Fig. 5.8(e') the path is traced out in such a way that there are 4 left turns and 4 right turns for a total weight of $\alpha^4\alpha^{-4} = 1$. The total weight then of the two clockwise paths is $-1 + 1 = 0$. It is clear that the two counterclockwise paths have weight 0 also.

Similarly you should be able to show that 5.8(b), 5.8(c), and 5.8(d) and all simple polygons give the correct contribution (remember that everything is multiplied by $-1/2$).

Method of Calculating partition Function

Let q denote our corrected value of q. Then

$$q = \sum_{l=1}^{\infty} \frac{\{\text{corrected } h(l)\}}{l} T^l$$

and

$$Q' = e^{Nq} - 2T \frac{dq}{dT} = -2 \sum_{l=1}^{\infty} \{\text{corrected } h(l)\} T^l$$

---

* Sherman, S., *J. Math. Phys.*, **1**, 202 (1960); **4**, 1213 (1963).

which is the sum over all paths leaving and returning to the origin of a complex "amplitude" or weight for each path. This amplitude is the product of amplitudes for each lattice point passed:

T if one continues forward:

$T\alpha$ if one turns counterclockwise: $\alpha = e^{i\pi/4}$;

$T\alpha^{-1}$ if one turns clockwise: $\alpha^{-1} = e^{-i\pi/4}$;

0 if one turns 180°.

In our rule for determining the amplitude for each lattice point passed, we have neglected to say how to deal with the first step, whose amplitude depends on the last step. In what follows, we will calculate the amplitude to start at the origin and arrive back at the origin from below, that is, moving upward. Thus, we will choose as our rule for the amplitude of the first step:

T if one starts upward;

$T\alpha$ if one starts to the left;

$T\alpha^{-1}$ if one starts to the right;

0 if one starts downward.

More generally, we will use the above rule for the first step and work with amplitudes for starting at the origin and arriving at any lattice point (x, y). We shall have to describe the amplitude by giving four components: $U(x, y)$, the amplitude to arrive at x, y moving upward; $D(x, y)$, the amplitude to arrive moving downward; $R(x, y)$, the amplitude to arrive *from* the right; and $L(x, y)$, the amplitude to arrive *from* the left. This, as mentioned before, is because the amplitude to make the next step depends on the arrival direction, and we shall have to keep track of it.

Roughly, what we will do is calculate the amplitudes for arriving at (x, y) from various directions in exactly n steps, and then sum over n. $U_n(x, y)$ is the amplitude to arrive at (x, y) in exactly n steps with the last step upward and similarly for D_n, R_n, and L_n. We assume that in the zeroth step we arrive at the origin moving upward. Thus $U_0(x, y) = \delta_{x,0}\delta_{y,0}$, $D_0 = R_0 = L_0 = 0$.

In other words, the amplitude to arrive in zero steps is one if we arrive upward at the origin and zero for any other point or any other direction of arrival. With these amplitudes for the zeroth step, we can get the amplitudes for any number of steps by recursion. We can write for the amplitude to arrive upward at (x, y) in $n + 1$ steps in terms of amplitudes to arrive from $(x, y - 1)$ in n steps.

$$U_{n+1}(x, y) = TU_n(x, y - 1) + T\alpha L_n(x, y - 1) + T\alpha^{-1}R_n(x, y - 1).$$

If we came to $x, y - 1$ going up, we have an additional amplitude factor T. If we came to $x, y - 1$ from the left, we need to turn counterclockwise and add a factor $T\alpha$, and so on. We cannot get to x, y going up if we previously entered $x, y - 1$ going down, for we do not allow paths that reverse themselves. Similar expressions for $R_n(x, y)$, $L_n(x, y)$, and $D_n(x, y)$ are easily written.

Then we see that our assumption for the amplitude of the zeroth step will give, for the first step, the correct amplitudes (such as T if one starts upwards). When summing over n, we will eventually have to subtract out the $n = 0$ term, because we do not consider a figure with zero side to be a polygon, even though it starts and ends at the origin.

To simplify the summation over n, we define the transform of $U_n(x, y)$

$$U_n(\xi, \eta) = \sum_{x=-\infty}^{\infty} \sum_{y=-\infty}^{\infty} U_n(x, y)e^{-i\xi x}e^{-i\eta y}. \qquad (5.22)$$

Then

$$U_n(x, y) = \int_0^{2\pi} \int_0^{2\pi} e^{i\xi x}e^{i\eta y}U_n(\xi, \eta)\frac{d\xi\, d\eta}{(2\pi)^2}.$$

For example, the transform of $U_n(x, y - 1)$ is $e^{-i\eta}u_n(\xi, \eta)$, and the transform of $U_0(x, y)$ is 1. The transformed equation then reads

$$U_{n+1}(\xi, \eta) = Te^{-i\eta}U_n(\xi, \eta) + 0 \cdot e^{-i\eta}D_n(\xi, \eta) + T\alpha e^{-i\eta}L_n(\xi, \eta)$$
$$+ T\alpha^{-1}e^{-i\eta}R_n(\xi, \eta). \qquad (5.23)$$

and similarly for $D_{n+1}(\xi, \eta)$, $L_{n+1}(\xi, \eta)$, and $R_{n+1}(\xi, \eta)$.

If we let ψ_n be a four-component symbol (column vector) with components (U_n, D_n, L_n, R_n), we can write the matrix equation

$$\psi_{n+1}(\xi, \eta) = TM(\xi, \eta)\psi_n(\xi, \eta) \qquad (5.24)$$

where $M(\xi, \eta)$ is the matrix

$$M(\xi, \eta) = \begin{vmatrix} e^{-i\eta} & 0 & \alpha e^{-i\eta} & \alpha^{-1}e^{-i\eta} \\ 0 & e^{i\eta} & \alpha^{-1}e^{i\eta} & \alpha e^{i\eta} \\ \alpha^{-1}e^{-i\xi} & \alpha e^{-i\xi} & e^{-i\xi} & 0 \\ \alpha e^{i\xi} & \alpha^{-1}e^{i\xi} & 0 & e^{i\xi} \end{vmatrix}. \qquad (5.25)$$

$$\psi_0(\xi, \eta) = (1, 0, 0, 0).$$
$$\psi_1(\xi, \eta) = TM\psi_0$$
$$\psi_2 = TM\psi_1 = (TM)^2\psi_0$$
$$\vdots$$
$$\psi_n = (TM)^n \psi_0.$$

Then

$$\psi = \sum_{n=0}^{\infty} (TM)^n \psi_0 = \frac{1}{1 - TM} \psi_0$$

is the transform of the amplitude to arrive in any number of steps. The amplitude to arrive at the origin moving upward is

$$\sum_{n=1}^{\infty} U_n(0, 0) = -1 + \sum_{n=0}^{\infty} U_n(0, 0) = -1 + \psi_0 \cdot \int_0^{2\pi} \int_0^{2\pi} e^{i\xi 0} e^{i\eta 0}$$

$$\times \frac{1}{1 - TM} \psi_0 \frac{d\xi \, d\eta}{(2\pi)^2}$$

$$= \int_0^{2\pi} \int_0^{2\pi} \left(\psi_0 \cdot \frac{1}{1 - TM} \psi_0 - 1 \right) \frac{d\xi \, d\eta}{(2\pi)^2} .$$

We could have arrived at the origin moving in any direction, so we must sum over all four values of ψ_0, that is $\psi_0 = (1, 0, 0, 0)$, $\psi_0 = (0, 1, 0, 0)$, $\psi_0 = (0, 0, 1, 0)$, $\psi_0 = (0, 0, 0, 1)$. The result is a trace. Thus,

$$-2T \frac{dq}{dT} = \int_0^{2\pi} \int_0^{2\pi} \text{Tr} \left[\frac{1}{1 - TM} - 1 \right] \frac{d\xi \, d\eta}{(2\pi)^2} \tag{5.26}$$

$$q = \int_0^{2\pi} \int_0^{2\pi} \frac{d\xi \, d\eta}{(2\pi)^2} \text{Tr} \left[\int_0^T - \frac{1}{2t(1 - tM)} + \frac{1}{2t} \right] dt$$

$$= \int_0^{2\pi} \int_0^{2\pi} \frac{d\xi \, d\eta}{(2\pi)^2} \text{Tr} \left[\int_0^T \frac{-M}{2(1 - tM)} \, dt \right]$$

$$= \frac{1}{2} \int_0^{2\pi} \int_0^{2\pi} \frac{d\xi \, d\eta}{(2\pi)^2} \text{Tr} \left[\log (1 - TM) \right]$$

$$= \frac{1}{2} \int_0^{2\pi} \int_0^{2\pi} \frac{d\xi \, d\eta}{(2\pi)^2} \log \det (1 - TM), \tag{5.27}$$

we have used the following theorem:

$$\text{Tr} \log A = \log \det A.$$

To prove this identity, note that

$$\det e^B = \lim_{N \to \infty} (\det e^{B/N})^N = \lim_{N \to \infty} \left[\det \left(1 + \frac{B}{N} \right) \right]^N$$

$$= \lim_{N \to \infty} \left(1 + \frac{1}{N} \text{Tr} \, B \right)^N = e^{\text{Tr} \, B}.$$

Therefore, $\log \det e^B = \operatorname{Tr} B$. Let $B = \log A$. Then $\operatorname{Tr} \log A = \log \det A$. M is given by Eq. (5.25), and one obtains

$$\det [1 - TM] = (T^2 + 1)^2 - 2T(1 - T^2)(\cos \xi + \cos \eta). \qquad (5.28)$$

Summarizing, we see that

$$BF/N = -\ln Q/N,$$

$$Q = 2^N(\cosh^{2N} H)Q',$$

$$Q' = e^{Nq}$$

$$q = \frac{1}{2} \int_0^{2\pi} \int_0^{2\pi} \frac{d\xi \, d\eta}{(2\pi)^2} \log \det (1 - TM),$$

$$\det (1 - TM) = (T^2 + 1)^2 - 2T(1 - T^2)(\cos \xi + \cos \eta)$$

$$T = \tanh H.$$

Putting all this together we obtain*

$$\frac{BF}{N} = -\ln 2 - \frac{1}{2} \int_0^{2\pi} \int_0^{2\pi} \frac{d\xi \, d\eta}{(2\pi)^2} \ln [\cosh^2 2H - \sinh 2H(\cos \xi + \cos \eta)].$$

$$(5.29)$$

5.5 MISCELLANEOUS COMMENTS

In the above we sometimes assumed that the lattice was finite (with N points) and we sometimes assumed it was infinite (when we neglected the boundary for example). Because N was assumed to be very large, it is plausible to suppose that our inconsistency made little difference. However, let us modify the above work to make it consistent.

Assume the lattice has the topology of a torus. If we pass through L points in one direction (or L' points in the perpendicular direction) we finish back where we started. Formerly, to compute the amplitude to arrive back at the origin moving upward, we took the sum

$$\sum_{n=1}^{\infty} U_n(0, 0).$$

Now, we must take the sum

$$\sum_{p=-\infty}^{\infty} \sum_{q=-\infty}^{\infty} \sum_{n=1}^{\infty} U_n(pL, qL').$$

---

* A slightly different treatment for the Onsager problem was given by N.V.Vdovichenko, *Soviet Phys. JETP* **20**, 2, 477 (1965); also see *Soviet Phys. JETP* **21**, 2, 350 (1965).

Continuing as before, we get in place of Eq. (5.29),

$$\frac{\beta F}{N} = -\ln 2 - \frac{1}{2} \int_0^{2\pi} \int_0^{2\pi} \frac{d\xi \, d\eta}{(2\pi)^2} \sum_{p,q} e^{ipL\xi} e^{iqL'\eta}$$

$$\times \ln \left[\cosh^2 2H - \sinh 2H(\cos \xi + \cos \eta)\right].$$

But

$$\sum_{p=-\infty}^{\infty} e^{ipL\xi} = \frac{2\pi}{L} \sum_{n=-\infty}^{\infty} \delta\left(\xi - \frac{2\pi n}{L}\right).$$

So

$$\frac{\beta F}{N} = -\ln 2 - \frac{1}{2} \sum_{m=0}^{L} \sum_{n=0}^{L'} \left(\frac{2\pi}{L}\right)\left(\frac{2\pi}{L'}\right)\frac{1}{(2\pi)^2}$$

$$\times \ln \left[\cosh^2 2H - \sinh 2H \left(\cos \frac{2\pi m}{L} + \cos \frac{2\pi n}{L'}\right)\right]. \quad (5.30)$$

As L and L' tend to infinity, Eq. (5.30) approaches Eq. (5.29).

A few comments about the form of Eq. (5.29) are in order. If we use the expression

$$\beta F = -\log \sum_{\text{states}} e^{-\beta E}$$

we can verify that Eq. (5.29) is correct for high temperatures ($H \to 0$) and low temperatures ($H \to \infty$). Another check of Eq. (5.29) is by way of Eq. (5.19). If we consider Q as a function of $y = \sinh 2H$ instead of as a function of H, Eq. (5.19) can be written in an especially simple form. Because $\sinh 2H = 2 \cosh H \sinh H = 2 \cosh^2 H \tanh H$,

$$Q(y) = 2^N \cosh^{2N} H \tanh^N HQ\left(\frac{1}{y}\right) = \sinh^N 2HQ\left(\frac{1}{y}\right) = y^N Q\left(\frac{1}{y}\right). \quad (5.19a)$$

Equation (5.29) can be written

$$\frac{\beta F}{N} = -\ln 2 - \frac{1}{2}\ln y - \frac{1}{2}\int_0^{2\pi}\int_0^{2\pi}\frac{d\xi \, d\eta}{(2\pi)^2}\ln\left[y + \frac{1}{y} - (\cos \xi + \cos \eta)\right].$$

$$(5.29a)$$

from which it can be seen that Eq. (5.19a) holds.

Kac made the conjecture that the solution for the three-dimensional Onsager problem (cubic lattice), is of the same form as Eq. (5.29) except that $\cos \xi + \cos \eta$ is replaced by $\cos \xi + \cos \eta + \cos \zeta$ and $d\xi \, d\eta/(2\pi)^2$ by $d\xi \, d\eta \, d\zeta/(2\pi)^3$. However, this conjecture is false.

Instead let us conjecture that the specific linkage of the Ising lattice will not greatly affect the character of the transition point in the specific heat curve. Also, an approximation starting from the transition point should prove very valuable. If a magnetic field is added to the Onsager problem, the problem of evaluating the partition function can again be converted to a path problem, which has not been solved. The problem here is to keep track of the area encircled as well as the length of the path.

We have done a lot of work just to evaluate the expression

$$Q = \sum_{\varepsilon_i = \pm 1} \exp\left(-H \sum_{\text{bonds}} \varepsilon_i \varepsilon_j\right).$$

One reason for going to all this trouble is that maybe you will be able to generalize the methods used, or apply them somewhere else. For example, we considered a statistical-mechanical problem for which we eventually had to evaluate a sum over all paths leaving and returning to the origin, with an amplitude depending on the preceding step. This is essentially a Markovian walk, with each step depending only on the one previous step. A Markovian-walk problem, however, can be handled when each step depends on a specific finite number of preceding steps. In a way analogous to the method used above, Q_n can be found for the case where the amplitude of a step depends on the n steps preceding it. The question is: Can one find an interesting problem whose partition function is Q_n?

CHAPTER 6

CREATION AND ANNIHILATION OPERATORS

6.1 A SIMPLE MATHEMATICAL PROBLEM

In this chapter we shall describe an operator formalism that has widespread applications in quantum mechanics, notably in dealing with harmonic oscillators and in describing many-particle systems.

We begin by formulating and solving the following simple problem: Suppose an operator a satisfies

$$[a, a^+] = 1. \tag{6.1}$$

The problem is to find the eigenvalues of the Hermitian operator a^+a, and to relate the eigenvectors. (Note: a^+ denotes the Hermitian conjugate of a, and $[A, B]$ is, of course, the commutator $AB - BA$.)

We first note that, if $|\alpha\rangle$ is a normalized eigenvector with

$$a^+a|\alpha\rangle = \alpha|\alpha\rangle, \tag{6.2}$$

then

$$\alpha = \langle\alpha|a^+a|\alpha\rangle = \|a|\alpha\rangle\|^2 \geq 0. \tag{6.3}$$

That is, the eigenvalues are all real and nonnegative. Using the identity $[AB, C] = A[B, C] + [A, C]B$, we observe that

$$[a^+a, a] = [a^+, a]a = -a, \tag{6.4}$$

$$[a^+a, a^+] + a^+[a, a^+] = a^+; \tag{6.5}$$

or, equivalently,

$$(a^+a)a = a(a^+a - 1), \tag{6.4'}$$

$$(a^+a)a^+ = a^+(a^+a + 1). \tag{6.5'}$$

From Eq. (6.4') we have, for an eigenvector $|\alpha\rangle$,

$$(a^+a)a|\alpha\rangle = a(a^+a - 1)|\alpha\rangle = a(\alpha - 1)|\alpha\rangle = (\alpha - 1)a|\alpha\rangle. \tag{6.6}$$

Therefore $a|\alpha\rangle$ is an eigenvector with eigenvalue $\alpha - 1$, unless $a|\alpha\rangle = 0$. Similarly $a^+|\alpha\rangle$ is an eigenvector with eigenvalue $\alpha + 1$, unless $a^+|\alpha\rangle = 0$. The norm of $a|\alpha\rangle$ is found from

$$\|a|\alpha\rangle\|^2 = \langle\alpha|a^+a|\alpha\rangle = \alpha\langle\alpha|\alpha\rangle = \alpha,$$

151

or

$$\| a|\alpha\rangle \| = \sqrt{\alpha}. \tag{6.7}$$

Similarly,

$$\| a^+|\alpha\rangle \| = \sqrt{\alpha + 1}. \tag{6.8}$$

Now, suppose that $a^n|\alpha\rangle \neq 0$ for all n. Then by repeated application of Eq. (6.6), $a^n|\alpha\rangle$ is an eigenvector of a^+a with eigenvalue $\alpha - n$. This contradicts Eq. (6.3), because $\alpha - n < 0$ for sufficiently large n. Therefore we must have

$$a^n|\alpha\rangle \neq 0 \quad \text{but} \quad a^{n+1}|\alpha\rangle = 0 \tag{6.9}$$

for some nonnegative integer n.

Let $|\alpha - n\rangle = a^n|n\rangle / \| a^n|n\rangle \|$, so that $|\alpha - n\rangle$ is a normalized eigenvector with eigenvalue $\alpha - n$. Then from Eqs. (6.7) and (6.9),

$$\sqrt{\alpha - n} = \| a|\alpha - n\rangle \| = 0,$$

and therefore $\alpha = n$. This shows that the eigenvalues of a^+a must be nonnegative integers, and that there is a "ground state" $|0\rangle$ such that

$$a|0\rangle = 0. \tag{6.10}$$

By repeatedly applying a^+ to the ground state we see that $(a^+)^n|0\rangle$ has the eigenvalue n and, because of Eq. (6.8), it is never zero. Thus the eigenvalues of a^+a are 0, 1, 2, 3, ...

If $|n\rangle$ is a normalized eigenvector with eigenvalue n, then, from Eq. (6.8),

$$|n - 1\rangle = (1/\sqrt{n})a|n\rangle$$

is a normalized eigenvector with eigenvalue $n - 1$. Also

$$a^+|n - 1\rangle = (1/\sqrt{n})a^+a|n\rangle = \sqrt{n}|n\rangle.$$

So applying a^+ to $|n - 1\rangle$ gives us back $|n\rangle$ (within a factor), rather than some other state with eigenvalue n.

We may then construct the eigenstates of a^+a as follows: First we find a state $|0\rangle$ such that

$$a|0\rangle = 0. \tag{6.11}$$

($|0\rangle$ may be unique; if not, we find other operators commuting with a and a^+, and classify the $|0\rangle$'s according to their eigenvalues.) Then we define

$$|1\rangle = a^+|0\rangle; \quad |2\rangle = \frac{1}{\sqrt{2}}a^+|1\rangle = \frac{1}{\sqrt{2}}(a^+)^2|0\rangle; \quad \cdots$$

and in general

$$|n\rangle = \frac{1}{\sqrt{n!}} (a^+)^n |0\rangle. \tag{6.12}$$

(Note that we could have included arbitrary phase factors in the definition of $|n\rangle$; our convention here is to make them unity.) With this definition, the $|n\rangle$ are orthonormal* and satisfy

$$a^+ |n\rangle = \sqrt{n+1} |n+1\rangle \tag{6.13}$$

$$a|n\rangle = \sqrt{n} |n-1\rangle \tag{6.14}$$

$$a^+ a|n\rangle = n|n\rangle. \tag{6.15}$$

Equations (6.11) through (6.15) form the answer to the problem posed at the beginning of this section.

The operators a^+ and a are called "raising" and "lowering" operators, respectively, because they raise and lower the eigenvalue of $a^+ a$. In later applications $a^+ a$ will be interpreted as the observable representing the number of particles of a certain kind, in which case a^+ and a are called "creation" and "annihilation" (destruction operators, or "emission" and "absorption" operators. Equations (6.13) and (6.14) may be alternatively expressed in terms of matrix elements:

$$\langle m|a^+|n\rangle = \sqrt{n+1}\, \delta_{m,n+1}, \tag{6.13'}$$

$$\langle m|a|n\rangle = \sqrt{n}\, \delta_{m,n-1}. \tag{6.14'}$$

---

* For, by (6.12) we have

$$\langle n|m\rangle = \langle 0|a^n (a^+)^m|0\rangle (1/\sqrt{n!m!}).$$

From Eq. (6.1) we easily obtain

$$[a,(a^+)^n] = n(a^+)^{n-1},$$

so that

$$\begin{aligned}
\langle 0|a^n (a^+)^m|0\rangle &= \langle 0|a^{n-1}(a^+)^m a|0\rangle + \langle 0|na^{n-1}(a^+)^{m-1}|0\rangle \\
&= n\langle 0|a^{n-1}(a^+)^{m-1}|0\rangle \\
&= n(n-1)\cdots(n-m+1)\langle 0|a^{n-m}|0\rangle \\
&= n!\, \delta_{nm}
\end{aligned}$$

and the orthonormality follows.

6.2 THE LINEAR HARMONIC OSCILLATOR

Our first application of the results of Section 6.1 will be to the one-dimensional harmonic oscillator, which has a Hamiltonian of the form

$$H = \frac{1}{2m} p^2 + \frac{m\omega^2}{2} x^2, \tag{6.16}$$

where x and p are the position and momentum operators for the particle and satisfy

$$[x, p] = i\hbar. \tag{6.17}$$

Our task is to find the eigenvalues and eigenstates of H.

Note that $\sqrt{(m\omega/\hbar)}x$ and $(1/\sqrt{m\omega\hbar})p$ are dimensionless. Let us define

$$a = \frac{1}{\sqrt{2}}\left(\sqrt{\frac{m\omega}{\hbar}} x + i \frac{1}{\sqrt{m\omega\hbar}} p\right). \tag{6.18}$$

Because x and p are Hermitian it follows that

$$a^+ = \frac{1}{\sqrt{2}}\left(\sqrt{\frac{m\omega}{\hbar}} x - i \frac{1}{\sqrt{m\omega\hbar}} p\right). \tag{6.19}$$

From Eq. (6.17) we obtain

$$[a, a^+] = 1. \tag{6.20}$$

Expressing x and p in terms of a and a^+, we have

$$x = \sqrt{\frac{\hbar}{m\omega}} \frac{a + a^+}{\sqrt{2}}, \tag{6.21}$$

$$p = \sqrt{m\omega\hbar} \frac{a - a^+}{i\sqrt{2}}. \tag{6.22}$$

We get, for the Hamiltonian,

$$H = \frac{\hbar\omega}{2} (a^+a + aa^+) = \hbar\omega(a^+a + \tfrac{1}{2}). \tag{6.23}$$

Thus, the eigenstates of H are those of a^+a. Now we can apply the results of Section 6.1, obtaining the eigenstates $|0\rangle, |1\rangle, |2\rangle, \ldots$ that satisfy

$$H|n\rangle = (n + \tfrac{1}{2})\hbar\omega|n\rangle. \tag{6.24}$$

The energy levels are thus $E_n = (n + \tfrac{1}{2})\hbar\omega$.

The eigenstates themselves are given by Eqs. (6.11) and (6.12). We can easily obtain the wave functions $\varphi_n(x) = \langle x|n \rangle$ as follows: from Eqs. (6.18) and (6.11),

$$0 = a|0\rangle = \sqrt{\frac{m\omega}{2\hbar}}\left(x + \frac{i}{m\omega}\,p\right)|0\rangle. \tag{6.25}$$

Applying $\langle x|$ and noticing that $\langle x|p|\varphi \rangle = -i\hbar(d\langle x|\varphi \rangle/dx)$, we get

$$0 = \sqrt{\frac{m\omega}{2\hbar}}\left(x + \frac{\hbar}{m\omega}\frac{d}{dx}\right)\langle x|0\rangle \tag{6.26}$$

(where x is now a number, rather than an operator.) Equation (6.26) is merely Eq. (6.11) in coordinate representation, in which it takes the form of a differential equation. Solving it, we get

$$\langle x|0\rangle = Ae^{-(m\omega/2\hbar)x^2},$$

where A is a constant. Normalization requires that

$$1 = \langle 0|0\rangle = \int_{-\infty}^{\infty}\langle 0|x\rangle\langle x|0\rangle\,dx = |A|^2\int_{-\infty}^{\infty}e^{-(m\omega/\hbar)x^2}\,dx$$

$$= |A|^2\sqrt{\frac{\pi\hbar}{m\omega}},$$

so

$$A = e^{i\theta}\left(\frac{m\omega}{\pi\hbar}\right)^{1/4}.$$

The phase θ of A is arbitrary, and we set it equal to zero. Then

$$A = \left(\frac{m\omega}{\pi\hbar}\right)^{1/4},$$

so

$$\langle x|0\rangle = \left(\frac{m\omega}{\pi\hbar}\right)^{1/4}e^{-(m\omega/2\hbar)x^2}. \tag{6.27}$$

We have thus found the wave function for the ground state. For the other states we apply a^+ according to Eq. (1.12):

$$\langle x|n\rangle = \frac{1}{\sqrt{n!}}\langle x|(a^+)^n|0\rangle. \tag{6.28}$$

Since

$$\langle x|a^+ = \sqrt{\frac{m\omega}{2\hbar}} \langle x| \left(x - \frac{i}{m\omega} p \right)$$

$$= \sqrt{\frac{m\omega}{2\hbar}} \left(x - \frac{\hbar}{m\omega} \frac{d}{dx} \right) \langle x|,$$

we have

$$\langle x|n \rangle = \frac{1}{\sqrt{n!}} \left(\frac{m\omega}{2\hbar} \right)^{n/2} \left(x - \frac{\hbar}{m\omega} \frac{d}{dx} \right)^n \langle x|0 \rangle$$

$$= \frac{1}{\sqrt{n!}} \left(\frac{m\omega}{\pi\hbar} \right)^{1/4} \left(\frac{m\omega}{2\hbar} \right)^{n/2} \left(x - \frac{\hbar}{m\omega} \frac{d}{dx} \right)^n e^{-(m\omega/2\hbar)x^2}. \quad (6.29)$$

The matrix elements of observables between harmonic oscillator states can be found without having to express the states in coordinate representation and integrating over x. We simply express the observable in terms of the raising and lowering operators. An example of this procedure is given in the following section.*

6.3 AN ANHARMONIC OSCILLATOR

Suppose a system has the Hamiltonian

$$H = \frac{p^2}{2m} + \frac{m\omega^2}{2} x^2 + \lambda x^4. \quad (6.30)$$

Assume that λ is small enough ($\ll \hbar\omega$) that we can use first-order perturbation theory, treating λx^4 as a perturbation of the Hamiltonian (Eq. (6.16)). Then the perturbed energy levels are

$$E_n \approx (n + \tfrac{1}{2})\hbar\omega + \Delta_n, \quad (6.31)$$

where

$$\Delta_n = \langle n|\lambda x^4|n \rangle. \quad (6.32)$$

From Eq. (6.21) we have

$$\Delta_n = \lambda \left(\frac{\hbar}{2m\omega} \right)^2 \langle n|(a + a^+)^4|n \rangle. \quad (6.33)$$

---

* *Problem:* Prove that

$$|x\rangle = \left(\frac{m\omega}{\pi\hbar} \right)^{1/4} \exp \left(\frac{m\omega}{2\hbar} x^2 \right) \exp \left[\left(-\frac{1}{2} \right) \left(a^+ - x \sqrt{\frac{2m\omega}{\hbar}} \right)^2 \right] |0\rangle,$$

where $f(a^+)$ is interpreted as $\sum a_n(a^+)^n$, when $f(x) = \sum_n a_n x^n$. From this formula, find a generating function for $\langle x|n \rangle$. (Hint: Prove first that $[a, f(a^+)] = f'(a^+)$.)

Expanding $(a + a^+)^4$ gives us 16 terms, but, thanks to the raising and lowering properties of a^+ and a, the only terms giving a nonzero expectation value are those with two a's and two a^+'s:

$$\langle n|(a + a^+)^4|n\rangle = \langle n|(a^+a^+aa + a^+aa^+a + a^+aaa^+$$
$$+ aa^+a^+a + aa^+aa^+ + aaa^+a^+)|n\rangle$$
$$= n(n - 1) + n^2 + n(n + 1)$$
$$+ n(n + 1) + (n + 1)^2 + (n + 1)(n + 2)$$
$$= 6n^2 + 6n + 3,$$

where we have used Eqs. (6.13) and (6.14) repeatedly. Therefore

$$\Delta_n = 3\lambda \left(\frac{\hbar}{2m\omega}\right)^2 (2n^2 + 2n + 1).$$

6.4 SYSTEMS OF HARMONIC OSCILLATORS

Suppose a system has the Hamiltonian

$$H = \sum_i \frac{1}{2m_i} P_i^2 + \sum_{i,j} V_{ij}Q_iQ_j \qquad (6.34)$$

where Q_i and P_i are canonical coordinates and momenta:

$$[Q_i, Q_j] = [P_i, P_j] = 0; \qquad [Q_i, P_j] = i\hbar\delta_{ij}, \qquad (6.35)$$

and $V_{ij} = V_{ji}$. To simplify the presentation a little let us make a change of scale, defining

$$q_i = \sqrt{m_i}\, Q_i; \qquad p_i = P_i/\sqrt{m_i} \qquad (6.36)$$

and

$$U_{ij} = \frac{2}{\sqrt{m_i m_j}}\, V_{ij}. \qquad (6.37)$$

Then q_i and p_i are also canonical:

$$[q_i, p_j] = i\hbar\delta_{ij}, \qquad (6.38)$$

and in terms of them the Hamiltonian is

$$H = \tfrac{1}{2}\sum_i p_i^2 + \tfrac{1}{2}\sum_{i,j} U_{ij}q_iq_j. \qquad (6.39)$$

We shall express H in terms of raising and lowering operators as we did for the one-dimensional oscillator. The procedure involves two steps: the first is finding a set of normal coordinates $\tilde{q}_\alpha$ with respect to which the potential is in diagonal form, and the second is expressing the coordinates and momenta in terms of raising and lowering operators.

Let the coordinates q_i and $\tilde{q}_\alpha$ be related by

$$\tilde{q}_\alpha = \sum_i C_{\alpha i} q_i. \tag{6.40}$$

Because (U_{ij}) is assumed to be real and symmetric, the transformation matrix $C_{\alpha i}$ that diagonalizes it is orthogonal:

$$\sum_i C_{\alpha i} C_{\beta i} = \delta_{\alpha\beta}; \qquad \sum_\alpha C_{\alpha i} C_{\alpha j} = \delta_{ij}. \tag{6.41}$$

The inverse transformation of Eq. (6.40) is then

$$q_i = \sum_\alpha C_{i\alpha} \tilde{q}_\alpha. \tag{6.42}$$

We further assume that the eigenvalues of (U_{ij}) are all positive, that is, that the matrix is positive definite (this ensures that $q_i = 0$ is a point of stable equilibrium). Denoting these eigenvalues by $\omega_\alpha^2(\omega_\alpha > 0)$, we have

$$\sum_{i,j} C_{\alpha i} C_{\beta j} U_{ij} = \omega_\alpha^2 \delta_{\alpha\beta},$$

and thus

$$\sum_{i,j} U_{ij} q_i q_j = \sum_\alpha \omega_\alpha^2 \tilde{q}_\alpha^2. \tag{6.43}$$

Finally, we define $\tilde{p}_\alpha$ in such a way as to preserve the canonical commutation relations:

$$\tilde{p}_\alpha = \sum_i C_{\alpha i} p_i; \tag{6.44}$$

$$[\tilde{q}_\alpha, \tilde{p}_\beta] = i\hbar \delta_{\alpha\beta}. \tag{6.45}$$

The result of our efforts is that

$$H = \tfrac{1}{2} \sum_\alpha (\tilde{p}_\alpha^2 + \omega_\alpha^2 \tilde{q}_\alpha^2), \tag{6.46}$$

which means we have a system of decoupled harmonic oscillators (one for each value of α).

Using the methods of Section 6.2, we form lowering and raising operators for each mode:

$$a_\alpha = \frac{1}{\sqrt{2\hbar}} \left(\sqrt{\omega_\alpha}\, \tilde{q}_\alpha + \frac{i}{\sqrt{\omega_\alpha}}\, \tilde{p}_\alpha \right), \tag{6.47}$$

$$a_\alpha^+ = \frac{1}{\sqrt{2\hbar}} \left(\sqrt{\omega_\alpha}\, \tilde{q}_\alpha - \frac{i}{\sqrt{\omega_\alpha}}\, \tilde{p}_\alpha \right), \tag{6.48}$$

$$\tilde{q}_\alpha = \sqrt{\frac{\hbar}{2\omega_\alpha}} \, (a_\alpha + a_\alpha^+), \tag{6.49}$$

$$\tilde{p}_\alpha = -i \sqrt{\frac{\hbar\omega_\alpha}{2}} \, (a_\alpha - a_\alpha^+). \tag{6.50}$$

Then,

$$[a_\alpha, a_\beta] = [a_\alpha^+, a_\beta^+] = 0, \tag{6.51}$$

$$[a_\alpha, a_\beta^+] = \delta_{\alpha\beta}, \tag{6.52}$$

$$H = \sum_\alpha \hbar\omega_\alpha(a_\alpha^+ a_\alpha + \tfrac{1}{2}). \tag{6.53}$$

The eigenstates of H are described by giving, for each α, the eigenvalue n_α of $a_\alpha^+ a_\alpha$. Thus,

$$H|n_1 n_2 n_3 \cdots \rangle = \sum_\alpha (n_\alpha + \tfrac{1}{2})\hbar\omega_\alpha |n_1 n_2 n_3 \cdots \rangle, \tag{7.54}$$

$$|n_1 n_2 n_3 \cdots \rangle = \left[\prod_\alpha \frac{(a_\alpha^+)^{n_\alpha}}{\sqrt{n_\alpha!}} \right] |000 \cdots \rangle, \tag{6.55}$$

where the ground state $|000 \cdots \rangle$ is defined by

$$a_\alpha|000 \cdots \rangle = 0 \text{ for all } \alpha. \tag{6.56}$$

Note that the energy of the ground state is $\sum_\alpha \tfrac{1}{2}\hbar\omega_\alpha$. For a system with infinitely many degrees of freedom (which we will consider shortly), this quantity will generally be infinite. Because the zero point of energy is a matter of definition (only the difference between levels being of physical importance), it is convenient to redefine the Hamiltonian of such a system so that the ground-state energy is zero. Thus, if we let

$$H = \tfrac{1}{2} \sum_\alpha (p_\alpha^2 + \omega_\alpha^2 q_\alpha^2 - \tfrac{1}{2}\hbar\omega_\alpha) \tag{6.57}$$

with a corresponding (but more complicated) expression in terms of the original coordinates q_i, then

$$H = \sum_\alpha \hbar\omega_\alpha a_\alpha^+ a_\alpha, \tag{6.58}$$

and

$$H|n_1 n_2 n_3 \cdots \rangle = \sum_\alpha n_\alpha \hbar\omega_\alpha |n_1 n_2 n_3 \cdots \rangle. \tag{6.59}$$

6.5 PHONONS

The states of the system considered in the preceding section can be given a simple interpretation in terms of "noninteracting phonons". Assume that the Hamiltonian is given by Eqs. (6.57) and (6.58), so that the energy of the ground state is zero. The ground state is then called the "vacuum state" and represents the state of the system in which there are no phonons. If the system is in the state $|n_1 n_2 n_3 \cdots \rangle$ we say that there are n_α phonons of type $\alpha(\alpha = 1, 2, 3, \ldots)$. The n_α are called "occupation numbers." Note that the energy of this state is $n_1 \hbar \omega_1 + n_2 \hbar \omega_2 + \cdots$, so that the energy of a single phonon of type α is $\hbar \omega_\alpha$, and the total energy is the sum of the energies of the individual phonons. In other words, the phonons are noninteracting.

Since

$$a_\alpha^+ |n_1 \cdots n_{\alpha-1} n_\alpha n_{\alpha+1} \cdots \rangle = \sqrt{n_\alpha + 1} |n, \ldots, n_{\alpha-1}, (n_\alpha + 1), n_{\alpha+1} \cdots \rangle, \tag{6.60}$$

$$a_\alpha |n_1 \cdots n_{\alpha-1} n_\alpha n_{\alpha+1} \cdots \rangle = \sqrt{n_\alpha} |n, \ldots, n_{\alpha-1}, (n_\alpha - 1), n_{\alpha+1} \cdots \rangle, \tag{6.61}$$

we may call a_α^+ and a_α creation and annihilation operators for phonons of type α. The operator for the number of phonons of type α is $a_\alpha^+ a_\alpha$, and the operator for the total number of phonons is

$$N = \sum_\alpha a_\alpha^+ a_\alpha. \tag{6.62}$$

Let the vacuum state be denoted by $|0\rangle$, and let

$$|\alpha\rangle = a_\alpha^+ |0\rangle = |0, \ldots, 0, 1, 0, \ldots \rangle, \tag{6.63}$$

be the state with $n_\alpha = 1$ and $n_{\alpha'} = 0$ for $a' \neq \alpha$. A phonon, then, is the system that is decribed by the states $|\alpha\rangle$ (i.e., the system whose quantum-mechanical Hilbert space is spanned by the $|\alpha\rangle$). If $|\alpha\rangle$ and $|\beta\rangle$ are (one-) phonon states, then

$$\langle \alpha | \beta \rangle = \langle 0 | a_\alpha a_\beta^+ | 0 \rangle$$

$$= \langle 0 | (a_\beta^+ a_\alpha + \delta_{\alpha\beta}) | 0 \rangle$$

$$= \delta_{\alpha\beta} \tag{6.64}$$

so that the states $|\alpha\rangle$ are orthonormal. To each normal mode of vibration of the original system of harmonic oscillators corresponds a one-phonon state (since they are both indexed by α).

We can also use a similar notation for states of two or more phonons, defining

$$|\alpha_1, \ldots, \alpha_n\rangle = a_{\alpha_1}^+ \cdots a_{\alpha_n}^+ |0\rangle. \tag{6.65}$$

These states are normalized as they stand if $\alpha_1, \ldots, \alpha_n$ are distinct; otherwise, they have a norm larger than 1. Assume for definiteness that α takes on the values 1, 2, 3, ... Then the state with n_1, of the α's equal to 1, n_2 α's equal to 2, and so on, is

$$|\underbrace{1, \ldots, 1}_{n_1}, \underbrace{2, \ldots, 2}_{n_2}, \ldots\rangle = (a_1^+)^{n_1}(a_2^+)^{n_2}\cdots|0\rangle$$

$$= \sqrt{n_1!\, n_2!\cdots}\,|n_1 n_2 \cdots\rangle \qquad (6.66)$$

so that its norm is $\sqrt{n_1!\, n_2!\cdots}$.

When we deal with only a few phonons at a time, it is usually more convenient to use the notation $|\alpha_1, \ldots, \alpha_n\rangle$ rather than the occupation-number description $|n_1 n_2 \cdots\rangle$, especially when α can take on a continuum of values. The effects of creation and destruction operators on $|\alpha_1, \ldots, \alpha_n\rangle$ are

$$a_\alpha^+|\alpha_1, \ldots, \alpha_n\rangle = |\alpha, \alpha_1, \ldots, \alpha_n\rangle \qquad (6.67)$$

$$a_\alpha|\alpha_1, \ldots, \alpha_n\rangle = \sum_{k=1}^{n} \delta_{\alpha\alpha_k}|\alpha_1, \ldots, \alpha_{k-1}, \alpha_{k+1}, \ldots, \alpha_n\rangle \qquad (6.68)$$

Equation (6.68) comes from Eq. (6.65) and the relation

$$[a_\alpha, a_{\alpha_1}^+ \cdots a_{\alpha_n}^+] = \sum_{k=1}^{n} \delta_{\alpha\alpha_k} a_{\alpha_1}^+ \cdots a_{\alpha_{k-1}}^+ a_{\alpha_{k+1}}^+ \cdots a_{\alpha_n}^+.$$

Note that phonons act like Bose particles (insofar as we can call them particles*), as an arbitrary number of them may be in any given state (i.e., $|\alpha, \ldots, \alpha\rangle$ exists for any number of α's). Their Bose nature is also reflected in the symmetry of the states (e.g., $|\alpha, \beta\rangle = |\beta, \alpha\rangle$). In Section 6.7 we shall show how the ordinary rules for quantum-mechanically describing systems of many Bose particles lead to a set of states and operators with the same form as those obtained here, so that the interpretation of the oscillator as a system of many Bose particles is correct.

We conclude this section by considering the qualitative effect of an anharmonic perturbation on the oscillator system. Suppose the perturbation has terms of the form

$$\sum_{i,j,k} \Gamma_{ijk} q_i q_j q_k \quad \text{and} \quad \sum_{i,j,k,l} \Gamma_{ijkl} q_i q_j q_k q_l.$$

In terms of creation and annihilation operators, the cubic terms are of the form

$$a_\alpha^+ a_\beta^+ a_\gamma^+, a_\alpha^+ a_\beta^+ a_\gamma, \ldots, a_\alpha a_\beta a_\gamma,$$

which always changes the number of phonons (e.g., the first term creates three

---

* The phonons will look more like particles (e.g., carrying momentum and energy) when the oscillator system is a field as described in the next section.

new phonons). Thus, if we start with a definite number of phonons and let the Hamiltonian drive the system forward in time,

$$|\psi(t)\rangle = e^{-iHt}|\psi(0)\rangle$$
$$= |\psi(0)\rangle - itH|\psi(0)\rangle + 0(t^2),$$

we will soon start finding different numbers of particles. The quartic terms similarly change the number of particles, except for terms like

$$a_\alpha^+ a_\beta^+ a_\gamma a_\delta, \; a_\alpha^+ a_\beta a_\gamma^+ a_\delta, \ldots,$$

which conserve the number of particles but act as a mutual interaction between them; that is, the particles are no longer independent. The description of a mutual interaction will be considered in more detail later.

Exercise: Verify that the number-of-phonons operator N, defined by Eq. (6.62), commutes with a product of creation and destruction operators if and only if the number of a^+'s equals the number of a's in the product.

6.6 FIELD QUANTIZATION

A notable example of a system with infinitely many degrees of freedom is a *field*. Examples are the amplitude of sound waves, drumhead vibrations, light, and so on. Consider a real scalar* field $\varphi(x)$ whose motion† is described by the Lagrangian‡

$$L(\varphi, \dot{\varphi}) = \tfrac{1}{2} \int d^3x \dot{\varphi}(x)\dot{\varphi}(x) - \tfrac{1}{2} \int d^3x \int d^3x' K(x - x')\varphi(x)\varphi(x'), \qquad (6.69)$$

where $K(x - x') = K(x' - x)$. The classical equations of motion, found by varying $\varphi(x)$, are

$$0 = \frac{\partial}{\partial t} \frac{\delta L}{\delta \dot{\varphi}(x)} - \frac{\delta L}{\delta \varphi(x)}$$

$$= \ddot{\varphi}(x) + \int d^3x' K(x - x')\varphi(x'). \qquad (6.70)$$

---

* The following procedure can be generalized for a multicomponent field by putting indices on everything.
† Classically $\varphi(x)$ depends on t, but (as with q_i in Section 7.4) we will not show it explicitly. Besides, in the Schrödinger picture, the operator $\varphi(x)$ is time-independent.
‡ We assume that the system is invariant under translations, so that K is only a function of $x - x'$.

Note how Eqs. (6.69) and (6.70) resemble the corresponding equations for the system of harmonic oscillators described in Section 6.4:

$$L = \tfrac{1}{2} \sum_i \dot{q}_i^2 - \tfrac{1}{2} \sum_{i,j} U_{ij} q_i q_j,$$

$$0 = \ddot{q}_i + \sum_j U_{ij} q_j.$$

Thus, we are justified in treating the field as a system of harmonic oscillators (at least formally): φ corresponds to the symbol "q" and x corresponds to i. $\varphi(x)$ can be thought of as a separate coordinate of the system for each x.

As an example, suppose that

$$K(x - x') = -c^2 \nabla^2 \delta^3(x - x'). \tag{6.71}$$

Then Eq. (6.69) becomes, after a few integrations by parts,

$$L = \tfrac{1}{2} \int d^3x [\dot{\varphi}(x)\dot{\varphi}(x) - c^2 \nabla\varphi(x) \cdot \nabla\varphi(x)], \tag{6.72}$$

and Eq. (6.70) becomes

$$\nabla^2 \varphi(x) - \frac{1}{c^2} \ddot{\varphi}(x) = 0, \tag{6.73}$$

which is the usual wave equation.

If we assume that $\varphi(x)$ is a coordinate of the system for each x, the conjugate momentum to $\varphi(x)$ is

$$\Pi(x) = \frac{\delta L}{\delta \dot{\varphi}(x)} = \dot{\varphi}(x). \tag{6.74}$$

The Hamiltonian is then

$$H = \int d^3x \Pi(x)\dot{\varphi}(x) - L$$

$$= \tfrac{1}{2} \int d^3x \Pi(x)\Pi(x) + \tfrac{1}{2} \int d^3x \int d^3x' K(x - x')\varphi(x)\varphi(x'). \tag{6.75}$$

To quantize the system we let $\varphi(x)$ and $\Pi(x)$ be Hermitian operators satisfying

$$[\varphi(x), \varphi(x')] = [\Pi(x), \Pi(x')] = 0, \tag{6.76}$$

$$[\varphi(x), \Pi(x')] = i\hbar\delta^3(x - x'), \tag{6.77}$$

and assume that the Hamiltonian is given by Eq. (6.75), except for a scalar term to make the ground-state energy zero.

We next express everything in terms of "normal modes." The situation turns out to be slightly different from that of Section 6.4 because it is convenient here to use "complex" (that is, non-hermitian) normal coordinates.

Because the system is translationally invariant, we expect that it might help to express the fields in "momentum" representation. Therefore we define

$$\tilde{\varphi}(k) = \int d^3x \varphi(x) e^{-ik \cdot x} \tag{6.78}$$

$$\tilde{\Pi}(k) = \int d^3x \Pi(x) e^{-ik \cdot x}. \tag{6.79}$$

The inverse transformation is*

$$\varphi(x) = \int \frac{d^3k}{(2\pi)^3} \tilde{\varphi}(k) e^{ik \cdot x} \tag{6.80}$$

with a similar expression for $\Pi(x)$. Since $\varphi(x)$ and $\Pi(x)$ are Hermitian, we have

$$\tilde{\varphi}^+(k) = \tilde{\varphi}(-k); \qquad \tilde{\Pi}^+(k) = \tilde{\Pi}(-k). \tag{6.81}$$

From Eqs. (6.76) and (6.77) we obtain

$$[\tilde{\varphi}(k), \tilde{\varphi}(k')] = [\tilde{\Pi}(k), \tilde{\Pi}(k')] = 0, \tag{6.82}$$

$$[\tilde{\varphi}(k), \tilde{\Pi}(k')] = i\hbar(2\pi)^3 \delta^3(k + k'). \tag{6.83}$$

Now let

$$\omega^2(k) = \int d^3x K(x) e^{-ik \cdot x}. \tag{6.84}$$

From $K(x) = K(-x) = K^*(x)$ it follows that

$$\omega^2(k) = \omega^2(k)^* = \omega^2(-k). \tag{6.85}$$

Rewriting the Hamiltonian of Eq. (6.75), we get

$$H = \frac{1}{2} \int \frac{d^3k}{(2\pi)^3} [\tilde{\Pi}(-k)\tilde{\Pi}(k) + \omega^2(k)\tilde{\varphi}(-k)\tilde{\varphi}(k)]$$

$$= \frac{1}{2} \int \frac{d^3k}{(2\pi)^3} [\tilde{\Pi}^+(k)\tilde{\Pi}(k) + \omega^2(k)\tilde{\varphi}^+(k)\tilde{\varphi}(k)]. \tag{6.86}$$

We assume that $\omega^2(k) > 0$, so that the Hamiltonian is positive definite. [Thus $\omega(k)$ is real, and we take $\omega(k) > 0$.]

In the example described in Eqs. (6.71), (6.72), and (6.73), $\omega(k) = c|k|$.

---

* Throughout all of this the following integrals will be useful:

$$\int d^3x e^{-ik \cdot x} = (2\pi)^3 \delta^3(k), \qquad \int \frac{d^3k}{(2\pi)^3} e^{ik \cdot x} = \delta^3(x).$$

Next we will define annihilation and creation operators (compare Eqs. (6.47) through (6.50)):

$$a(k) = \frac{1}{\sqrt{2\hbar}} \left[\sqrt{\omega(k)}\, \tilde{\varphi}(k) + \frac{i}{\sqrt{\omega(k)}}\, \tilde{\Pi}(k) \right], \tag{6.87}$$

$$a^+(k) = \frac{1}{\sqrt{2\hbar}} \left[\sqrt{\omega(k)}\, \tilde{\varphi}(-k) - \frac{i}{\sqrt{\omega(k)}}\, \tilde{\Pi}(-k) \right], \tag{6.88}$$

or

$$\tilde{\varphi}(k) = \sqrt{\frac{\hbar}{2\omega(k)}}\, [a(k) + a^+(-k)], \tag{6.89}$$

$$\tilde{\Pi}(k) = -i \sqrt{\frac{\hbar\omega(k)}{2}}\, [a(k) - a^+(-k)]. \tag{6.90}$$

The commutation relations are, from Eqs. (6.82) and (6.83),

$$[a(k), a(k')] = [a^+(k), a^+(k')] = 0, \tag{6.91}$$

$$[a(k), a^+(k')] = (2\pi)^3 \delta^3(k - k'). \tag{6.92}$$

If we write H in terms of a and a^+, making the change of variable $k \to -k$ when necessary, we obtain

$$H = \frac{1}{2} \int \frac{d^3k}{(2\pi)^3}\, \hbar\omega(k)[a^+(k)a(k) + a(k)a^+(k)]$$

plus a correction term to make the vacuum energy zero. The corrected Hamiltonian is evidently

$$H = \int \frac{d^3k}{(2\pi)^3}\, \hbar\omega(k)a^+(k)a(k). \tag{6.93}$$

[Note that the correction term is the infinite quantity $-\frac{1}{2}\int d^3k\,\hbar\omega(k)\delta^3(0)$]. Finally, we express the original field variables in terms of the creation and destruction operators, using Eqs. (6.80), (6.89), and (6.90):

$$\varphi(x) = \int \frac{d^3k}{(2\pi)^3} \sqrt{\frac{\hbar}{2\omega(k)}}\, [a(k)e^{ik\cdot x} + a^+(k)e^{-ik\cdot x}] \tag{6.94}$$

$$\Pi(x) = \int \frac{d^3k}{(2\pi)^3} \sqrt{\frac{\hbar\omega(k)}{2}}\, [-ia(k)e^{ik\cdot x} + ia^+(k)e^{-ik\cdot x}]. \tag{6.95}$$

Equations (6.91) through (6.95) are the important results of the quantization procedure.

The commutation relation of Eq. (6.92) may appear strange, in that

$[a(k), a^+(k)]$ is infinite (instead of unity), so that the analysis of Section 6.1 does not apply directly to $a(k)$. Suppose, however, that we use a less singular representation. Choose a complete orthonormal set of functions $\psi_\alpha(k)$, where α is a discrete index:

$$\int \frac{d^3k}{(2\pi)^3} \, \psi_\alpha^*(k)\psi_\beta(k) = \delta_{\alpha\beta}, \tag{6.96}$$

$$\sum_\alpha \psi_\alpha(k)\psi_\alpha^*(k') = (2\pi)^3\delta^3(k - k') \tag{6.97}$$

and define

$$a_\alpha = \int \frac{d^3k}{(2\pi)^3} \, \psi_\alpha^*(k)a(k). \tag{6.98}$$

Then

$$[a_\alpha, a_\beta^+] = \delta_{\alpha\beta}, \tag{6.99}$$

so that we can apply previous results and construct $|n_1 \cdots n_\alpha \cdots \rangle$. But these states may not be eigenstates of H. The states

$$|k\rangle = a^+(k)|0\rangle, \qquad |k, k'\rangle = a^+(k)a^+(k')|0\rangle,$$

and so forth, though unnormalizable, are eigenstates of H.

What kind of phonons do that unnormalizable states represent? The state $|k\rangle$ is a phonon of energy $\hbar\omega(k)$, and we may also say that it has momentum $\hbar k$. To discover the reason for this, consider the operator

$$P = \int \frac{d^3k}{(2\pi)^3} \, \hbar k a^+(k)a(k), \tag{6.100}$$

which satisfies

$$[P, a^+(k)] = \hbar k a^+(\mathrm{k}), \qquad [P, a(k)] = -\hbar k a(k) \tag{6.101}$$

so that

$$P|k_1, k_2, \ldots \rangle = (\hbar k + \hbar k_2 + \cdots)|k_1, k_2, \ldots, \rangle. \tag{6.102}$$

Now, from Eqs. (6.94) and (6.101) we obtain

$$[P, \varphi(x)] = i\hbar\nabla\varphi(x). \tag{6.103}$$

One can then show that

$$e^{a \cdot P/i\hbar}\varphi(x)e^{-a \cdot P/i\hbar} = \varphi(x + a), \tag{6.104}$$

so that P generates translations and is therefore the momentum operator.

In Sections 6.7 and 6.8 we will describe further the relation between the operators and the states they create and destroy, as well as how other operators, such as the Hamiltonian, can be written in terms of the creation and destruction operators using any basis of states.

Note: In relativistic quantum mechanics, when one quantizes a free-particle field with $\omega(k) = \sqrt{k^2 + m^2}$ ($\hbar = c = 1$), a different normalization and summation convention is commonly used for momentum states. Everything is written in terms of $\int d^3k/(2\pi)^3 2\omega(k)$ and $(2\pi)^3 2\omega(k)\delta^3(k - k')$, which happen to be relativistically invariant; to accomplish this change of normalization one uses a "relativistic" $a(k)$ equal to $\sqrt{2\omega(k)}$ times our $a(k)$. Equations (6.92), (6.93), and (6.94) then become

$$[a(k), a^+(k')] = (2\pi)^3 2\omega(k)\delta^3(k - k');$$

$$H = \int \frac{d^3k}{(2\pi)^3 2\omega(k)} \omega(k)a^+(k)a(k);$$

$$\varphi(x) = \int \frac{d^3k}{(2\pi)^3 2\omega(k)} [a(k)e^{ik\cdot x} + a^+(k)e^{-ik\cdot x}].$$

In some texts the $(2\pi)^3$ is also treated differently. We will not use the relativistic normalization here, but it is mentioned in case the reader finds it elsewhere and wants to reconcile it with our notation.

6.7 SYSTEMS OF INDISTINGUISHABLE PARTICLES

In the preceding sections we considered the quantum states of an oscillator system as being states of various numbers of a "particle" called a *phonon*. We identified certain states as one-phonon states, and others as states containing more than one phonon.

In this section we shall follow a different line of reasoning. We will start with a space of states describing a *single* particle, either Bose or Fermi, and construct the multiparticle states according to standard methods. For the Bose case we will arrive at a system of states and operators that is mathematically the same as that found previously for a system of oscillators, thereby showing that the interpretation of oscillator states as many-phonon states is consistent with the usual description of many-particle systems. In the meantime we will have also developed a formalism for dealing with Fermi particles, for which the states do not resemble those of a harmonic-oscillator system.

We will treat the Bose and Fermi cases simultaneously, distinguishing them by the number ζ;

$$\zeta = \begin{cases} +1 & \text{if the particles are Bose} \\ -1 & \text{if the particles are Fermi.} \end{cases} \tag{6.105}$$

We will use the symbol ζ^P (where P is a permutation) to denote 1 for the Bose case and $(-1)^P$ for the Fermi case.

Consider first the case of distinguishable particles. If $|\psi_1\rangle, \ldots, |\psi_n\rangle$ are one-particle states, then

$$|\psi\rangle = |\psi_1\rangle|\psi_2\rangle \cdots |\psi_n\rangle \qquad (6.106)$$

describes the n-particle state with the ith particle in state $|\psi_i\rangle$. If $|\varphi\rangle = |\varphi_1\rangle|\varphi_2\rangle \cdots |\varphi_n\rangle$, then we write

$$\langle\varphi|\psi\rangle = ((\langle\varphi_1|\langle\varphi_2| \cdots \langle\varphi_n|)(|\psi_1\rangle|\psi_2\rangle \cdots |\psi_n\rangle))$$

$$= \langle\varphi_1|\psi_1\rangle\langle\varphi_2|\psi_2\rangle \cdots \langle\varphi_n|\psi_n\rangle, \qquad (6.107)$$

which defines the inner product $\langle\varphi|\psi\rangle$. The Hilbert space describing the n-particle system is that spanned by all nth-rank tensors with the form of Eq. (6.106).

The state in which the ith particle is localized at the point x_i is $|x_1\rangle|x_2\rangle \cdots |x_n\rangle$. As each x_i runs over all space, the resulting states form a complete orthonormal set for the n-particle space (ignoring spin and other variables):

$$((\langle x_1| \cdots \langle x_n|)(|y_1\rangle \cdots |y_n\rangle)) = \delta^3(x_1 - y_1) \cdots \delta^3(x_n - y_n), \qquad (6.108)$$

$$\int d^3x_1 \cdots \int d^3x_n(|x_1\rangle \cdots |x_n\rangle)(\langle x_1| \cdots \langle x_n|) = 1. \qquad (6.109)$$

Using this basis we can express the n-particle states in coordinate representation:

$$\psi(x_1, \ldots, x_n) = ((\langle x_1| \cdots \langle x_n|)|\psi\rangle. \qquad (6.110)$$

For the particle $|\psi\rangle$ of Eq. (6.106), we have, using Eq. (6.107),

$$\psi(x_1, \ldots, x_n) = \psi_1(x_1)\psi_n \cdots (x_n), \qquad (6.111)$$

where $\psi_i(x_i) = \langle x_i|\psi_i\rangle$.

Next, let us consider indistinguishable particles. We assume that the particles obey Bose or Fermi statistics, which means that we must symmetrize or anti-symmetrize, respectively, the states obtained in Eq. (6.106). We therefore define

$$|\psi_1\rangle \times |\psi_2\rangle \times \cdots \times |\psi_n\rangle = (1/\sqrt{n!}) \sum_P \zeta^P |\psi_{P(1)}\rangle|\psi_{P(2)}\rangle \cdots |\psi_{P(n)}\rangle, \qquad (6.112)$$

where P runs through all permutations of n objects. It will often be convenient to write $|\psi_1, \psi_2, \ldots, \psi_n\rangle$ for $|\psi_1\rangle \times |\psi_2\rangle \times \cdots \times |\psi_n\rangle$.

The space of n-particle states is that spanned by all "products" of the form of Eq. (6.112). Note that $|\psi_1\rangle \times \cdots \times |\psi_n\rangle$ is totally symmetric in the Bose case and totally antisymmetric in the Fermi case, as it should be.

Example: Let $|a\rangle$ and $|b\rangle$ be two single-particle states. If $\zeta = +1$ (Bose particles),

$$|a\rangle \times |b\rangle = |a, b\rangle = \frac{1}{\sqrt{2}}(|a\rangle|b\rangle + |b\rangle|a\rangle),$$

$$|a\rangle \times |a\rangle = |a, a\rangle = \sqrt{2}|a\rangle|a\rangle.$$

If $\zeta = -1$ (Fermi particles)

$$|a, b\rangle = \frac{1}{\sqrt{2}}(|a\rangle|b\rangle - |b\rangle|a\rangle),$$

$$|a, a\rangle = 0.$$

Thus, we have the expected result that two Fermi particles cannot be in the same state.

What is the inner product of two of these n-particle states? The answer is given by the following theorem*

$$\langle \varphi_1, \ldots, \varphi_n | \psi_1, \ldots, \psi_n \rangle = \begin{vmatrix} \langle \varphi_1 | \psi_1 \rangle & \cdots & \langle \varphi_1 | \psi_n \rangle \\ \langle \varphi_n | \psi_1 \rangle & \cdots & \langle \varphi_n | \psi_n \rangle \end{vmatrix}_\zeta, \qquad (6.113)$$

where, for any $n \times n$ matrix $A = (A_{ij})$,

$$|A|_\zeta \equiv \sum_P \zeta^P A_{1P(1)} \cdots A_{nP(n)}. \qquad (6.114)$$

That is, $|A|_-$ is the determinant of A, and $|A|_+$ is what is often called the *permanent* of A.

Proof:

$$\langle \varphi_1, \ldots, \varphi_n | \psi_1, \ldots, \psi_n \rangle = \frac{1}{n!} \sum_P \sum_Q \zeta^P \zeta^Q (\langle \varphi_{P(1)} | \cdots \langle \varphi_{P(n)} |)(|\psi_{Q(1)}\rangle \cdots |\psi_{Q(n)}\rangle)$$

$$= \frac{1}{n!} \sum_P \sum_Q \zeta^P \zeta^Q \langle \varphi_{P(1)} | \psi_{Q(1)} \rangle \cdots \langle \varphi_{P(n)} | \psi_{Q(n)} \rangle$$

$$= \frac{1}{n!} \sum_P \sum_Q \zeta^P \zeta^Q \langle \varphi_1 | \psi_{QP^{-1}(1)} \rangle \cdots \langle \varphi_n | \psi_{QP^{-1}(n)} \rangle$$

(where we have permuted the factors by P)

$$= \frac{1}{n!} \sum_P \sum_Q \zeta^{QP^{-1}} \langle \varphi_1 | \psi_{QP^{-1}(1)} \rangle \cdots \langle \varphi_n | \psi_{QP^{-1}(n)} \rangle$$

---

* Compare this theorem with the well-known formula for vectors in 3-space,

$$(a \times b) \cdot (c \times d) = \begin{vmatrix} a \cdot c & a \cdot d \\ b \cdot c & b \cdot d \end{vmatrix}.$$

(since $\zeta^P = \zeta^{P^{-1}}$ and $\zeta^Q \zeta^{P^{-1}} = \zeta^{QP^{-1}}$)

$$= \frac{1}{n!} \sum_P \sum_R \zeta^R \langle \varphi_1 | \psi_{R(1)} \rangle \cdots \langle \varphi_n | \psi_{R(n)} \rangle$$

(letting $R = QP^{-1}$)

$$= \sum_R \zeta^R \langle \varphi_1 | \psi_{R(1)} \rangle \cdots \langle \varphi_n | \psi_{R(n)} \rangle$$

$$= |(\langle \varphi_i | \psi_j \rangle)|_\zeta,$$

which is the desired result.

Now let $\{|1\rangle, |2\rangle, \dots \}$ be a complete orthornormal set of states:

$$\langle \alpha | \beta \rangle = \delta_{\alpha\beta}; \qquad \sum_\alpha |\alpha\rangle\langle\alpha| = 1. \tag{6.115}$$

A complete set of n-particle states consists of $|\alpha_1, \alpha_2, \dots, \alpha_n\rangle$, where $\alpha_1 \leq \cdots \leq \alpha_n$ in the Bose case, and $\alpha_1 < \cdots < \alpha_n$ in the Fermi case. These states are orthogonal to one another, but not always normalized. The reader can show, using Eq. (6.113), that for a complete orthonormal set of states we may take

$$\frac{|\alpha_1, \dots, \alpha_n\rangle}{\sqrt{n_1! \, n_2! \cdots}} \, (\alpha_1 \leq \cdots \leq \alpha_n) \quad \text{for bosons}$$

$$|\alpha_1, \dots, \alpha_n\rangle (\alpha_1 < \cdots < \alpha_n) \quad \text{for fermions},$$

where n_α is the number of times that α occurs in the sequence $\alpha_1, \dots, \alpha_n$ (for Fermi particles, $n_\alpha = 0$ or 1).

For either case the completeness relation can be written in the following convenient form:

$$\frac{1}{n!} \sum_{\alpha_1} \cdots \sum_{\alpha_n} |\alpha_1, \dots, \alpha_n\rangle\langle\alpha_1, \dots, \alpha_n| = 1. \tag{6.116}$$

Here the range of each α_i is unrestricted, duplication of states being taken care of by the $1/n!$ and the normalization. In the Fermi case, the terms with non-distinct α_i are, of course, zero. The 1 on the right-hand side of Eq. (6.116) means the unit operator on the space of (properly symmetrized) n-particle states. Equation (6.116) can be verified by applying the left side to a state $|\beta, \dots, \beta_n\rangle$ and using Eqs. (6.113) and (6.114).

The case $n = 0$ may require some explanation. The zero-particle states are tensors of rank zero, that is, scalars (complex numbers). They form a one-dimensional space, all of whose elements are proportional to the number 1. The "state" 1 will be denoted by $|vac\rangle$ (or sometimes by $|0\rangle$) and called the "vacuum state." The zero-particle states are thus spanned by the state $|vac\rangle$.

We have constructed, for each n, a Hilbert space that describes a system of n particles; thus we have an infinite sequence of spaces. In many processes the number of particles is not constant: particles can be created and destroyed. To describe such processes we need a Hilbert space that contains states of varying numbers of particles. To get such a space we simply combine all the n-particle spaces into one big space that we may call the *multiparticle space*. A general state in the multiparticle space is of the form

$$|\psi\rangle = |\psi^{(0)}\rangle + |\psi^{(1)}\rangle + |\psi^{(2)}\rangle + |\psi^{(3)}\rangle + \cdots, \qquad (6.117)$$

where $|\psi^{(n)}\rangle$ is an n-particle state.

We define states of different numbers of particles to be orthogonal to each other, so that if $|\varphi\rangle$ is another multiparticle state and is expressed in the manner of Eq. (6.117), then

$$\langle\varphi|\psi\rangle \equiv \langle\varphi^{(0)}|\psi^{(0)}\rangle + \langle\varphi^{(1)}|\psi^{(1)}\rangle + \cdots. \qquad (6.118)$$

If $\{|\alpha\rangle\}$ is a complete orthonormal set of states, so that Eq. (6.115) holds, then using Eqs. (6.113), (6.115), and (6.118) we may summarize orthogonality by

$$\langle\alpha_1,\ldots,\alpha_n|\beta_1,\ldots,\beta_m\rangle = \delta_{nm}\begin{vmatrix}\delta_{\alpha_1\beta_1} & \cdots & \delta_{\alpha_1\beta_n}\\ \delta_{\alpha_n\beta_1} & \cdots & \delta_{\alpha_n\beta_n}\end{vmatrix}_\zeta. \qquad (6.119)$$

From Eq. (6.116) we also have the completeness relation

$$\sum_{n=0}^{\infty}\frac{1}{n!}\sum_{\alpha_1,\ldots,\alpha_n}|\alpha_1,\ldots,\alpha_n\rangle\langle\alpha_1,\ldots,\alpha_n| = 1. \qquad (6.120)$$

In this equation "1" means the unit operator on the whole multiparticle space.

As an example, suppose we describe the states in coordinate representation. The (unnormalizable) state $|x_1,\ldots,x_n\rangle$ describes the situation in which there is one particle each at points $x_1,\ldots,x_n$. Then Eqs. (6.119) and (6.120) become

$$\langle x_1,\ldots,x_n|y_1,\ldots,y_m\rangle = \delta_{nm}\begin{vmatrix}\delta^3(x_1-y_1) & \cdots & \delta^3(x_1-y_n)\\ \delta^3(x_n-y_1) & \cdots & \delta^3(x_n-y_n)\end{vmatrix}_\zeta, \qquad (6.121)$$

$$\sum_{n=0}^{\infty}\frac{1}{n!}\int d^3x_1\cdots\int d^3x_n|x_1,\ldots,x_n\rangle\langle x_1,\ldots,x_n| = 1 \qquad (6.122)$$

We may expand an arbitrary multiparticle state $|\psi\rangle$ as follows, using Eq. (6.122):

$$|\psi\rangle = \sum_{n=0}^{\infty}\frac{1}{n!}\int d^3x_1\cdots\int d^3x_n|x_1,\ldots,x_n\rangle\psi^{(n)}(x_1,\ldots,x_n). \qquad (6.123)$$

Here

$$\psi^{(n)}(x_1,\ldots,x_n) = \langle x_1,\ldots,x_n|\psi\rangle \qquad (6.124)$$

is (if $|\psi\rangle$ is normalized) the amplitude for the state $|\psi\rangle$ to have n particles, one at each x_i. Note that $\psi^{(n)}(x_1, \ldots, x_n)$ is symmetric or antisymmetric according to the statistics. Note also that if $|\psi\rangle$ is an n-particle state and is of the form $|\psi_1, \ldots, \psi_n\rangle$, where each $|\psi_i\rangle$ is a one-particle state, then

$$\psi^{(m)}(x_1, \ldots, x_n) = 0 \quad \text{unless} \quad m = n;$$

$$\psi^{(n)}(x_1, \ldots, x_n) = \begin{vmatrix} \psi_1(x_1) & \cdots & \psi_1(x_n) \\ \psi_n(x_1) & \cdots & \psi_n(x_n) \end{vmatrix}_\zeta, \tag{6.125}$$

where $\psi_i(x_j) = \langle x_j | \psi_i \rangle$. Equation (6.125) follows from Eqs. (6.124) and (6.113). In the Fermi case the determinant is called the "Slater determinant."

We are now ready to define creation and destruction operators. These operators are fundamental for two reasons; first, we constructed the multiparticle states so that we could describe changing numbers of particles, and we need some operators that can effect this change, and second, other operators, such as the total energy, will turn out to be simply expressible in terms of the creation and destruction operators.

Let $|\varphi\rangle$ be any one-particle state. We define $a^+(\varphi)$ to be that linear operator which satisfies

$$a^+(\varphi)|\psi_1, \ldots, \psi_n\rangle = |\varphi, \psi_1, \ldots, \psi_n\rangle \tag{6.126}$$

for any n-particle state $|\psi_1, \ldots, \psi_n\rangle$. For $n = 0$ this is understood to mean $a^+(\varphi)|\text{vac}\rangle = |\varphi\rangle$. We call $a^+(\varphi)$ the *creation operator* for the state $|\varphi\rangle$, and its adjoint $a(\varphi)$ the *destruction operator*.

A creation operator clearly converts an n-particle state into an $(n + 1)$-particle state. It is easily seen that a destruction operator turns an n-particle state into an $(n - 1)$-particle state and annihilates the vacuum state. To find the effect of $a(\varphi)$ on an n-particle state $|\psi_1 \cdots \psi_n\rangle$ we multiply on the left by an arbitrary $(n - 1)$-particle state $\langle \chi_1 \cdots \chi_{n-1}|$.

$$\langle \chi_1 \cdots \chi_{n-1} | a(\varphi) | \psi_1 \cdots \psi_n \rangle$$

$$= \langle \psi_1 \cdots \psi_n | a^+(\varphi) | \chi_1 \cdots \chi_{n-1} \rangle^*$$

$$= \langle \psi_1 \cdots \psi_n | \varphi, \chi_1 \cdots \chi_{n-1} \rangle^*$$

$$= \begin{vmatrix} \langle \psi_1 | \varphi \rangle \langle \psi_1 | \chi_1 \rangle & \cdots & \langle \psi_1 | \chi_{n-1} \rangle \\ \langle \psi_n | \varphi \rangle \langle \psi_n | \chi_1 \rangle & \cdots & \langle \psi_n | \chi_{n-1} \rangle \end{vmatrix}_\zeta^*$$

$$= \left\{ \sum_{k=1}^{n} \zeta^{k-1} \langle \psi_k | \varphi \rangle \begin{vmatrix} \langle \psi_1 | \chi_1 \rangle & \cdots & \langle \psi_1 | \chi_{n-1} \rangle \\ & (\text{no } \psi_k) & \\ \langle \psi_n | \chi_1 \rangle & \cdots & \langle \psi_n | \chi_{n-1} \rangle \end{vmatrix}_\zeta \right\}^*$$

(an expansion by minors)

$$= \sum_{k=1}^{n} \zeta^{k-1} \langle \psi_k | \varphi \rangle^* \langle \psi_1 \cdots (\text{no } \psi_k) \cdots \psi_n | \chi_1 \cdots \chi_{n-1} \rangle^*$$

$$= \sum_{k=1}^{n} \zeta^{k-1} \langle \varphi | \psi_k \rangle \langle \chi_1 \cdots \chi_{n-1} | \psi_1 \cdots (\text{no } \psi_k) \cdots \psi_n \rangle.$$

Because this is for arbitrary $\langle \chi_1 \cdots \chi_{n-1} |$ we have finally,

$$a(\varphi)| \psi_1 \cdots \psi_n \rangle = \sum_{k=1}^{n} \zeta^{k-1} \langle \varphi | \psi_k \rangle | \psi_1 \cdots (\text{no } \psi_k) \cdots \psi_n \rangle. \quad (6.127)$$

Thus the destruction operator removes the states $| \psi_i \rangle$, one at a time, leaving a sum of $(n - 1)$-particle states. In the Bose case ($\zeta = 1$) the terms all have a $+$ sign, whereas in the Fermi case ($\zeta = -1$) they alternate in sign.

Equations (6.126) and (6.127) describe the action of creation and destruction operators on many-particle states. From Eq. (6.126) it follows that

$$a^+(\varphi_1) a^+(\varphi_2) = \zeta a^+(\varphi_2) a^+(\varphi_1),$$

or

$$[a^+(\varphi_1), a^+(\varphi_2)]_{-\zeta} = 0, \quad (6.128)$$

where $[A, B]_{-\zeta} \equiv AB - \zeta BA$; that is, $[AB]_-(\zeta = +1)$ is the commutator and $[A, B]_+(\zeta = -1)$ is the anticommutator.

Taking the adjoint of Eq. (6.128) we obtain the further result

$$[a(\varphi_1), a(\varphi_2)]_{-\zeta} = 0. \quad (6.129)$$

Thus, the creation operators commute for Bose particles and anticommute for Fermi particles, and similarly for destruction operators.

Now, what is $[a(\varphi_1), a^+(\varphi_2)]_{-\zeta}$? Does it (or any similar expression) reduce to anything simple? We first calculate

$$a(\varphi_1) a^+(\varphi_2) | \psi_1 \cdots \psi_n \rangle$$

$$= a(\varphi_1) | (\varphi_2, \psi_1 \cdots \psi_n \rangle$$

$$= \langle \varphi_1 | \varphi_2 \rangle | \psi_1 \cdots \psi_n \rangle + \sum_{k=1}^{n} \zeta^k \langle \varphi_1 | \psi_k \rangle | \varphi_2, \psi_1 \cdots (\text{no } \psi_k) \cdots \psi_n \rangle, \quad (6.130)$$

and then

$$a^+(\varphi_2) a(\varphi_1) | \psi_1 \cdots \psi_n \rangle = \sum_{k=1}^{n} \zeta^{k-1} \langle \varphi_1 | \psi_k \rangle a^+(\varphi_2) | \psi_1 \cdots (\text{no } \psi_k) \cdots \psi_n \rangle$$

$$= \sum_{k=1}^{n} \zeta^{k-1} \langle \varphi_1 | \psi_{k,} | \varphi_2, \psi_1 \cdots (\text{no } \psi_k) \cdots \psi_n \rangle.$$

$$(6.131)$$

Multiplying Eq. (6.131) by ζ and subtracting it from Eq. (6.130) we see that

$$[a(\varphi_1), a^+(\varphi_2)]_{-\zeta} = \langle \varphi_1 | \varphi_2 \rangle. \tag{6.132}$$

Equations (6.128), (6.129), and (6.132) are the fundamental "commutation" relations for creation and destruction operators.

The relations we have derived are usually stated in terms of an orthonormal basis, and we shall now do that. Let $\{|\alpha\rangle\} = \{|1\rangle, |2\rangle, \dots\}$ be a complete orthonormal set of one-particle states. It is usual to let $a_\alpha = a(\alpha)$. Then, since $\langle \alpha | \beta \rangle = \delta_{\alpha\beta}$, we have $[a_\alpha, a_\beta^+]_{-\zeta} = \delta_{\alpha\beta}$. We consider the Bose and Fermi cases separately.

Bose Case

Let

$$|n_1 n_2 \cdots \rangle = \frac{|1, \dots, 1, 2, \dots, 2, \dots \rangle}{\sqrt{n_1! \, n_2! \cdots}}, \tag{6.133}$$

where n_α is the number of times α appears in the ket on the right. Then the $|n_1 n_2 \cdots \rangle$ (each $n_\alpha = 0, 1, 2, 3, \dots$) form an orthonormal basis for the whole multiparticle space. From Eqs. (6.133), (6.126), and (6.127) we find

$$a_\alpha^+ |n_1 n_2 \cdots n_\alpha \cdots \rangle = \sqrt{n_\alpha + 1} \, |n_1 n_2 \cdots n_\alpha + 1 \cdots \rangle, \tag{6.134}$$

$$a_\alpha |n_1 n_2 \cdots n_\alpha \cdots \rangle = \sqrt{n_\alpha} \, |n_1 n_2 \cdots n_{\alpha-1} \cdots \rangle. \tag{6.135}$$

The commutation relations are

$$[a_\alpha, a_\beta] = [a_\alpha^+, a_\beta^+] = 0; \qquad [a_\alpha, a_\beta^+] = \delta_{\alpha\beta}. \tag{6.136}$$

Equations (6.134), (6.135), and (6.136) are identical to Eqs. (6.60), (6.61), (6.51), and (6.52) for raising and lowering operators for a system of harmonic oscillators. The operator for the number of particles in the state $|\alpha\rangle$ is

$$N_\alpha = a_\alpha^+ a_\alpha.$$

The notation of Eq. (6.133) (in terms of "occupation numbers") is generally not the most convenient. It is more natural, in fact, to continue using the notation we have been using all along in this section, that is, the notation $|\alpha_1, \alpha_2, \dots, \alpha_n\rangle$. This notation was discussed in Section 6.5 in connection with phonons. Note that Eqs. (6.126) and (6.127) of this section, when applied to states of the form $|\alpha_1, \dots, \alpha_n\rangle$, become identical to Eqs. (6.67) and (6.68). Thus, creation and destruction operators for a system of Bose particles look just like those for what we called a "phonon" system.

Fermi Case

Using the notation $|\alpha_1, \ldots, \alpha_n\rangle$, we have

$$a_\alpha^+|\alpha_1, \ldots, \alpha_n\rangle = |\alpha, \alpha_1, \ldots, \alpha_n\rangle \tag{6.137}$$

and

$$a_\alpha|\alpha_1, \ldots, \alpha_n\rangle = \sum_{k=1}^{n} (-1)^{k-1}\delta_{\alpha\alpha_k}|\alpha_1, \ldots, \alpha_{k-1}, \alpha_{k+1}, \ldots, \alpha_n\rangle \tag{6.138}$$

We could also use the occupation-number notation

$$|n_1 n_2 \cdots\rangle = |\alpha_1, \alpha_2 \cdots\rangle$$

where $\alpha_1 < \alpha_2 < \cdots$, and n_α is the number of times α occurs ($n_\alpha = 0$ or 1) in this sequence. If $n_\alpha = 0$, then a_α^+ changes it to 1, whereas a_α annihilates the state. If $n_\alpha = 1$, then a_α changes it to 0, whereas a_α^+ annihilates the state. There are also factors of ± 1 involved, depending on what other states are occupied. It is easiest merely to remember Eqs. (6.137) and (6.138).

Note that $a(\varphi)^2 = a^+(\varphi)^2 = 0$ for any one-particle state $|\varphi\rangle$. This statement follows from Eqs. (6.128) and (6.129) (with $\zeta = -1$ and $\varphi_1 = \varphi_2 = \varphi$), and it is also equivalent to the fact that two fermions cannot be in the same state, that is, $|\varphi, \varphi\rangle = 0$.

One could also derive Eqs. (6.137) and (6.138) directly from the anti-commutation relations

$$[a_\alpha, a_\beta]_+ = [a_\alpha^+, a_\beta^+]_+ = 0; \qquad [a_\alpha, a_\beta^+]_+ = \delta_{\alpha\beta} \tag{6.139}$$

as was done in previous sections for the a_α's of the harmonic-oscillator system. But for the Fermi case there does not appear to be any *a priori* reason for postulating Eq. (6.139). (Remember, for oscillators the corresponding commutation rules followed from the canonical quantization procedure.) One may rather consider Eq. (6.139) as derived from the antisymmetrization postulate for fermions.

Let us return to the general case where Eqs. (6.126) through (6.132) apply. One advantage of deriving them in such a general form is that we are not tied down to a particular basis of states. Suppose we use a basis of momentum eigenstates, $|p\rangle$. Because $\langle p|p'\rangle = (2\pi)^3\delta^3(p - p')$ we have

$$[a(p), a^+(p')]_{-\zeta} = (2\pi)^3\delta^3(p - p'),$$
$$[a(p), a(p')]_{-\zeta} = [a^+(p), a^+(p')]_{-\zeta} = 0. \tag{6.140}$$

From the vacuum state we can construct the other states by

$$|p_1, \ldots, p_n\rangle = a^+(p_1) \cdots a^+(p_n)|\text{vac}\rangle. \tag{6.141}$$

If we use a basis of position eigenstates $|x\rangle$, then since $\langle x|x'\rangle = \delta^3(x - x')$,

$$[a(x), a^+(x')]_{-\zeta} = \delta^3(x - x'), \tag{6.142}$$

$$|x_1, \ldots, x_n\rangle = a^+(x_1) \cdots a^+(x_n)|\text{vac}\rangle. \tag{6.143}$$

If we use a basis of hydrogen-atom energy eigenstates $|nlm\rangle$, then $[a(nlm), a^+(n'l'm')]_{-\zeta} = \delta_{nn'} \delta_{ll'} \delta_{mm'}$, and so on.

How do the creation and destruction operators change when we make a change of basis? This question is easily answered by noting that if

$$|\chi\rangle = \alpha|\psi\rangle + \beta|\varphi\rangle, \tag{6.144}$$

then

$$a^+(\chi) = \alpha a^+(\psi) + \beta a^+(\varphi),$$

$$a(\chi) = \alpha^* a(\psi) + \beta^* a(\varphi), \tag{6.145}$$

This means that creation operators "transform" like kets, whereas destruction operators "transform" like bras (because $\langle \chi | = \alpha^* \langle \psi | + \beta^* \langle \varphi |$). Equation (6.145) are readily generalized to infinite series and integrals. Now if we change, for example, from position to momentum representation, so that

$$|p\rangle = \int d^3x |x\rangle\langle x|p\rangle = \int d^3x |x\rangle e^{ip \cdot x},$$

$$|x\rangle = \int \frac{d^3p}{(2\pi)^3} |p\rangle\langle p|x\rangle = \int \frac{d^3p}{(2\pi)^3} |p\rangle e^{-ip \cdot x}, \tag{6.146}$$

the creation operators are related by

$$a^+(p) = \int d^3x\, a^+(x) e^{ip \cdot x},$$

$$a^+(x) = \int \frac{d^3p}{(2\pi)^3} a^+(p) e^{-ip \cdot x}. \tag{6.147}$$

To relate the destruction operators $a(p)$ and $a(x)$, simply take the Hermitian adjoint of Eq. (6.147). One proceeds in a similar way for any other change of basis.

Exercise: Suppose we have a complete orthonormal set of states $|\alpha\rangle$, and we let the "wave functions" of these states be $\langle x|\alpha\rangle = u_\alpha(x)$. Write down the formulas for $a^+(x)$ and $a(x)$ in terms of a_α^+ and a_α, and vice versa.

6.8 THE HAMILTONIAN AND OTHER OPERATORS

In the last section we developed a method for describing systems containing many Bose or Fermi particles, and we defined creation and destruction operators for the particle states. We will now show that these operators have other uses than merely creating and destroying states.

Suppose $A^{(1)}$ is an operator that acts only on one-particle states. We wish to find an operator A that represents the "sum of $A^{(1)}$ over all of the particles." That is, for any n-particle state

$$|\psi\rangle = |\psi_1, \ldots, \psi_n\rangle = |\psi_1\rangle \times \cdots \times |\psi_n\rangle \qquad (6.148)$$

we want $A|\psi\rangle$ to satisfy

$$A|\psi\rangle = A^{(1)}|\psi_1\rangle \times |\psi_2\rangle \times \cdots \times |\psi_n\rangle + |\psi_1\rangle \times A^{(1)}|\psi_2\rangle \times \cdots \times |\psi_n\rangle$$
$$+ \cdots + |\psi_1\rangle \times |\psi_2\rangle \times \cdots \times A^{(1)}|\psi_n\rangle. \qquad (6.149)$$

To see what this means, suppose each $|\psi_i\rangle$ is an eigenstate of $A^{(1)}$ with eigenvalue a_i. Then Eq. (6.149) implies that

$$A|\psi\rangle = (a_1 + \cdots + a_n)|\psi\rangle. \qquad (6.150)$$

For example, if $A^{(1)}$ is the single-particle Hamiltonian, then A is the total energy (ignoring mutual interactions, which we shall consider later in this section). If $A^{(1)}$ is the momentum operator for a single particle, then A is the total momentum. If $A^{(1)} = 1^{(1)}$ (the unit operator on one-particle states), then $A = N$, the "number-of-particles operator."

The desired operator A is easy to find. We first find it for the special case $A^{(1)} = |\alpha\rangle\langle\beta|$. In this case Eq. (6.149) becomes

$$A|\psi\rangle = \langle\beta|\psi_1\rangle|\alpha, \psi_2, \ldots, \psi_n\rangle + \langle\beta|\psi_2\rangle|\psi_1, \alpha, \ldots, \psi_n\rangle$$
$$+ \cdots + \langle\beta|\psi_n\rangle|\psi_1, \psi_2, \ldots, \alpha\rangle. \qquad (6.151)$$

Now look back at Eq. (6.131) and notice that when $\varphi_2 = \alpha$ and $\varphi_1 = \beta$ we have

$$a^+(\alpha)a(\beta)|\psi\rangle = \sum_{k=1}^{n} \zeta^{k-1}\langle\beta|\psi_k\rangle|\alpha_1\psi_1, \ldots, (\text{no } \psi_k), \ldots, \psi_n\rangle. \qquad (6.152)$$

But

$$\zeta^{k-1}|\alpha, \psi_1, \ldots, (\text{no } \psi_k), \ldots, \psi_n\rangle = |\psi_1, \ldots, \psi_{k-1}, \alpha, \psi_{k+1}, \ldots, \psi_n\rangle$$

due to the symmetry property of the n-particle state. Using this equation in Eq. (6.152) and comparing it with Eq. (6.151) we find

$$A = a^+(\alpha)a(\beta) \qquad \text{when} \qquad A^{(1)} = |\alpha\rangle\langle\beta|. \qquad (6.153)$$

The generalization of Eq. (6.153) for an arbitrary one-particle operator $A^{(1)}$ is immediate. We choose a basis—any basis—of one-particle state $|\alpha\rangle$, and write

$$A^{(1)} = \sum_{\alpha,\beta} |\alpha\rangle\langle\alpha|A^{(1)}|\beta\rangle\langle\beta| = \sum_{\alpha,\beta} A^{(1)}_{\alpha\beta}|\alpha\rangle\langle\beta|. \qquad (6.154)$$

Then, from Eq. (6.153) and linearity,

$$A = \sum_{\alpha,\beta} A^{(1)}_{\alpha\beta} a^+(\alpha)a(\beta).$$

(6.155)

As a first example we consider $A^{(1)} = 1^{(1)}$, so that $A = N$, the operator for the number of particles. Using various bases we have

$$1^{(1)} = \sum_{\alpha} |\alpha\rangle\langle\alpha|$$

$$= \int d^3x |x\rangle\langle x|$$

$$= \int \frac{d^3p}{(2\pi)^3} |p\rangle\langle p|$$

(6.156)

from which we can immediately write

$$N = \sum_{\alpha} a^+_{\alpha} a_{\alpha}$$

$$= \int d^3x a^+(x)a(x)$$

$$= \int \frac{d^3p}{(2\pi)^3} a^+(p)a(p).$$

(6.157)

Next, consider the momentum operator. Because

$$P^{(1)} = \int \frac{d^3p}{(2\pi)^3} p |p\rangle\langle p|$$

$$= \int d^3x |x\rangle \frac{1}{i} \nabla\langle x|$$

(6.158)

we have for the total momentum

$$P = \int \frac{d^3p}{(2\pi)^3} p a^+(p)a(p)$$

$$= \int d^3x a^+(x) \frac{1}{i} \nabla a(x).$$

(6.159)

(Compare the first of these expressions for P with Eq. (6.100), another expression for P.)

Finally, suppose the Hamiltonian for a single particle is

$$H^{(1)} = \frac{P^2}{2m} + V(x),$$

(6.160)

where x is the position operator. In coordinate representation,

$$\langle x|H^{(1)}|x'\rangle = -\frac{1}{2m}\nabla^2\delta^3(x - x') + V(x)\delta^3(x - x'), \qquad (6.161)$$

so that

$$H = \int d^3x \int d^3x' \left[-\frac{1}{2m}\nabla^2\delta^3(x - x') + V(x)\delta^3(x - x') \right] a^+(x)a(x')$$

$$= \int d^3x\, a^+(x)\left[-\frac{\nabla^2}{2m} + V(x) \right] a(x). \qquad (6.162)$$

In momentum representation*

$$\langle p|H^{(1)}|p'\rangle = \frac{p^2}{2m}(2\pi)^3\delta^3(p - p') + \int d^3x\, e^{-ip\cdot x}V(x)e^{ip\cdot x'}$$

$$= \frac{p^2}{2m}(2\pi)^3\delta^3(p - p') + \tilde{V}(p - p'), \qquad (6.163)$$

where

$$\tilde{V}(q) = \int d^3x\, V(x)e^{-iq\cdot x}. \qquad (6.164)$$

Therefore

$$H = \int \frac{d^3p}{(2\pi)^3}\frac{p^2}{2m}a^+(p)a(p) + \int \frac{d^3p}{(2\pi)^3}\int \frac{d^3p'}{(2\pi)^3}\tilde{V}(p - p')a^+(p)a(p')$$

$$= \int \frac{d^3p}{(2\pi)^3}\frac{p^2}{2m}a^+(p)a(p) + \int \frac{d^3p}{(2\pi)^3}\int \frac{d^3q}{(2\pi)^3}\tilde{V}(q)a^+(p + q)a(p). \qquad (6.165)$$

The term $\tilde{V}(q)a^+(p + q)a(p)$ annihilates a particle of momentum p and recreates it with momentum $p + q$, the amplitude for this process being $\tilde{V}(q)$. If we use a basis of eigenstate $|\alpha\rangle$ of $H^{(1)}$, so that

$$\langle \alpha|H^{(1)}|\beta\rangle = E_\alpha\delta_{\alpha\beta} \qquad (6.166)$$

then

$$H = \sum_\alpha E_\alpha a_\alpha^+ a_\alpha, \qquad (6.167)$$

which is what we had for phonons with $E_\alpha = \hbar\omega_\alpha$.

---

* The second term is obtained by noting that

$$V(X_p) = \int d^3x\, V(x)|x\rangle\langle x|$$

and putting this between states $\langle p|$ and $|p'\rangle$.

All of these expressions for H can be derived from each other using the formulas relating $a^+(x)$ and $a^+(p)$, and so on, described at the end of Section 6.7. But it is often simplest to obtain such expressions directly from the single-particle Hamiltonians, as we have done here.

The density of particles (that is, number per unit volume) at the point x is given by the operator

$$\rho(x) = a^+(x)a(x) \tag{6.168}$$

(which corresponds to the one-particle operator $|x\rangle\langle x|$). Thus the number-of-particles operator in Eq. (6.157) may be written as

$$N = \int d^3x\rho(x),$$

and the potential-energy term in Eq. (6.162) as

$$V = \int d^3xV(x)\rho(x).$$

This last equation represents the integral of the potential energy weighted by the density.

So far we have described a system of *independent* particles, each particle being (possibly) acted on by an external potential but no two particles influencing each other. Suppose, however, that there is an additional potential $V^{(2)}(x_i, x_j)$ between any two particles at x_i and x_j (giving rise to a "two-body" force). We assume that $V^{(2)}(x_i, x_j) = V^{(2)}(x_j, x_i)$. On two-particle states the operator is then

$$V^{(2)} = \tfrac{1}{2}\int d^3x \int d^3y|x, y\rangle V^{(2)}(x, y)\langle x, y|, \tag{6.169}$$

as can be verified by applying it to a two-particle state $|x_1, x_2\rangle$. We now want an operator V on the whole multiparticle space such that

$$V|x_1, \ldots, x_n\rangle = \sum_{i<j} V^{(2)}(x_i, x_j)|x_1, \ldots, x_n\rangle$$

$$= \tfrac{1}{2}\sum_{i\neq j} V^{(2)}(x_i, x_j)|x_1, \ldots, x_n\rangle. \tag{6.170}$$

Looking at Eq. (6.169) and noticing that $a^+(x)a^+(y)$ creates the state $|x, y\rangle$, whereas $a(y)a(x)$ destroys the same state, we might guess that

$$V = \tfrac{1}{2}\int d^3x \int d^3ya^+(x)a^+(y)V^{(2)}(x, y)a(y)a(x). \tag{6.171}$$

This is in fact correct, as can be verified by applying V to $|\bar{x}_1, \ldots, x_n\rangle$. Using Eq. (6.127) twice, we have

$$a(y)a(x)|x_1 \cdots x_n\rangle$$

$$= a(y) \sum_{k=1}^{n} \zeta^{k-1} \delta^3(x - x_k)|x_1 \cdots (\text{no } x_k) \cdots x_n\rangle$$

$$= \sum_{k=1}^{n} \zeta^{k-1} \delta^3(x - x_k) \sum_{\substack{j=1 \\ j \neq k}}^{n} \eta_{jk} \delta^3(y - x_j)|x_1 \cdots (\text{no } x_k, x_j) \cdots x_n\rangle,$$

where

$$\eta_{jk} = \begin{cases} \zeta^{j-1} & \text{if } j < k \\ \zeta^{j} & \text{if } j > k \end{cases}.$$

Then

$$a^+(x)a^+(y)a(y)a(x)|x_1, \ldots, x_n\rangle$$

$$= \sum_{j \neq k} \zeta^{k-1} \eta_{jk} \delta^3(x - x_k) \delta^3(y - x_j)|x, y, x_1, \ldots, (\text{no } x_k, x_j), \ldots, x_n\rangle$$

$$= \sum_{j \neq k} \zeta^{k-1} \eta_{jk} \delta^3(x - x_k) \delta^3(y - x_j)|x_k, x_j, \ldots, (\text{no } x_k, x_j), \ldots, x_n\rangle$$

$$= \sum_{j \neq k} \delta^3(x - x_k) \delta^3(y - x_j)|x_1, \ldots, x_n\rangle.$$

Multiplying by $\frac{1}{2}V^{(2)}(x, y)$ and integrating over x and y, we find that V as given by Eq. (6.171) indeed satisfies Eq. (6.170).

In view of Eq. (6.168) and the remarks following it, we might expect that the mutual interaction could also be described in terms of the particle density by the operator

$$V' = \frac{1}{2} \int d^3x \int d^3y V^{(2)}(x, y)\rho(x)\rho(y). \tag{6.172}$$

However, V' is not quite the same as V. To see the difference we write

$$\rho(x)\rho(y) = a^+(x)a(x)a^+(y)a(y)$$
$$= \zeta a^+(x)a^+(y)a(x)a(y) + \delta^3(x - y)a^+(x)a(y)$$
$$= a^+(x)a^+(y)a(y)a(x) + \delta^3(x - y)a^+(x)a(x),$$

so that

$$V' = V + \frac{1}{2} \int d^3x V^{(2)}(x, x)\rho(x). \tag{6.173}$$

Thus V' contains an extra term, which may be interpreted as a self-energy; it contributes even when there is only one particle present. The true mutual interaction V is zero unless there are two or more particles. We want only the

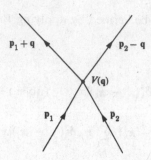

Fig. 6.1 Operator V adds momentum q to one particle and subtracts it from the other.

mutual interaction, because any self-energy (if it exists in nature) can be included in the Hamiltonian of Eq. (6.162). Besides, for many potentials (for example, the Coulomb potential), V' is infinite and is not what we would consider to be the true energy.

If we express V in momentum representation, using Eqs. (6.171) and (6.147), and assume

$$V^{(2)}(x, y) = V(x - y) = V(y - x), \tag{6.174}$$

we arrive at

$$V = \frac{1}{2} \int \frac{d^3q}{(2\pi)^3} \int \frac{d^3p}{(2\pi)^3} \int \frac{d^3p'}{(2\pi)^3} \tilde{V}(q)a^+(p + q)a^+(p' - q)a(p')a(p). \tag{6.175}$$

Here $\tilde{V}(q)$ is as defined in Eq. (6.164) (in proving Eq. (6.175) we use $\tilde{V}(q) = \tilde{V}(-q)$, because $V(x) = V(-x)$). Note that

$$V|p_1, p_2\rangle = \int \frac{d^3q}{(2\pi)^3} \tilde{V}(q)|p_1 + q, p_2 - q\rangle,$$

which says that V adds momentum q to one particle and subtracts it from the other with amplitude $V(q)$. This process is denoted by the diagram in Fig. 6.1. The total momentum is conserved, as we would expect, because Eq. (6.174) implies that the mutual interaction is invariant under translations.

The Hamiltonian and mutual interactions described here all conserve the total number of particles (see the exercise at the end of Section 6.5).

6.9 GROUND STATE FOR A FERMION SYSTEM

In this section we will consider fermions, so that the results of Sections 6.7 and 6.8 apply with $\zeta = -1$. Suppose we know the eigenstates and eigenvalues of the single-particle Hamiltonians, $H^{(1)}$:

$$H^{(1)}|\alpha\rangle = E_\alpha|\alpha\rangle, \qquad (\alpha = 1, 2, 3, \ldots). \tag{6.176}$$

Then, ignoring mutual interactions, we have for the multiparticle Hamiltonian:

$$H = \sum_\alpha E_\alpha a_\alpha^+ a_\alpha. \tag{6.177}$$

Every state may be built up from the vacuum state by applying creation operators:

$$|\alpha_1, \ldots, \alpha_n\rangle = a_\alpha^+, \ldots, a_{\alpha_n}^+ |vac\rangle. \tag{6.178}$$

(Both sides vanish unless the α_i are all distinct.)

It is often inconvenient to refer everything to the vacuum state, as in Eq. (6.178). In practice we may be considering states that differ from some "ground" state only by the presence or absence of a few particles. Suppose there are G particles present. Assume that the energy levels are ordered such that

$$E_1 \le E_2 \le E_3 \le \cdots.$$

Then the state of lowest energy is

$$|gnd\rangle = |1, \ldots, G\rangle = a_1^+ \cdots a_G^+ |vac\rangle, \tag{6.179}$$

which we call the *ground state*; its energy is

$$E_{gnd} = E_1 + \cdots + E_G. \tag{6.180}$$

Any other G-particle state will have some of the levels $1, \ldots, G$ unoccupied and some higher levels occupied. It is convenient to use $|gnd\rangle$ as a reference, describing the removal of a particle from $1, \ldots, G$ as the creation of a "hole." Particles in the levels $G + 1, G + 2, \ldots$ are still called "particles." If a particle is excited from the level α to the level $\beta(\alpha \le G \le \beta)$, then we say that a hole with energy $-E_\alpha$ has been created as well as a particle with energy E_β.

The concept of holes may be formulated mathematically as follows: Define

$$b_\alpha = a_\alpha^+ \qquad \text{for} \quad \alpha \le G. \tag{6.181}$$

From the anticommutation relations in Eq. (6.139) we have, if $\alpha, \alpha' > G$ and $\beta, \beta' \le G$,

$$[a_\alpha, a_{\alpha'}]_+ = [a_\alpha, b_\beta]_+ = [b_\beta, b_{\beta'}]_+ = 0,$$
$$[a_\alpha, a_{\alpha'}^+]_+ = \delta_{\alpha\alpha'}; \; [b_\beta, b_{\beta'}^+]_+ = \delta_{\beta\beta'}, \tag{6.182}$$
$$[a_\alpha, b_\beta^+]_+ = 0.$$

Thus the operators $a_\alpha^+ (\alpha > G)$ and $b_\alpha^+ (\alpha \le G)$ behave like creation operators. We can now express the states in the form

$$|\alpha_1, \ldots, \alpha_m, \beta_1, \ldots, \beta_n, gnd\rangle \equiv a_{\alpha_1}^+ \cdots a_\alpha^+ b_{\beta_1}^+ \cdots b_{\beta_n}^+ |gnd\rangle \tag{6.183}$$

$(\alpha_i > G, \beta_i \leq G)$. We call this a state with m particles and n holes. The ground state acts like a vacuum state in that

$$a_\alpha|\text{gnd}\rangle = b_\beta|\text{gnd}\rangle = 0. \tag{6.184}$$

The Hamiltonian of Eq. (6.177) may be written in the form

$$
\begin{aligned}
H &= \sum_{\alpha > G} E_\alpha a_\alpha^+ a_\alpha + \sum_{\alpha \leq G} E_\alpha b_\alpha b_\alpha^+ \\
&= \sum_{\alpha > G} E_\alpha a_\alpha^+ a_\alpha + \sum_{\alpha \leq G} E_\alpha(-b_\alpha^+ b_\alpha + 1) \\
&= E_{\text{gnd}} + \sum_{\alpha > G} E_\alpha a_\alpha^+ a_\alpha - \sum_{\alpha \leq G} E_\alpha b_\alpha^+ b_\alpha, \tag{6.185}
\end{aligned}
$$

so that the energy of a hole in state α is $-E_\alpha$. In other words, it takes an energy $-E_\alpha$ to create the hole. The number-of-particles operator is

$$
\begin{aligned}
N &= \sum_\alpha a_\alpha^+ a_\alpha \\
&= \sum_{\alpha > G} a_\alpha^+ a_\alpha + \sum_{\alpha \leq G} (-b_\alpha^+ b_\alpha + 1) \\
&= G + \sum_{\alpha > G} a_\alpha^+ a_\alpha - \sum_{\alpha \leq G} b_\alpha^+ b_\alpha; \tag{6.186}
\end{aligned}
$$

thus a hole counts as -1. The number of particles and holes outside the ground state (that is, the number $m + n$ in Eq. (6.183)) is given by the operator

$$N' = \sum_{\alpha > G} a_\alpha^+ a_\alpha + \sum_{\alpha \leq G} b_\alpha^+ b_\alpha, \tag{6.187}$$

which counts a hole as $+1$ and the ground state as nothing.

Suppose now that a perturbation is applied to the system in the form of an external potential

$$U = \sum_{\alpha, \beta} U_{\alpha\beta}^{(1)} a_\alpha^+ a_\beta, \tag{6.188}$$

where we assume that the single-particle potential $U_{\alpha\beta}^{(1)}$ may be nondiagonal. Then

$$
\begin{aligned}
U &= \sum_{\substack{\alpha > G \\ \beta > G}} U_{\alpha\beta}^{(1)} a_\alpha^+ a_\beta + \sum_{\substack{\alpha > G \\ \beta \leq G}} U_{\alpha\beta}^{(1)} a_\alpha^+ b_\beta^+ \\
&\quad + \sum_{\substack{\alpha \leq G \\ \beta > G}} U_{\alpha\beta}^{(1)} b_\alpha a_\beta - \sum_{\substack{\alpha \leq G \\ \beta \leq G}} U_{\alpha\beta}^{(1)} b_\beta^+ b_\alpha + \sum_{\alpha \leq G} U_{\alpha\alpha}^{(1)}. \tag{6.189}
\end{aligned}
$$

The first and fourth terms of this equation modify the energies of the particles and holes, respectively, and the fifth term modifies the ground-state energy. The second term creates particle–hole pairs, and the third term destroys them. Note that N, as given by Eq. (6.186), is conserved, whereas N', as defined in Eq. (6.187), is not.

Suppose further that there is a mutual interaction, a two-body potential of the form described in Section 6.8:

$$V = \sum_{\alpha,\beta,\gamma,\delta} V^{(2)}_{\alpha\beta,\gamma\delta} a^+_\alpha a^+_\beta a_\delta a_\gamma. \tag{6.190}$$

As in Eq. (6.189), we can express V in terms of a_α ($\alpha > G$) and b_α ($\alpha \le G$), and then use the anticommutation rules of Eq. (6.182) to write V in a form such that in every term all of the creation operators are to the left of all of the destruction operators. Such terms are called "normal products." The result will be a number of terms involving four operators, plus other terms that can be lumped into U or added to E_{gnd}.

6.10 HAMILTONIAN FOR A PHONON–ELECTRON SYSTEM

Now we will consider, as an application of the creation and destruction operator formalism, the interaction between the electrons and lattice vibrations in a crystal. It is this interaction (as well as the presence of crystal impurities and imperfections) that accounts for the finite conductivity of metals under most conditions. (One might otherwise expect from the band theory of metals that, once an electron got into an unfilled band, it would move unhindered, resulting in infinite conductivity.) We will show how the Hamiltonian is derived and written in terms of creation and destruction operators.

We start with the electron in a lattice with no vibrations. Let $N = n_1 a_1 + n_2 a_2 + n_3 a_3$ (where n_1, n_2, n_3 are integers) describe the positions of the nuclei. The potential $V(x)$ felt by an electron in the lattice is periodic and has the form

$$V_1(x) = \sum_N V_0(x - N). \tag{6.191}$$

The Hamiltonian for a system of independent electrons in the lattice is, from Section 6.8,

$$H_{\text{el}} = \int d^3x \, a^+(x) \left[-\frac{\hbar^2}{2m} \nabla^2 + V_1(x) \right] a(x). \tag{6.192}$$

Now, suppose we have solved this part of the problem and know the one-particle eigenstates $|\alpha\rangle$ and eigenvalues E_α. Denoting $\langle x|\alpha\rangle$ by $\varphi_\alpha(x)$ we have

$$\left[-\frac{\hbar^2}{2m} \nabla^2 + V_1(x) \right] \varphi_\alpha(x) = E_\alpha \varphi_\alpha(x). \tag{6.193}$$

Then we can express the electron Hamiltonian as

$$H_{\text{el}} = \sum_\alpha E_\alpha a^+_\alpha a_\alpha. \tag{6.194}$$

Because

$$|x\rangle = \sum_\alpha |\alpha\rangle\langle\alpha|x\rangle = \sum_\alpha |\alpha\rangle \varphi^*_\alpha(x)$$

we have

$$a^+(x) = \sum_\alpha a_\alpha^+ \varphi_\alpha^*(x) \qquad (6.195)$$

and similarly

$$a_\alpha^+ = \int d^3x\, a^+(x)\varphi_\alpha(x). \qquad (6.196)$$

To obtain relations between the destruction operators, simply take the Hermitian adjoints of Eqs. (6.195) and (6.196).

If, as in Section 6.9, we refer all states to a ground state (called the "electron sea"), which is normally filled, then in the notation of that section,

$$H_{el} = E_{gnd} + \sum_{\alpha>G} E_\alpha a_\alpha^+ a_\alpha - \sum_{\alpha\le G} E_\alpha b_\alpha^+ b_\alpha. \qquad (6.197)$$

In what follows we shall ignore the ground-state energy, so that the electron Hamiltonian becomes H_{el}':

$$H_{el}' = \sum_{\alpha>G} E_\alpha a_\alpha^+ a_\alpha - \sum_{\alpha\le G} E_\alpha b_\alpha^+ b_\alpha. \qquad (6.198)$$

Note that, in terms of electron and hole operators, Eq. (6.195) and its adjoint become

$$\begin{aligned} a^+(x) &= \sum_{\alpha>G} a_\alpha^+ \varphi_\alpha^*(x) + \sum_{\alpha\le G} b_\alpha \varphi_\alpha^*(x), \\ a(x) &= \sum_{\alpha>G} a_\alpha \varphi_\alpha(x) + \sum_{\alpha\le G} b_\alpha^+ \varphi_\alpha(x). \end{aligned} \qquad (6.199)$$

Consider next the lattice vibrations. For each N let Z_N be the displacement of the corresponding nucleus from its equilibrium position N. The Z_N form a set of coordinates for a system of harmonic oscillators. The procedure for finding the normal modes and quantizing this system is similar to the field quantization of Section 6.6, except that Fourier transforms are replaced by Fourier series, with the "momentum" k of the phonons running over a limited region. The normal coordinates $q_{k,a}$ are related to the Z_N by equations of the following form (assuming for simplicity that "a" runs from 1 to 3 as in the case of one atom per unit cell of the crystal):

$$\begin{aligned} \frac{q_{k,a}}{\sqrt{V}} &= \sum_N e^{-ik\cdot N} e_{k,a}^* \cdot Z_N, \\ Z_N &= \int_K^V \frac{d^3k}{(2\pi)^3} \sum_{a=1}^3 e_{k,a} e^{ik\cdot N} \frac{q_{k,a}}{\sqrt{V}}. \end{aligned} \qquad (6.200)$$

Here the region of integration is $K = \{k|-\pi \le k\cdot a_i \le \pi;\ i = 1, 2, 3\}$; $V = |a_1 \times a_2 \cdot a_3|$ is the volume of the unit cell of the crystal lattice (recall that the a_i are vectors describing the periodicity of the lattice), and $e_{k,a}$ ($a = 1, 2, 3$)

form an orthonormal basis of 3-space for each k (in that $e_{k,a}^* \cdot e_{k,b} = \delta_{ab}$), chosen so that the mode (k, a) is a normal mode.

You should check the consistency of Eqs. (6.200) as an exercise. It is convenient to use the relation

$$V \int_K \frac{d^3 k}{(2\pi)^3} \, e^{iK \cdot (N - N')} = \delta_{N,N'},$$

and the easiest way to verify it is to make a change of variables $r_i = K \cdot a_i$. $N - N'$ can be written as

$$\sum_{i=1}^{3} m_i a_i,$$

where m_i are integers, so that $K \cdot (N - N') = m \cdot r$, and the Jacobian of the transformation is

$$\left\| \frac{\partial k}{\partial r} \right\| = \frac{1}{\left\| \frac{\partial r}{\partial k} \right\|} = \frac{1}{V}.$$

Equations 6.200 are essentially a special case of Eqs. (1.28) and (1.29). The change of notation is

Chapter 1	Chapter 6
$Q_r(k)$	$\dfrac{q_{k,r}}{\sqrt{V}}$
$a_\alpha^r(k)$	The αth component of $e_{k,r}$
$Z_{\alpha,N}$	The αth component of Z_N
$\dfrac{V}{n}$	V

The normalization conventions used in Eq. (6.200) are convenient because, in the limit that the lattice approaches a continuum (i.e., $a_i \to 0$), K becomes all of momentum space and, letting $x = N$, we have

$$V \sum_N \to \int d^3 x \quad \text{and} \quad \frac{1}{V} \delta_{NN'} \to \delta^3(x - x'), \qquad (6.201)$$

Thus we have obtained the normalization of Section 6.6 for a field $\varphi(x)$, where

$$\sqrt{M/V} \, Z_N \to \varphi(x).$$

Following the procedure of Section 6.6 we have an expression for Z_N in terms of creation and destruction operators:

$$Z_N = \int_K \frac{d^3k}{(2\pi)^3} \sum_a \sqrt{\frac{\hbar V}{2M\omega(k, a)}} \left[A(k, a)e^{ik\cdot N}e_{k,a} + A^+(k, a)e^{-ik\cdot N}e^*_{k,a}\right],$$

$$(6.202)$$

where

$$[A(k, a), A^+(k', a')] = (2\pi)^3\delta^3(k - k')\delta_{aa'}. \qquad (6.203)$$

M is the mass of the vibrating atom, and $\omega(k, a)$ is the frequency of the corresponding mode. The Hamiltonian for the oscillator system (apart from a constant term) is

$$H_{\text{osc}} = \int_K \frac{d^3k}{(2\pi)^3} \sum_a \hbar\omega(k, a)A^+(k, a)A(k, a). \qquad (6.204)$$

(We have used A for the phonon operators* to avoid confusion with the electron creation and destruction operators.)

Having written down Hamiltonians for the electrons and the phonons (lattice-vibration states), we now turn to the interaction between them. The potential energy of an electron in an undisturbed lattice was given by Eq. (6.191). If the nucleus at N is displaced by an amount Z_N, then the potential energy changes to

$$V_1(x) + \Delta V_1(x) = \sum_N V_0(x - N - Z_N), \qquad (6.205)$$

where we have assumed that each atom in the crystal acts like a rigid body when it is displaced, so that the potential V_0 arising from it is simply displaced by Z_N. (In practice, not all the electron shells around the nucleus move by the same amount, so that the potential changes its shape as well as being displaced; however, we shall ignore this fact.) The potential $V_1(x)$ was already included in H_{el}. Writing

$$V_0(x - N - Z_N) \approx V_0(x - N) - Z_N \cdot \nabla V_0(x - N)$$

we have for the interaction energy (that is, the extra energy of the system due to displacement of the lattice):

$$\Delta V_1(x) = -\sum_N Z_N \cdot \nabla V_0(x - N). \qquad (6.206)$$

---

* Note that $A^+(k, a)$ (as well as the state it creates) has the dimensions of (length)$^{3/2}$, as do the corresponding operators for other particles in momentum representation.

For the many electron system the interaction Hamiltonian is

$$H_{int} = \int d^3 x \, \Delta V_1(x) a^+(x) a(x)$$

$$= \sum_{\alpha,\beta} V_{\alpha\beta} a_\alpha^+ a_\beta, \tag{6.207}$$

where

$$V_{\alpha\beta} = \int d^3 x \varphi_\alpha^*(x) \, \Delta V_1(x) \varphi_\beta(x)$$

$$= - \sum_N Z_N \cdot \int d^3 x \varphi_\alpha^*(x) \nabla V_0(x - N) \varphi_\beta(x). \tag{6.208}$$

Now $V_{\alpha\beta}$ is also an operator on phonon states, because Z_N is an operator. Using Eq. (6.202) we get

$$V_{\alpha\beta} = \int_K \frac{d^3 k}{(2\pi)^3} \sum_a [C_{\alpha\beta}(k, a) A^+(k, a) + C_{\beta\alpha}^*(k, a) A(k, a)], \tag{6.209}$$

where

$$C_{\alpha\beta}(k, a) = - \sqrt{\frac{\hbar V}{2M\omega(k, a)}} \, e_{k,a} \cdot \sum_N e^{ik \cdot N} \int d^3 x \varphi_\alpha^*(x) \nabla V_0(x - N) \varphi_\beta(x). \tag{6.210}$$

Therefore,

$$H_{int} = \int_K \frac{d^3 k}{(2\pi)^3} \sum_a \sum_{\alpha,\beta} [C_{\alpha\beta}(k, a) A^+(k, a) + C_{\beta\alpha}^*(k, a) A(k, a)] a_\alpha^+ a_\beta. \tag{6.211}$$

The total Hamiltonian of the system is

$$H = H_{el} + H_{osc} + H_{int}. \tag{6.212}$$

Note: The states of our system are of the form

$$|\alpha_1, \ldots, \alpha_n; k_1, a_1, \ldots, k_m, a_m\rangle \qquad (n \text{ electrons and } m \text{ phonons}),$$

or, if we use the hole notation,

$$|\alpha_1, \ldots, \alpha_m; \beta_1, \ldots, \beta_n; k_1, a_1, \ldots, k_p, a_p\rangle$$

$$(n \text{ electrons, } m \text{ holes, and } p \text{ phonons}).$$

The effect of a creation operator $A^+(k, a)$ on such a state may be defined as

$$A^+(k, a)|\text{electrons, holes, } k_1, a_1, \ldots \rangle = |k, a, \text{electrons, holes, } a_1, \ldots \rangle,$$

which is in turn defined to be

$$|\text{electrons, holes, } k, a, k_1, a_1, \ldots \rangle.$$

The result of this definition is that all phonon operators commute with all electron (and hole) operators. (In general, it is conventional to say that creation and destruction operators for different particles always commute unless the particles are both fermions. In the latter case it is convenient to define the operators to anticommute; then everything is consistent if we decide to call the particles different states of the same particle, as with the proton and neutron.)

If we write Eq. (6.211) in terms of electron and hole operators, we get an expression four times as long, involving terms of the form

$$A^+a^+a, \qquad A^+a^+b^+, \qquad A^+ba, \qquad A^+bb^+,$$
$$A\ a^+a, \qquad A\ a^+b^+, \qquad A\ ba, \qquad A\ bb^+.$$

These terms represent the transition of an electron or hole from one state to another, or the creation or annihilation of an electron–hole pair. In each process a phonon is emitted or absorbed.

The foregoing derivation assumed that there is only one atom per unit cell of the crystal. However, the results are similar if there are more atoms (say A) per unit cell, the only difference in the final result being that there are more phonon modes for a given k (a running from 1 to 3A). As a simplification it may turn out that some of the modes do not "couple" to the electrons, that is, they do not influence the potential felt by the electrons; these modes are independent of the others and may be ignored. Such is the case in the Polaron Problem (Chapter 8), where only one phonon mode contributes for each k.

6.11 PHOTON–ELECTRON INTERACTIONS

Suppose we shine light on the crystal of Section 6.10. What is the Hamiltonian now? To H we must add a term H_γ for the free electromagnetic field (γ refers to phonons) and a term $H_{e\gamma}$ for the interaction between electrons and phonons.

For the free electromagnetic field, the classical Lagrangian density is

$$
\begin{aligned}
\mathcal{L} &= \tfrac{1}{2}(E^2 - c^2B^2) \\
&= \tfrac{1}{2}[\dot{A}^2 - c^2(\nabla \times A)^2] \\
&= \tfrac{1}{2}\left[\dot{A}^2 - c^2 \sum_{i,j=1}^{3} (\nabla_i A_j \nabla_i A_j - \nabla_i A_j \nabla_j A_i)\right],
\end{aligned}
\tag{6.213}
$$

where $A(x)$ satisfies the subsidiary condition*)

$$\nabla \cdot A = 0. \tag{6.214}$$

---

* We assume also that the scalar potential is zero. We are using "rationalized units," but with $\varepsilon_0 = 1$; thus $e^2/4\pi\hbar c \approx 1/137$ and $\nabla \cdot E = \rho$, $\nabla \times B = (1/c^2)(j + \partial E/\partial t)$, etc.

The equation of motion resulting from the Lagrangian is

$$0 = c^2[\nabla^2 A - \nabla(\nabla \cdot A)] - \ddot{A} = c^2\left[\nabla^2 - \frac{1}{c^2}\frac{\partial^2}{\partial t^2}\right]A. \quad (6.215)$$

Carrying out the quantization procedure of Section 6.6, we find $\omega(k) = c|k| = ck$, and

$$A(x) = \int \frac{d^3k}{(2\pi)^3} \sum_{r=1}^{2} \sqrt{\frac{\hbar}{2ck}} \left[C(k, r)e^{ik \cdot x} + C^+(k, r)e^{-ik \cdot x}\right]e_{k,r}. \quad (6.216)$$

Here $e_{k,r}$ ($r = 1, 2$) are two unit vectors, perpendicular to each other and to k, and $C^+(k, r)$ is the creation operator for a photon with momentum $\hbar k$ and polarization $e_{k,r}$:

$$[C(k, r), C^+(k', r')] = (2\pi)^3\delta^3(k - k')\delta_{rr'}. \quad (6.217)$$

The Hamiltonian is

$$H_\gamma = \int \frac{d^3k}{(2\pi)^3} \sum_{r=1}^{2} \hbar ck C^+(k, r)C(k, r). \quad (6.218)$$

To find the photon–electron interaction we replace the operator* P by $P + eA(x)$ (the charge of the electron being $-e$) in the single-particle Hamiltonian

$$H_{el}^{(1)} = \frac{P^2}{2m} + V_1,$$

so that

$$H_{el}^{(1)} + H_{e\gamma}^{(1)} = \frac{(P + eA)^2}{2m} + V_1$$

$$= H_{el}^{(1)} + \frac{e}{2m}(P \cdot A + A \cdot P) + \frac{e^2}{2m}A^2. \quad (6.219)$$

Writing

$$A(X) = \int d^3x A(x)|x\rangle\langle x|, \quad (6.219')$$

we have

$$H_{e\gamma}^{(1)} = -\int d^3x A(x) \cdot j^{(1)}(x) + \frac{e^2}{2m}\int d^3x A(x)^2|x\rangle\langle x|, \quad (6.220)$$

---

* P and X are the electron momentum and position operators, and $A(x)$ operates on photon states, so that $A(X)$ (see Eq. (6.219')) operates on both photons and electrons.

where

$$j^{(1)}(x) = -\frac{e}{2m}(P|x\rangle\langle x| + |x\rangle\langle x|P)$$

$$= -\frac{e\hbar}{2m}i[(\nabla|x\rangle)\langle x| - |x\rangle(\nabla\langle x|)]. \qquad (6.221)$$

Note that $j^{(1)}(x)$ is the charge $-e$ multiplied by the probability current-density operator. The expectation value in any one-electron state $|\psi\rangle$ is

$$\langle\psi|j^{(1)}(x)|\psi\rangle = -\frac{e\hbar}{2m}i[\psi(x)\nabla\psi^*(x) - \psi^*(x)\nabla\psi(x)].$$

By inspection of Eq. (6.220) we have, for the interaction between photons and systems of arbitrary numbers of electrons,

$$H_{e\gamma} = -\int d^3x A(x)\cdot j(x) + \frac{e^2}{2m}\int d^3x A(x)^2 a^+(x)a(x), \qquad (6.222)$$

where

$$j(x) = -\frac{e\hbar}{2m}i[\nabla a^+(x)a(x) - a^+(x)\nabla a(x)] \qquad (6.223)$$

is the electromagnetic current-density operator.

In Eq. (6.222), $A(x)$ is itself an operator for each x, given by Eq. (6.216). If we use Eq. (6.216) in Eq. (6.222), and also express $a^+(x)$ in terms of a_α^+ using Eq. (6.195), we get terms involving

$$C^+(k, r)a_\alpha^+ a_\beta, \quad C(k, r)a_\alpha^+ a_\beta, \quad C(k, r)C(k', r')a_\alpha^+ a_\beta, \quad C(k, r)C^+(k', r')a_\alpha^+ a_\beta,$$

and so forth, which have interpretations similar to those of the phonon–electron interaction.

The electromagnetic current defined in Eq. (6.223) may be expressed in momentum representation as follows:

$$j(x) = -e\int\frac{d^3p'}{(2\pi\hbar)^3}\int\frac{d^3p}{(2\pi\hbar)^3}\frac{p'+p}{2m}a^+(p')a(p)e^{-i(p'-p)\cdot x/\hbar}. \qquad (6.224)$$

This expression will be useful in Section 6.12.

6.12 FEYNMAN DIAGRAMS

A graphical method employing what have come to be called "Feynman diagrams" has proven to be very convenient in dealing with perturbation solutions of complicated Hamiltonians. These diagrams serve as a "bookkeeping" device to keep track of all the perturbation terms and a guide in writing down the value of each term. (The diagrams assume their full power in the relativistic case, with which we are not concerned here.) To show how the method works we will consider the Hamiltonian for a single, otherwise free, electron interacting with the electromagnetic field:

$$H = H_{\text{free}} + H_{\text{int}},$$

where

$$H_{\text{free}} = \int \frac{d^3p}{(2\pi)^3} \frac{p^2}{2m} a^+(p)a(p) + \int \frac{d^3k}{(2\pi)^3} \sum_{r=1}^{2} ck A^+(k, r)A(k, r) \qquad (6.225)$$

and

$$H_{\text{int}} = -\int d^3x j(x) \cdot A(x) + \int d^3x \frac{e^2}{2m} a^+(x)a(x)A(x)^2. \qquad (6.226)$$

Here we denote the photon-creation operator by $A^+(k, r)$ instead of by $C^+(k, r)$:

$$A(x) = \int \frac{d^3k}{(2\pi)^3} \sum_{r=1}^{2} \sqrt{\frac{1}{2ck}} \left[A(k, r)e^{ik \cdot x} + A^+(k, r)e^{-ik \cdot x} \right] e_{k,r}, \qquad (6.227)$$

and $j(x)$ is given by Eqs. (6.223) and (6.224). In this section, $\hbar = 1$.

The states under consideration are of the form

$$|p; k_1, r_1; \ldots ; k_n, r_n\rangle \qquad \text{(one electron and } n \text{ photons)}. \qquad (6.228)$$

According to standard quantum-mechanical techniques, the transition amplitude M_{fi} for a transition from an initial state $|\text{i}\rangle$ to a final state $|\text{f}\rangle$ can be expressed in terms of the Hamiltonian (to second order) as follows:

$$(2\pi)^3 \delta^3(p_{\text{f}} - p_{\text{i}})M_{\text{fi}} = \langle \text{f}|H|\text{i}\rangle + \sum_n \frac{\langle \text{f}|H|n\rangle \langle n|H|\text{i}\rangle}{E_{\text{i}} - E_n + i\varepsilon} + O(H^3). \qquad (6.229)$$

Cross sections and transition rates are proportional to the absolute square of the amplitude:

$$\text{Rate} = \sum_{\text{f}} |M_{\text{fi}}|^2 (2\pi)^4 \delta^3(p_{\text{f}} - p_{\text{i}})\delta(E_{\text{f}} - E_{\text{i}}).$$

In practice, the states $|\text{i}\rangle$ and $|\text{f}\rangle$ are usually states of the form given in Eq. (6.228), that is, states of particles of definite momentum. It is therefore convenient to express H_{int} in terms of the $A(k, r)$, substituting Eq. (6.227) in Eq. (6.226), and to use Eq. (6.224) for $j(x)$.

$$H_{\text{int}} = H_1 + H_2, \qquad (6.230)$$

where

$$H_1 = \int \frac{d^3p'}{(2\pi)^3} \int \frac{d^3p}{(2\pi)^3} \int \frac{d^3k}{(2\pi)^3} \sum_r e \frac{p' + p}{2m} \cdot e_{k,r} \frac{1}{\sqrt{2ck}}$$

$$\times \; [(2\pi)^3 \delta^3(p' - p - k)a^+(p')a(p)A(k, r)$$

$$+ (2\pi)^3 \delta^3(p' + k - p)a^+(p')a(p)A^+(k, r)], \qquad (6.231)$$

and

$$H_2 = \int \frac{d^3p'}{(2\pi)^3} \int \frac{d^3p}{(2\pi)^3} \int \frac{d^3k'}{(2\pi)^3} \int \frac{d^3k}{(2\pi)^3} \sum_{r',r} \frac{e^2}{2m} \frac{1}{\sqrt{2ck'}\sqrt{2ck}} e_{k',r'} \cdot e_{k,r}$$

$$\times \; [(2\pi)^3 \delta^3(p' - p - k' - k)a^+(p')a(p)A(k', r')A(k, r)$$

$$+ (2\pi)^3 \delta^3(p' + k' - p - k)a^+(p')a(p)A^+(k', r')A(k, r)$$

$$+ (2\pi)^3 \delta^3(p' + k - p - k')a^+(p')a(p)A(k', r')A^+(k, r)$$

$$+ (2\pi)^3 \delta^3(p' + k' + k - p)a^+(p')a(p)A^+(k', r')A^+(k, r)]. \qquad (6.232)$$

Note that the total momentum is conserved.

Each of the terms in Eqs. (6.231) and (6.232) is represented by a diagram in which a straight line denotes an electron and a wavy line a photon. For H_1 the diagrams are given in Fig. 6.2a and 6.2b, and the amplitude for each is

$$e \frac{p' + p}{2m} \cdot e_{k,r} \frac{1}{\sqrt{2ck}}.$$

For H_2 the diagrams are those shown in Fig. 6.3a through 6.3d, and the amplitude for each is

$$\frac{e^2}{2m} \frac{1}{\sqrt{2ck'}\sqrt{2ck}} e_{k',r'} \cdot e_{k,r}.$$

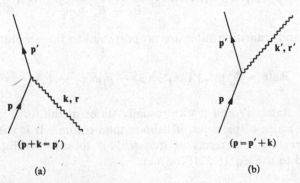

Fig. 6.2 Feynman diagrams for H_1.

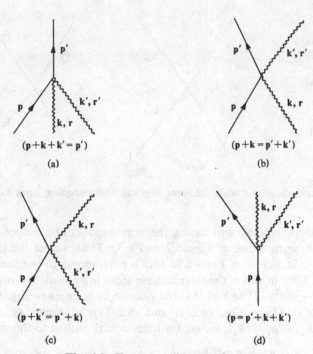

Fig. 6.3 Feynman diagrams for H_2.

Note that parts b and c of Fig. 6.3 represent essentially the same process. In fact, after the integrations are performed, the second and third terms of Eq. (6.232) are equal;* so in any process involving such terms we need only calculate for the case shown in Fig. 6.3b, say, and multiply by 2.

Now suppose we have a definite amplitude to calculate, for example, that for Compton scattering. The whole process is denoted by Fig. 6.4. To find the

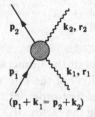

Fig. 6.4 Feynman diagrams for Compton scattering.

---

* We ignore the infinite self-energy of the electron that comes from rearranging the A and A^+ of the third term in Eq. (6.232).

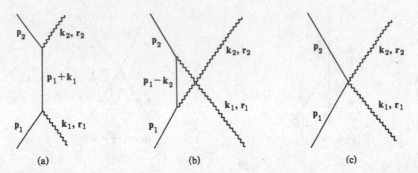

Fig. 6.5 Feynman diagrams with the same ingoing and outgoing lines as Fig. 6.4.

amplitude we draw all diagrams having the same ingoing and outgoing lines as that of Fig. 6.4, using those of Figs. 6.2 and 6.3. Thus we get the three cases shown in Fig. 6.5a, b, and c. Parts 6.5a and b correspond to the second-order term of Eq. (6.229), in which the intermediate state $|n\rangle$ consists of one electron and $H = H_1 + 0(e^2)$. (The fact that the photon lines cross in Fig. 6.5b is of no significance.) Using Eqs. (6.229) and (6.231) and removing the factor $(2\pi)^3\delta^3(p_2 + k_2 - p_1 - k_1)$, we get for these contributions to the amplitude

$$M^{(a)} = \frac{\left(\dfrac{e}{\sqrt{2ck_2}}\dfrac{p_2 + p_1 + k_1}{2m}\cdot e_{k_2,r_2}\right)\left(\dfrac{e}{\sqrt{2ck_1}}\dfrac{p_1 + k_1 + p_1}{2m}\cdot e_{k_1,r_1}\right)}{\dfrac{p_1^2}{2m} + ck_1 - \dfrac{(p_1 + k_1)^2}{2m}}$$

(6.233)

and

$$M^{(b)} = \frac{\left(\dfrac{e}{\sqrt{2ck_1}}\dfrac{p_2 + p_1 - k_2}{2m}\cdot e_{k_1,r_1}\right)\left(\dfrac{e}{\sqrt{2ck_2}}\dfrac{p_1 - k_2 + p_1}{2m}\cdot e_{k_2,r_2}\right)}{\dfrac{p_1^2}{2m} + ck_1 - \dfrac{(p_1 - k_2)^2}{2m}}\cdot$$

(6.234)

We may associate the two factors in the numerator with each vertex, and the denominator with the "propagation" of the intermediate electron.

Figure 6.5c corresponds to the first-order terms of Eq. (6.229) in which only H_2 can contribute. Remembering the factor of 2 due to the two ways in which this process occurs in Eq. (6.232), we get

$$M^{(c)} = 2\cdot\frac{e^2}{2m}\frac{1}{\sqrt{2ck_2}\sqrt{2ck_1}}e_{k_2,r_2}\cdot e_{k_1,r_1}.$$

(6.235)

The total amplitude for Compton scattering is

$$M = M^{(a)} + M^{(b)} + M^{(c)} + 0(e^4) \tag{6.236}$$

In equations (6.233) and (6.234)

$$p_1 + k_1 = p_2 + k_2 \quad \text{and} \quad k_1 \cdot e_{k_1, r_1} = k_2 \cdot e_{k_2, r_2} = 0.$$

In fact, if the initial electron is at rest (i.e., $p_1 = 0$), then $M^{(a)}$ and $M^{(b)}$ both vanish.

The results we have derived are satisfactory for low photon energies ($ck_1 \ll mc^2$). For high energies we would have to use a relativistic theory and also take into account the spin of the electron.

CHAPTER 7

SPIN WAVES

7.1 SPIN-SPIN INTERACTIONS

In this section we will consider interactions that give rise to a coupling between the spins of two objects of the form

$$H \propto S_1 \cdot S_2. \tag{7.1}$$

In later sections we will find out what happens when such a coupling appears in a lattice full of spins.

An interaction of the form of Eq. (7.1) can be due, first of all, to the magnetic interaction between two magnetic dipoles. The interaction between two magnetic dipoles μ_1 and μ_2 with separation r is

$$H = \left[\frac{\mu_1 \cdot \mu_2}{r^3} - 3 \frac{(\mu_1 \cdot r)(\mu_2 \cdot r)}{r^5} \right] - \frac{8\pi}{3} \mu_1 \cdot \mu_2 \delta^3(r) = H' + H'', \tag{7.2}$$

where H' is the term in brackets. (The term H'' with $\delta^3(r)$ arises from a corresponding term in the magnetic field of a dipole, which appears when one evaluates $\nabla \times (\mu \times r)/r^3$).

Suppose the two particles are orbiting around each other (as in an atom) and we consider the matrix element of H between the two states of orbital angular momentum l and l'. As a function of the direction (θ, φ) of r, H' is a combination of spherical harmonics of order 2; therefore the matrix element is zero unless $|l - 2| \leq l' \leq l + 2$. In particular, the first-order splitting of the S-state ($l = l' = 0$) is zero. (This argument is just another way of saying that S states are spherically symmetric and H' averaged over all directions is zero). However, H'' contributes only for S states, because only those have nonzero wave functions at $r = 0$ (a state with angular momentum l varies as r^l near the origin). Thus, for S states we have an interaction proportional to $\mu_1 \cdot \mu_2$, which is proportional to $S_1 \cdot S_2$ because

$$\mu_i = g_i \frac{q_i}{2m_i c} S_i \tag{7.3}$$

for a particle with mass m_i, charge q_i and g-factor g_i (Eq. (7.3) in fact defines g_i)).

In addition to direct magnetic interactions, a term of the form of Eq. (7.1) can appear indirectly as a result of spin-independent interactions combined with the exclusion principle. As an example, consider the hydrogen molecule; take the nuclei to be fixed and ignore their spins. The two electrons may be described by their positions x_1 and x_2 and some component of their spins S_1 and S_2. If ψ_a and ψ_b are energy eigenfunctions for a single electron with energies E_a and E_b, then

$$\psi_{ab}(x_1, x_2) = \psi_a(x_1)\psi_b(x_2) \tag{7.4}$$

has energy $E_a + E_b$, neglecting mutual interactions between the electrons. The state $\psi_{ba}(x_1, x_2) = \psi_b(x_1)\psi_b(x_2)$ has the same energy.

If we add a mutual interaction $V(x_1, x_2) = V(x_2, x_1)$ which is spin-independent between the electrons, then the two-electron eigenfunctions become

$$\psi_S = \frac{1}{\sqrt{2}}(\psi_{ab} + \psi_{ba}),$$

$$\psi_A = \frac{1}{\sqrt{2}}(\psi_{ab} - \psi_{ba}), \tag{7.5}$$

with their respective energies

$$E_S = E_a + E_b + I - J$$
$$E_A = E_a + E_b + I + J. \tag{7.6}$$

Here

$$I = \int \psi_{ab}^* V \psi_{ab}\, d^3x_1\, d^3x_2 = \int \psi_{ba}^* V \psi_{ba}\, d^3x_1\, d^3x_2$$

and

$$J = -\int \psi_{ab}^* V \psi_{ba}\, d^3x_1\, d^3x_2 = -\int \psi_{ba}^* V \psi_{ab}\, d^3x_1\, dx_2. \tag{7.7}$$

As implied by the notation, ψ_S is symmetric and ψ_A is antisymmetric under exchange of x_1 and x_2.

We have not yet reckoned with the electron spins or the exclusion principle. We must multiply our spatial wave functions by spinors that describe the possible spins of the electrons, and the total wave function must be antisymmetric under the simultaneous exchanges of coordinates and spins. Now for two spin-$\frac{1}{2}$ particles, the symmetric spin states have total spin 1, and the antisymmetric states have spin 0 (see Section 7.2). The total wave functions are then of the form

$$\psi_S = \psi_S \chi^{(0)}, \qquad \psi_A = \psi_A \chi^{(1)} \tag{7.8}$$

where $\chi^{(s)}$ is any two-electron spin state with total spin s. (Note that S and A still refer to only the spatial symmetry). In other words, the energy eigenstates are also eigenstates of the total electron spin. We can then write the energy in terms of the spin as follows: If S is the total spin, $S \cdot S$ has eigenvalues $s(s + 1) = 0$ and 2. The energy of a state with total spin s can be written in the form

$$E = E_S + (E_A - E_S) \frac{s(s + 1)}{2} \tag{7.9}$$

because it equals E_S when $s = 0$ and E_A when $s = 1$. Therefore we can write the mutual-interaction Hamiltonian for the system in the form

$$H = E_S + (E_A - E_S) \frac{S \cdot S}{2}. \tag{7.10}$$

Now,

$$S \cdot S = (S_1 + S_2) \cdot (S_1 + S_2) = \tfrac{3}{2} + 2S_1 \cdot S_2, \tag{7.11}$$

because

$$S_i \cdot S_i = \tfrac{1}{2}(\tfrac{1}{2} + 1) = \tfrac{3}{4}.$$

Therefore

$$\begin{aligned} H &= E_S + (E_A - E_S)(\tfrac{3}{4} + S_1 \cdot S_2) \\ &= \tfrac{1}{4}(E_S + 3E_A) + (E_A - E_S)S_1 \cdot S_2 \\ &= E_a + E_b + I + \tfrac{1}{2}J + 2JS_1 \cdot S_2. \end{aligned} \tag{7.12}$$

In other words, the splitting of the energy levels is described by

$$\Delta H = 2JS_1 \cdot S_2, \tag{7.13}$$

which is of the form of Eq. (7.1). We have thus taken a spin-independent interaction (e.g., a Coulomb potential) between the electrons and made it look like a spin–spin interaction. Note that the splitting is proportional to the "exchange integral" J defined in Eq. (7.7), which depends on how much the wave functions $\psi_a(x)$ and $\psi_b(x)$ overlap. Note also that the derivation of Eq. (7.12) depends crucially on the fact that electrons have spin $\tfrac{1}{2}$; for higher spins we would get a polynomial in $S_1 \cdot S_2$ if we went through a similar procedure.

The interaction of Eq. (7.13), due to the exchange integral, turns out to be very strong compared to the usual magnetic-dipole interactions.

7.2 THE PAULI SPIN ALGEBRA

It is convenient and conventional to use the Pauli spin matrices when dealing with a particle of spin $\tfrac{1}{2}$:

$$\sigma_x = \begin{pmatrix} 0 & 1 \\ 1 & 0 \end{pmatrix}, \quad \sigma_y = \begin{pmatrix} 0 & -i \\ i & 0 \end{pmatrix}, \quad \sigma_z = \begin{pmatrix} 1 & 0 \\ 0 & -1 \end{pmatrix}. \tag{7.14}$$

The spin is then

$$S = \tfrac{1}{2}\sigma \tag{7.15}$$

where

$$\sigma = \sigma_x e_x + \sigma_y e_y + \sigma_z e_z \tag{7.16}$$

and where we take $\hbar = 1$ (otherwise $S = \hbar\sigma/2$). Equation (7.14) implies that we are using a basis of states in which the z-component of the spin is diagonalized; the states

$$|\alpha\rangle = \begin{pmatrix} 1 \\ 0 \end{pmatrix} \quad \text{and} \quad |\beta\rangle = \begin{pmatrix} 0 \\ 1 \end{pmatrix} \tag{7.17}$$

are states of spin up ($S_z = \frac{1}{2}$) and spin down ($S_z = -\frac{1}{2}$), respectively. From Eq. (7.14), we have

$$\sigma_x|\alpha\rangle = |\beta\rangle, \quad \sigma_y|\alpha\rangle = i|\beta\rangle, \quad \sigma_z|\alpha\rangle = |\alpha\rangle,$$
$$\sigma_x|\beta\rangle = |\alpha\rangle, \quad \sigma_y|\beta\rangle = -i|\alpha\rangle, \quad \sigma_z|\beta\rangle = -|\beta\rangle. \tag{7.18}$$

The algebra of the σ-matrices is as follows:

$$\sigma_x\sigma_y = -\sigma_y\sigma_x = i\sigma_z, \tag{7.19}$$

and cyclically in x, y, z;

$$\sigma_x\sigma_x = \sigma_y\sigma_y = \sigma_z\sigma_z = 1. \tag{7.20}$$

Alternatively one may use the following elegant form of Eq. (7.20):

$$(a \cdot \sigma)(b \cdot \sigma) = a \cdot b + ia \times b \cdot \sigma, \tag{7.21}$$

where the a_i and b_i are numbers (or other operators, provided that they commute with σ).

Consider next a system of two spin-$\frac{1}{2}$ particles (ignoring spatial and other degrees of freedom). As a basis of states we may take

$$|\alpha\alpha\rangle, \quad |\alpha\beta\rangle, \quad |\beta\alpha\rangle, \quad |\beta\beta\rangle, \tag{7.22}$$

where, for example, $|\alpha\beta\rangle$ is the state in which the first particle has its spin up and the second has its spin down. We then define σ_1 to act on the first spin and σ_2 on the second. That is, for example,

$$\sigma_{1y}|\alpha\alpha\rangle = i|\beta\alpha\rangle,$$
$$\sigma_{2y}|\alpha\alpha\rangle = i|\alpha\beta\rangle.$$

Note that $\sigma_{1i}\sigma_{2j} = \sigma_{2j}\sigma_{1i}$, which is to say that operators pertaining to different particles commute.

We have previously considered (and will consider further in the section to follow) the operator $S_1 \cdot S_2 = \frac{1}{4}\sigma_1 \cdot \sigma_2$. By explicit calculation, we find

$$\sigma_1 \cdot \sigma_2|\alpha\alpha\rangle = |\alpha\alpha\rangle,$$
$$\sigma_1 \cdot \sigma_2|\alpha\beta\rangle = 2|\beta\alpha\rangle - |\alpha\beta\rangle,$$
$$\sigma_1 \cdot \sigma_2|\beta\alpha\rangle = 2|\alpha\beta\rangle - |\beta\alpha\rangle,$$
$$\sigma_1 \cdot \sigma_2|\beta\beta\rangle = |\beta\beta\rangle. \tag{7.23}$$

Writing $|\alpha\alpha\rangle = 2|\alpha\alpha\rangle - |\alpha\alpha\rangle$ and similarly for $|\beta\beta\rangle$, we immediately come upon the useful relation

$$\sigma_1 \cdot \sigma_2 = 2p^{1,2} - 1, \tag{7.24}$$

where $p^{1,2}$ is the operator that exchanges the spins of the particles; that is,

$$p^{1,2}|\alpha\alpha\rangle = |\alpha\alpha\rangle, \qquad p^{1,2}|\alpha\beta\rangle = |\beta\alpha\rangle, \quad \text{and so on.}$$

The symmetric states

$$|1, 1\rangle = |\alpha\alpha\rangle, |1, 0\rangle = \frac{|\alpha\beta\rangle + |\beta\alpha\rangle}{\sqrt{2}}, |1, -1\rangle = |\beta\beta\rangle \tag{7.25}$$

all satisfy $p^{1,2}|1, m\rangle = |1, m\rangle$ and therefore $\sigma_1 \cdot \sigma_2|1, m\rangle = |1, m\rangle$ ($m = 1, 0, -1$). The antisymmetric state

$$|0, 0\rangle = \frac{|\alpha\beta\rangle - |\beta\alpha\rangle}{\sqrt{2}} \tag{7.26}$$

satisfies $p^{1,2}|0, 0\rangle = -|0, 0\rangle$ and thus $\sigma_1 \cdot \sigma_2|0, 0\rangle = -3|0, 0\rangle$. Therefore the eigenvalues of $\sigma_1 \cdot \sigma_2$ are 1 and -3. Now the total spin is

$$S = \tfrac{1}{2}\sigma_1 + \tfrac{1}{2}\sigma_2 \tag{7.27}$$

and its square is

$$S \cdot S = \tfrac{3}{2} + \tfrac{1}{2}\sigma_1 \cdot \sigma_2. \tag{7.28}$$

Therefore

$$S \cdot S|1, m\rangle = 2|1, m\rangle, \qquad S \cdot S|0, 0\rangle = 0,$$

so that (as we anticipated in the notation) the states $|1, m\rangle$ have spin 1 and $|0, 0\rangle$ spin 0. The fact that the symmetric states have spin 1 and the antisymmetric ones have spin 0 may also be seen directly from the relation

$$S \cdot S = p^{1,2} + 1, \tag{7.29}$$

which is derivable from Eqs. (7.27) and (7.28).

7.3 SPIN WAVE IN A LATTICE

Consider a lattice, the ith member of which has spin σ_i. The interaction between the spins is due to the exchange effect, and the Hamiltonian is given by

$$H = \sum_{i,j} A_{i,j}\sigma_i \cdot \sigma_j. \tag{7.30}$$

Consider the lattice to be cubic, and let each point on the lattice be denoted by the integer vector $N = a(n_x e_x + n_y e_y + n_z e_z)$, where a is the lattice spacing. Then

$$H = \sum_{N,N'} A_{N,N'}\sigma_N \cdot \sigma_{N'}. \tag{7.31}$$

It is clear that the interaction depends only on the distance $N - N' = M$, and if in addition it is assumed that only nearest-neighbor interactions are important,

$$H = \sum_N A_1(\sigma_N \cdot \sigma_{N+ae_x} + \sigma_N \cdot \sigma_{N+ae_y} + \sigma_N \cdot \sigma_{N+ae_z}). \tag{7.32}$$

Occasionally next-nearest-neighbor interactions are important. They may be even more important than the nearest-neighbor ones, but such a situation can be analyzed in the same way that we use here.

We now return to Eq. (7.32) and consider the one-dimensional case. The extension to three dimensions will be immediate. We also are interested in the ferromagnetic case ($A_1 < 0$), and we can neglect boundary effects by considering the system to be a closed loop. Let there be N lattice points; $\sigma_{N+1} \equiv \sigma_1$. Thus,

$$H = -A \sum_n \sigma_n \cdot \sigma_{n+1} \tag{7.33}$$

where $A = -A_1$. The minus sign in front of A is included to emphasize the fact that in the case of ferromagnets, $-A < 0$. That is, the energy is most negative when the spins are parallel. The (rare) instances when $-A > 0$ are due to a double-exchange effect. Recall that we used α to represent spin up and β spin down, and we defined the spin exchange operator $p^{1,2} = \frac{1}{2}(1 + \sigma_1 \cdot \sigma_2)$, satisfying

$$\begin{aligned} p^{1,2}|\alpha\alpha\rangle &= |\alpha\alpha\rangle, \\ p^{1,2}|\beta\beta\rangle &= |\beta\beta\rangle, \\ p^{1,2}|\alpha\beta\rangle &= |\beta\alpha\rangle, \\ p^{1,2}|\beta\alpha\rangle &= |\alpha\beta\rangle. \end{aligned} \tag{7.34}$$

Then

$$H = -A \sum_{n=1}^{N} (2p^{n,n+1} - 1) = NA - 2A \sum_{n=1}^{N} p^{n,n+1}. \tag{7.35}$$

To solve Schrödinger's equation with the above Hamiltonian, we must find the eigenstates and eigenvalues of H. To do so, notice first that Eq. (7.34) implies that H operating on a state with a fixed number of β's gives a state with the same number of β's. Since all states can be represented as linear combinations of those with a fixed number of β's, we have a method of partially diagonalizing H.

Rather than trying to completely diagonalize H, let us merely find eigenstates with low energy.

One suspects that the lowest state (most negative energy) occurs when all the spins are parallel. For example $|\alpha\alpha \cdots \alpha\rangle$ is such a state. More generally, when we say all spins are parallel what we really mean is that the total spin is as large in magnitude as possible, and it can be shown that the states that satisfy this condition are completely symmetric states. The completely symmetric states are invariant under $p^{n,n+1}$; so $p^{n,n+1}$ can be replaced by 1 in computing the energy of those states. Thus, we expect that

$$E_{gnd} = \langle \text{completely symmetric spin state } |H| \text{ completely symmetric state} \rangle$$

$$= NA - 2A \sum_{n=1}^{N} 1 = -NA$$

is the lowest-energy state. That this is the lowest energy is obvious. The eigenvalues of $p^{n,n+1}$ are ± 1; thus the minimum conceivable energy can occur only when for all n we get the maximum possible expectation value for $p^{n,n+1}$, that is, when we can replace $p^{n,n+1}$ with $+1$.

Problem: If $-A > 0$, what is the lowest state and what is the lowest energy?

In order to take the ground state as the zero of the energy spectrum, define

$$H' = H + NA = -2A \sum_{n=1}^{N} (p^{n,n+1} - 1).$$

One may suspect that the next-to-lowest states has all but one spin parallel. But which one? It is clear that

$$|\psi_1\rangle = |\beta\alpha\alpha \cdots \alpha\rangle, \ |\psi_2\rangle = |\alpha\beta\alpha \cdots \alpha\rangle,$$
$$|\psi_{n\prime} = |\alpha \cdots \alpha\beta\alpha \cdots \alpha\rangle \qquad (\beta \text{ in } n\text{th place}) \tag{7.36}$$

are equivalent and that none of them can be an eigenfunction. We look for a superposition of the $|\psi_n\rangle$. Let

$$|\psi\rangle = \sum a_n |\psi_n\rangle. \tag{7.37}$$

To solve the equation $H'|\psi\rangle = E|\psi\rangle$,

$$\sum_n (Ea_n)|\psi_n\rangle = E|\psi\rangle = H'|\psi\rangle$$

$$= -2A \sum_{n=1}^{N} a_n \sum_{m=1}^{N} (p^{m,m+1} - 1)|\psi_n\rangle$$

$$= -2A \sum_{n=1}^{N} a_n [(p^{n,n+1} - 1) + (p^{n-1,n} - 1)]|\psi_n\rangle$$

$$= -2A \sum_{n=1}^{N} a_n [|\psi_{n+1}\rangle + |\psi_{n-1}\rangle - 2|\psi_n\rangle].$$

Because the $|\psi\rangle$ are orthogonal, we can equate coefficients of $|\psi\rangle$ to get

$$Ea_n = -2A(a_{n+1} + a_{n-1} - 2a_n). \tag{7.38}$$

To solve this set of equations, let $a_n = e^{i\delta n}$ and substitute into Eq. (7.38). We thus obtain

$$E = -2A(e^{i\delta} - 2 + e^{-i\delta}) = 4A(1 - \cos \delta) = 8A \sin^2 \frac{\delta}{2}. \tag{7.39}$$

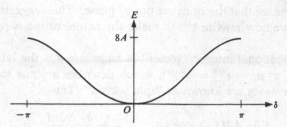

Fig. 7.1 Energy as a function of δ.

Neglecting boundary conditions as we have done is tantamount to assuming that the N lattice points are arranged in a circle, or that the $N + 1$ point is the first point. In other words,

$$a_{N+1} = a_1, \quad \text{or} \quad e^{i\delta(N+1)} = e^{i\delta}, \quad \text{or} \quad e^{iN\delta} = 1.$$

This means that $\delta = (2\pi/N)l$; l an integer;

$$-N/2 \leq l \leq N/2. \tag{7.40}$$

We have, from Eqs. (7.39) and (7.40):

$$E = 8A \sin^2 \frac{\delta}{2} = 8A \sin^2 \frac{2\pi}{N} \frac{l}{2}. \tag{7.41}$$

This function is plotted in Fig. 7.1. There is, in other words, a band of energies, the minimum energy occurring at $\delta = 0$. If $\delta = 0$ we have a completely symmetric state, which we have already concluded has the lowest possible energy, $E = 0$. Another way of seeing that the state with $\delta = 0$ has the same energy as $|\alpha\alpha \cdots \alpha\rangle$ is as follows. Let each spin be tipped from its "up" position by an angle θ, where θ is very small. Such a state is given by

$$|\chi\rangle = \left|\left(\cos\frac{\theta}{2}\alpha + \sin\frac{\theta}{2}\beta\right)\left(\cos\frac{\theta}{2}\alpha + \sin\frac{\theta}{2}\beta\right)\cdots\right\rangle.$$

Because $\theta/2$ is small, $\cos\theta/2 \approx 1$ and $\sin\theta/2 \approx \eta \ll 1$ ($\eta = \theta/2$). Then

$$|\chi\rangle = |(\alpha + \eta\beta)(\alpha + \eta\beta \cdots)\rangle.$$

Neglecting terms in η^2 or higher powers of η, we have

$$|\chi\rangle = |\alpha\alpha \cdots\rangle + \eta(|\beta\alpha\alpha \cdots \alpha\rangle + |\alpha\beta\alpha \cdots \alpha\rangle + \cdots).$$

Subtracting out the ground-state term, we obtain $|\chi'\rangle = \sum_n \eta|\psi_n\rangle$ where $|\psi_n\rangle$ is given by Eq. (7.36). Since $|\chi\rangle$ must be an eigenvector of H with the same eigenvalue (energy) as $|\alpha, \alpha, \ldots, \alpha\rangle$,* so must $|\chi'\rangle$, and $(1/\eta)|\chi'\rangle = \sum_n |\psi_n\rangle$.

---

* Since in the absence of external interaction there could be no change in energy when *all* of the spins are rotated in a certain angle.

For $\delta \neq 0$, we see that the spins are out of phase. This suggests the concept of a wave, and we now rewrite $e^{i\delta n}$ to make the nature of the wave somewhat clearer.

The one-dimensional integer "vector" is na, where a is the lattice spacing. We can write a_n as $a_n = e^{i\delta n} = e^{ik(an)}$, which describes a plane wave of wave number k. Such waves are known as "spin waves." Thus,

$$E = 4A(1 - \cos ka), \qquad k = \frac{\delta}{a} = \frac{2\pi l}{Na}. \tag{7.42}$$

For small ka, which means long waves, $\cos ka \approx 1 - k^2 a^2/2$, and

$$E \approx 2Aa^2 k^2 \text{ (long wavelength)}. \tag{7.43}$$

The extension to three dimensions gives us no trouble: The lowest state again has all spins parallel. Subtracting out the ground energy, we obtain

$$H' = - \sum_{N,M} A_M(\sigma_N \cdot \sigma_{N+M} - 1)$$

and, in a manner similar to that used in the one-dimensional case,

$$|\psi\rangle = \sum_N e^{iK \cdot Na}|\psi_N\rangle \quad \text{and} \quad E = - \sum_M A_M(e^{iK \cdot M} - 1).$$

For nearest-neighbor interactions (in a cubic crystal),

$$E = -2A\{(e^{iK_x a} - 1) + (e^{-iK_x a} - 1) + (e^{iK_y a} - 1) + \cdots\}$$

$$= 8A \left\{ \sin^2 \frac{K_x a}{2} + \sin^2 \frac{K_y a}{2} + \sin^2 \frac{K_z a}{2} \right\}. \tag{7.44}$$

For long waves

$$E \approx 2Aa^2 K^2 = K^2/2\mu, \tag{7.45}$$

where

$$1/2\mu = 2Aa^2.$$

To find the next band of energies, we must consider the case when all the spins but two are parallel. But first we will digress to discuss the semiclassical interpretation of spin waves. Problems dealing only with spins have often an immediate semiclassical interpretation. Let us look at the spin wave case.

7.4 SEMICLASSICAL INTERPRETATION OF SPIN WAVES

The Hamiltonian is

$$H = -2A \sum_{n=1}^{N} (p^{n,n+1} - 1) = -A \sum_{n=1}^{N} (\sigma_n \cdot \sigma_{n+1} - 1). \tag{7.46}$$

The Heisenberg equation of motion for the nth spin is

$$\dot{\sigma}_n = \frac{i}{\hbar}(H\sigma_n - \sigma_n H). \tag{7.47}$$

We transform this using the commutation relations:

$$\sigma_x \sigma_y - \sigma_y \sigma_x = 2i\sigma_z \quad \text{etc.} \tag{7.48}$$

so that

$$
\begin{aligned}
(\sigma_2 \cdot \sigma_1)\sigma_{1x} - \sigma_{1x}(\sigma_2 \cdot \sigma_1) &= (\sigma_{2x}\sigma_{1x} + \sigma_{2y}\sigma_{1y} + \sigma_{2z}\sigma_{1z})\sigma_{1x} - \sigma_{1x}(\cdots) \\
&= 2i(\sigma_{1y}\sigma_{2y} - \sigma_{1z}\sigma_{2y}) \\
&= 2i(\sigma_1 \times \sigma_2)_x. \tag{7.49}
\end{aligned}
$$

Using Eq. (7.49) in Eq. (7.47), we can write

$$\hbar\dot{\sigma}_n = 2A(\sigma_n \times \sigma_{n+1} - \sigma_{n-1} \times \sigma_n). \tag{7.50}$$

We regard this as the classical equation of motion for the vector σ_n. This is a nonlinear equation, but it can be linearized when we consider σ_{nz} to be close to 1, in which case we can approximate thus:

$$
\begin{cases}
\hbar\dot{\sigma}_{n,x} \approx 2A(2\sigma_{n,y} - \sigma_{n+1,y} - \sigma_{n-1,y}) \\
\hbar\dot{\sigma}_{n,y} \approx -2A(2\sigma_{n,x} - \sigma_{n+1,x} - \sigma_{n-1,x}).
\end{cases} \tag{7.51}
$$

We expect the vector σ_n to rotate around the z-axis, and we also expect the solution to be of the form

$$
\begin{aligned}
\sigma_{n,x} &\approx c \sin \omega t e^{inak}, \\
\sigma_{n,y} &\approx c \cos \omega t e^{inak},
\end{aligned} \tag{7.52}
$$

where a is the lattice constant. When we use Eq. (7.52) we see that

$$\hbar\omega = 4A(1 - \cos ak). \tag{7.53}$$

7.5 TWO SPIN WAVES

When the spins at positions n_1 and n_2 point downward we define the wave function φ_{n_1,n_2}:

$$|\varphi_{n_1,n_2}\rangle = |\alpha \cdots \alpha\beta\alpha \cdots \alpha\beta\alpha \cdots \alpha\rangle, \tag{7.54}$$

where the B's are in positions n_1 and n_2. The Hamiltonian is the same as that of Eq. 7.46, and the Schrödinger equation is

$$H|\psi^{(2)}\rangle = 4A\varepsilon^{(2)}|\psi^{(2)}\rangle; \qquad \text{Energy} = 4A\varepsilon^{(2)} \tag{7.55}$$

where

$$|\psi^{(2)}\rangle = \sum_{n_1 < n_2} a_{n_1,n_2}|\varphi_{n_1,n_2}\rangle. \tag{7.56}$$

When Eq. (7.56) is used in Eq. (7.55) and the coefficients of φ_{n_1,n_2} are compared we can distinguish two cases. When $n_2 \neq n_1 + 1$, in other words when the two down spins at n_1 and n_2 are nonadjacent, we have

$$2\varepsilon a_{n_1,n_2} = (a_{n_1,n_2} - a_{n_1-1,n_2}) + (a_{n_1,n_2} - a_{n_1+1,n_2})$$
$$+ (a_{n_1,n_2} - a_{n_1,n_2-1}) + (a_{n_1,n_2} - a_{n_1,n_2+1}). \tag{7.57}$$

On the other hand, when $n_2 = n_1 + 1$ the two down spins are next to each other and Eq. (7.57) does not hold. We then have

$$2\varepsilon a_{n,n+1} = (a_{n,n+1} - a_{n-1,n+1}) + (a_{n,n+1} - a_{n,n+2}). \tag{7.58}$$

Before solving Eqs. (7.57) and (7.58) rigorously let us derive some approximate results. Because the case $n_2 = n_1 + 1$ is only one out of N (the number of lattice points), for most of the time Eq. (7.57) holds. Therefore, the first crude approximation should correspond to two noninteracting waves:

$$a_{n_1 n_2} = e^{ik_1 n_1 a} e^{ik_2 n_2 a}. \tag{7.59}$$

When we use Eq. (7.59) in Eq. (7.57), we see that

$$\varepsilon^{(2)} = (1 - \cos k_1 a) + (1 - \cos k_2 a)$$
$$= \varepsilon(k_1) + \varepsilon(k_2), \tag{7.60}$$

where $4A\varepsilon(k_1)$ is the energy for one spin wave of wave number k_1.

At low temperatures, only the lowest energy modes are excited; thus for one spin wave, the energy might be,

$$E = \frac{k^2}{2\mu} \quad \left(\mu = \frac{1}{4Aa^2}\right),$$

and for two spin waves

$$E = \frac{k_1^2}{2\mu} + \frac{k_2^2}{2\mu}.$$

We can form wave packets with group velocity $\partial\omega/\partial k$, where $E = \hbar\omega$ and the phase velocity is ω/k. The statements above bring out strikingly the similarity of spin waves to crystal vibrations or phonons. By analogy to the concepts of "phonon," "photon," and so on, we call a spin-wave excitation a "magnon." Since the approximate system discussed above is made up of independent Bose particles, we can use the expressions that we discussed before. The free energy F is

$$F = kT \int \ln\left(1 - e^{-\beta E(k)}\right) \frac{d^3 k}{(2\pi)^3}. \tag{7.61}$$

For low temperatures only the $E(k)$ near the bottom are significant, so that we may approximate

$$E(k) \approx k^2/2\mu,$$

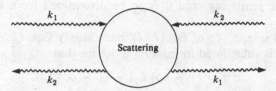

Fig. 7.2 Two spin waves, or the scattering of two quasi-particles.

and

$$F = kT \int_0^\infty \ln\left(1 - e^{-(k^2/2\mu kT)}\right) \frac{4\pi k^2 \, dk}{(2\pi)^3}. \qquad (7.62)$$

The energy is

$$U = \int_0^\infty \frac{k^2 \, 2\mu}{e^{k^2/2\mu k_b T} - 1} \frac{4\pi k^2 \, dk}{(2\pi)^3}. \qquad (7.63)$$

This energy depends on T to the 5/2 power:

$$U \propto T^{5/2}.$$

Therefore the specific heat is proportional to the 3/2 power of T:

$$C \propto T^{3/2}.$$

This result for spin waves was derived by Bloch.

Although the system is made up of ideal Bose particles, Bose–Einstein condensation does not occur because the number of particles is not fixed.

Problem: When there is an external magnetic field, derive the energy of the spin waves. Find the magnetic susceptibility.

7.6 TWO SPIN WAVES (RIGOROUS TREATMENT)

Bethe treated the two spin waves in a linear system rigorously.* When there are two spin waves in a system, the problem can be looked upon as a scattering of two quasiparticles, as in Fig. 7.2. For the scattering process the energy and the momentum are conserved, and the eigenstate is a superposition of the incoming and outgoing waves:

$$a_{n_1, n_2} = \alpha e^{i(k_1 n_1 a + k_2 n_2 a)} + \beta e^{i(k_2 n_1 a + k_1 n_2 a)}. \qquad (7.64)$$

From this point onward we will let $a = 1$. That is, for convenience we are either assuming unit lattice spacing or absorbing the factor a in k. As mentioned earlier, $n_1 < n_2$. The ratio α/β of the coefficients is related to the phase shift

* See A. Sommerfeld and H. Bethe, *Handbuch der Physik*, Vol. XXIV/2, p. 604 (1933).

resulting from the scattering, and it is to be determined from the scattering equation.

The left-hand side, a_{n_1,n_2}, of Eq. (7.64), must satisfy Eqs. (7.57) and (7.58). First, when a_{n_1,n_2} is substituted in Eq. (7.57), we see that

$$\varepsilon^{(2)} = (1 - \cos k_1) + (1 - \cos k_2). \tag{7.65}$$

This equation has the same form as Eq. (7.60), although we expect k_1 now satisfies a different relation from that found in Section 7.1.

Next we can use Eq. (7.64) in Eq. (7.58). Rather than doing that immediately, notice that Eq. (7.58) is a special case of Eq. (7.57) if we define $a_{n,n}$ (which was left unspecified before) by

$$2a_{n,n+1} = a_{n,n} + a_{n+1,n+1} \tag{7.66}$$

Then, substituting Eq. (7.64) in Eq. (7.66) is equivalent to substituting Eq. (7.64) in Eq. (7.58). From Eqs. (7.66) and (7.64) we have

$$2(\alpha e^{ik_2} + \beta e^{ik_1}) = \alpha + \beta + \alpha e^{i(k_1+k_2)} + \beta e^{i(k_1+k_2)}.$$

It follows that

$$
\begin{aligned}
\frac{\alpha}{\beta} &= -\frac{1 + e^{i(k_1+k_2)} - 2e^{ik_1}}{1 + e^{i(k_1+k_2)} - 2e^{ik_2}} \\
&= -\frac{\cos(k_1 + k_2)/2 - e^{i(k_1-k_2)/2}}{\cos(k_1 + k_2)/2 - e^{-i(k_1-k_2)/2}}.
\end{aligned} \tag{7.67}
$$

If k_1 and k_2 are real,

$$\frac{\alpha}{\beta} = \frac{z}{z^*}, \qquad \text{where} \qquad z = i\cos\frac{k_1 + k_2}{2} - ie^{i(k_1-k_2)/2}.$$

Therefore $|\alpha/\beta| = 1$, and in fact

$$\alpha/\beta = e^{i\varphi}, \tag{7.68}$$

where $\varphi/2$ is the phase of z:

$$
\begin{aligned}
\cot\frac{\varphi}{2} = \frac{\operatorname{Re} z}{\operatorname{Im} z} &= \frac{\sin(k_1 - k_2)/2}{\cos(k_1 + k_2)/2 - \cos(k_1 - k_2)/2} \\
&= \frac{1}{2}\left(\cot\frac{k_1}{2} - \cot\frac{k_2}{2}\right),
\end{aligned}
$$

or

$$2\cot\frac{\varphi}{2} = \cot\frac{k_1}{2} - \cot\frac{k_2}{2}. \tag{7.69}$$

We will also consider cases in which k_1 and k_2 are complex. Then $|\alpha/\beta|$ is no longer 1, but φ may still be defined by Eq. (7.68) (φ will now be complex).

Equation (7.69) holds, for such cases, as an algebraic identity or by analytic continuation from the case of real φ.

For scattering problems we assume k_1 and k_2 to be real. Then $|\alpha/\beta| = 1$, as is expected, and φ, which is real in this case, is the phase shift after the scattering. One special solution is

$$\alpha = e^{i\varphi/2}$$
$$\beta = e^{-i\varphi/2} \tag{7.70}$$

From the substitution of Eq. (7.70) in Eq. (7.64), this becomes

$$a_{n_1,n_2} = e^{i(k_1 n_1 + k_2 n_2 + \varphi/2)} + e^{i(k_2 n_1 + k_1 n_2 - \varphi/2)}. \tag{7.71}$$

When we impose the periodicity requirement with the period of N, we see that

$$a_{n_2,n_1+N} = a_{n_1,n_2}. \tag{7.72}$$

(Because $n_1 + N > n_2$, we do not write $a_{n_1+N,n_2} = a_{n_1,n_2}$.) Equation (7.72) is satisfied when, from Eq. (7.71),

$$\begin{cases} Nk_1 = 2\pi m_1 + \varphi \\ Nk_2 = 2\pi m_2 - \varphi, \end{cases} \tag{7.73}$$

where m_1 and m_2 are integers. It must be noticed here that Nk_1 is not equal to $2\pi m_1$, as was the case for one spin wave, but is changed slightly by the amount of φ. Using Eqs. (7.69) and (7.73) we can eliminate k_1 and k_2, and determine φ as a function of m_1 and m_2.

It can be shown that the real values of k do not exhaust the possibilities. We can examine the case of complex wave number by putting

$$k_1 = u + iv,$$
$$k_2 = u - iv. \tag{7.74}$$

Equations (7.70) through (7.73) still hold (although φ is complex). With Eqs. (7.74) we can rewrite Eq. (7.73) in terms of u and v:

$$u = \frac{\pi}{N}(m_1 + m_2), \tag{7.75}$$

$$\varphi = \pi(m_2 - m_1) + iNv. \tag{7.76}$$

From Eq. (7.75) it follows that u is real. Using Eqs. (7.74) in Eq. (7.67), and (7.76) in Eq. (7.68), we obtain a relation between u and v:

$$e^{i\pi(m_2 - m_1)}e^{-Nv} = -\frac{\cos u - e^{-v}}{\cos u - e^v}.$$

Now, if we hold v fixed and let $N \to \infty$, the left hand side $\to 0$; so we have for large N

$$e^{-v} \approx \cos u, \tag{7.77}$$

so that v is approximately real. (We have assumed that $\mathrm{Re}\, v > 0$, but the following result will be the same if $\mathrm{Re}\, v < 0$.) The energy is derived from Eqs. (7.65) and (7.74); use of Eq. (7.77) leads to

$$
\begin{aligned}
\varepsilon^{(2)} &= 2 - \cos (u + iv) - \cos (u - iv) \\
&= 2(1 - \cos u \cosh v) \\
&\approx \sin^2 u = \tfrac{1}{2}(1 - \cos k),
\end{aligned}
\tag{7.78}
$$

where we have defined

$$
k = 2u = k_1 + k_2.
\tag{7.79}
$$

To get a_{n_1, n_2} we use Eqs. (7.74) and (7.76) in Eq. (7.71). The result is (with a change in the normalization)

$$
a_{n_1, n_2} = e^{i(n_1 + n_2)k/2} \cosh v \left(\frac{N}{2} + n_1 - n_2 \right)
$$

or

$$
a_{n_1, n_2} = e^{i(n_1 + n_2)k/2} \sinh v \left(\frac{N}{2} + n_1 - n_2 \right),
\tag{7.80}
$$

according to whether $m_1 + m_2$ is even or odd. a_{n_1, n_2} has a sharp maximum when $n_1 = n_2$ (or $n_1 = n_2 - N$) and decreases as $n_2 - n_1$ increases (remember we assumed $n_1 \leqq n_2$).

Our solution represents a bound state. What is the binding energy for this quasiparticle? When there are two independent spin waves both having wave number K, the energy is

$$
\varepsilon = 2(1 - \cos K) = K^2 - K^4/12 + \cdots
\tag{7.81}
$$

For a quasiparticle of wave number $2K$, we find from Eq. (7.78) the energy

$$
\varepsilon = \tfrac{1}{2}(1 - \cos 2K) = K^2 - K^4/3 + \cdots.
\tag{7.82}
$$

Equation 7.82 gives a lower energy than Eq. (7.81) does, but the difference is very small, namely, $K^4/4$. In other words the binding is very weak, and it depends on K in such a way that the long-wavelength spin waves do not interact much. This is the reason why the approximation of non-interacting spin waves is a good approximation for low temperatures.

7.7 SCATTERING OF TWO SPIN WAVES

In this section and the one that follows we will approach the two-spin-wave system in a slightly different way. In Section 7.6 we found the eigenstates of the Hamiltonian containing two spin waves. Here we will consider the simple waves found in Section 7.5, which are not eigenstates of H but describe pairs of independent spin waves. Then we will apply perturbation theory to calculate the amplitude for scattering of these waves.

The general Hamiltonian for spin waves in three dimensions is written as

$$H = - \sum_{M,N} 2A_M(p^{N,N-M} - 1);$$ (7.83)

where A_M is a constant. For the one-dimensional, nearest-neighbor exchange interaction case, we may write

$$H = -2A \sum_N (p^{n,n+1} - 1).$$ (7.84)

The scattering of two spin waves will be described in this section for the one-dimensional case, although a similar analysis holds for the case of three dimensions.

We know that a ground state of Eq. (7.84) is

$$|\varphi^{(0)}\rangle \equiv |\alpha\alpha\cdots\alpha\rangle,$$ (7.85)

for which all spins point up. When one of the spins points down,

$$|\psi_k\rangle = \frac{1}{\sqrt{N}} \sum e^{ikn}|\varphi_n\rangle,$$ (7.86)

is an eigenfunction of Eq. (7.84) with the eigenvalue

$$E_k = 4A(1 - \cos k).$$ (7.87)

When there are two spins pointing down,

$$|\psi_{k_1,k_2}\rangle = \frac{1}{\sqrt{N^2}} \sum_{n_1 n_2} e^{ik_1 n_1} e^{ik_2 n_2}|\varphi_{n_1 n_2}\rangle$$ (7.88)

describes a state in which two spin waves exist. However, Eq. (7.88) is not an eigenfunction of Eq. (7.84), as we saw above, and the two spin waves scatter each other. Let us examine how the two spin waves interact. By definition, $|\varphi_{n_1 n_2}\rangle = |\varphi_{n_2 n_1}\rangle$ is the state where spins are down at n_1 and n_2, and $|\varphi_{n_1 n_1}\rangle = 0$. We want to evaluate

$$\sqrt{N^2} H|\psi_{k_1 k_2}\rangle = \sum_{n_1 n_2} e^{ik_1 n_k} e^{ik_2 n_2} H|\varphi_{n_1 n_2}\rangle.$$

Here $H|\varphi_{n_1 n_2}\rangle$ is calculated for four cases:

 i) n_1 and n_2 are separated, that is $n_2 \neq n_1$, $n_2 \neq n_1 + 1$, $n_2 \neq n_1 - 1$.

 ii) $n_2 = n_1 + 1$.

 iii) $n_2 = n_1 - 1$.

 iv) $n_2 = n_1$. In this case $|\varphi_{n_1 n_1}\rangle = 0$.

For case (i),

$$H|\varphi_{n_1 n_2}\rangle = -2A[(|\varphi_{n_1+1,n_2}\rangle - |\varphi_{n_1 n_2}\rangle) + (|\varphi_{n_1-1,n_2}\rangle - |\varphi_{n_1 n_2}\rangle)$$
$$+ (|\varphi_{n_1,n_2+1}\rangle - |\varphi_{n_1 n_2}\rangle) + (|\varphi_{n_1,n_2-1}\rangle - |\varphi_{n_1 n_2}\rangle)].$$ (7.89)

For case (ii),

$$H|\varphi_{n,n+1}\rangle = -2A[(|\varphi_{n-1,n+1}\rangle - |\varphi_{n,n+1}\rangle) + (|\varphi_{n,n+2}\rangle - |\varphi_{n,n+1}\rangle)].$$

In order to conform with Eq. (7.89) we will write the equation above in the form

$$H|\varphi_{n,n+1}\rangle = -2A[-2|\varphi_{n,n+1}\rangle + (|\varphi_{n-1,n+1}\rangle - |\varphi_{n,n+1}\rangle) \\ + (|\varphi_{n,n+2}\rangle - |\varphi_{n,n+1}\rangle)] - 4A|\varphi_{n,n+1}\rangle, \quad (7.90)$$

where the term in square brackets is Eq. (7.89) with $n_2 = n_1 + 1$. Similarly, for case (iii),

$$H|\varphi_{n,n-1}\rangle = -2A[-2|\varphi_{n,n-1}\rangle + (|\varphi_{n,n-2}\rangle - |\varphi_{n,n-1}\rangle) \\ + (|\varphi_{n+1,n-1}\rangle - |\varphi_{n,n-1}\rangle)] - 4A|\varphi_{n,n-1}\rangle. \quad (7.91)$$

For case (iv),

$$H|\varphi_{n,n}\rangle = 0 \\ = -2A[|\varphi_{n+1,n}\rangle + |\varphi_{n-1,n}\rangle + |\varphi_{n,n+1}\rangle + |\varphi_{n,n-1}\rangle] \\ + 4A[|\varphi_{n,n+1}\rangle + |\varphi_{n,n-1}\rangle]. \quad (7.92)$$

Here the first term is Eq. (7.89) with $n_1 = n_2$.

When we use Eqs. (7.89) through (7.92) to evaluate $\sqrt{N^2}\,H|\psi_{k_1 k_2}\rangle$, the first terms of Eqs. (7.90), (7.91), and (7.92) combined with Eq. (7.89) lead to the sum of Eq. (7.92) over n_1 and n_2 without restriction; thus

$$\sqrt{N^2}\,H\psi_{k_1 k_2} = -2A \sum_{n_1 n_2} e^{i(k_1 n_1 + k_2 n_2)}|\varphi_{n_1 n_2}\rangle[(e^{ik_1} - 1) + (e^{-ik_1} - 1) \\ + (e^{ik_2} - 1) + (e^{-ik_2} - 1)] \\ - 4A \sum_N [e^{i(k_1 n + k_2(n+1))}|\varphi_{n,n+1}\rangle + e^{i(k_1 n + k_2(n-1))}|\varphi_{n,n-1}\rangle \\ - e^{i(k_1 n + k_2 n)}(\varphi_{n,n+1} + \varphi_{n,n-1})] \\ = 4A[(1 - \cos k_1) + (1 - \cos k_2)]\sqrt{N^2}\,\psi_{k_1 k_2} + \sqrt{N^2}|x\rangle. \\ \quad (7.93)$$

The "remainder" $|x\rangle$ is written as

$$\sqrt{N^2}\,|x\rangle = -4A \sum_n e^{i(k_1 + k_2)n}(e^{ik_2} + e^{ik_1} - 1 - e^{i(k_1 + k_2)})|\varphi_{n,n+1}\rangle \quad (7.94)$$

The first term of Eq. (7.93) is, as seen in Eq. (7.88), a double sum over n, whereas Eq. (7.94) is a single sum over n. Therefore, the second term of Eq. (7.93) can be regarded as expressing the scattering of the two spin waves that appear in the first term of Eq. (7.93).

Notice that when H operates on $|\psi_{k_1 k_2}\rangle$ a new state emerges, but the new state is still a combination of states with two flipped spins. In other words, when two spin waves encounter, two spin waves come out.

With the help of Eq. (7.87) we may write Eq. (7.93) as

$$H|\psi_{k_1k_2}\rangle = (E_{k_1} + E_{k_2})|\psi_{k_1k_2}\rangle + |x\rangle, \qquad (7.95)$$

where

$$|x\rangle = -\frac{4A}{\sqrt{N^2}} \sum_n e^{i(k_1+k_2)n}\xi|\varphi_{n,n+1}\rangle, \qquad (7.96)$$

with

$$\xi = e^{ik_2} + e^{ik_1} - 1 - e^{i(k_1+k_2)}. \qquad (7.97)$$

7.8 NON-ORTHOGONALITY

The general theory of scattering tells us that the cross section σ satisfies

$$\sigma v = \frac{2\pi}{h} |M_{fi}|^2 \times \frac{(\text{number of final states})}{\Delta E}, \qquad (7.98)$$

where v is the velocity of the colliding particles. From perturbation theory,

$$M_{fi} = H'_{fi} + \sum_{n \neq i} \frac{H'_{fn}H'_{ni}}{E_i - E_n} + \cdots \qquad (7.99)$$

H'_{fi} is the matrix element of the perturbation Hamiltonian between the initial and final states. In this section the discussion will be limited to the first term of Eq. (7.99).

We are interested in the scattering from the initial state $|\psi_{k_1,k_2}\rangle$ to the final state $|\psi_{L_1,L_2}\rangle$ where

$$E_{k_1} + E_{k_2} = E_{L_1} + E_{L_2}. \qquad (7.100)$$

Then we cannot use Eq. (7.99) as such, because the states $|\psi_{k_1k_2}\rangle$ are not orthogonal to each other. This is shown as follows. We compute the inner product, using Eq. (7.88):

$$\langle\psi_{L_1,L_2}|\psi_{k_1,k_2}\rangle = \frac{1}{N^2} \sum_{n_1n_2} \sum_{n'_1n'_2} e^{i(k_1n_1 + k_2n_2 - L_1n'_1 - L_2n'_2)}\langle\varphi_{n'_1n'_2}|\varphi_{n_1n_2}\rangle. \qquad (7.101)$$

From the definition of $|\varphi_{n_1n_2}\rangle$, we see that

$$\langle\varphi_{n'_1n'_2}|\varphi_{n_1n_2}\rangle = 1 \text{ when } n'_1 = n_1 \neq n_2 = n'_2 \quad \text{or} \quad n'_1 = n_2 \neq n_1 = n'_2,$$

$$= 0 \text{ otherwise, including the case } n'_1 = n_1 = n_2 = n'_2.$$
$$(7.102)$$

Thus, we can write

$$\langle\psi_{L_1,L_2}|\psi_{k_1k_2}\rangle = \frac{1}{N^2} \sum_{n_1n_2} \{e^{i[(k_1-L_1)n_1 + (k_2-L_2)n_2]} + e^{i[(k_1-L_2)n_1 + (k_2-L_1)n_2]}\}$$

$$- \frac{2}{N^2} \sum_N e^{i(k_1+k_2-L_1-L_2)n}. \qquad (7.103)$$

In the first sum of Eq. (7.103), n_1 and n_2 are summed without constraints, so that we see

$$\frac{1}{N^2} \sum_{n_1 n_2} e^{i[(k_1 - L_1)n_1 + (k_2 - L_2)n_2]} = 1 \text{ when } k_1 = L_1 \text{ and } k_2 = L_2$$

$$= 0 \text{ otherwise.} \tag{7.104}$$

If the second term of Eq. (7.103) were absent, that equation would show the orthogonality of $\psi_{k_1 k_2}$. If k_1, k_2 are not equal to L_1, L_2 but

$$(k_1 + k_2) - (L_1 + L_2) = \text{``0''} \tag{7.105}$$

where "0" denotes any integral multiple of 2π. Then, although the first sum of Eq. (7.103) vanishes, the last term contributes $-2/N$. Thus, $|\psi_{k_1 k_2}\rangle$ are not strictly orthogonal to each other.

To calculate the scattering amplitude we consider the first term of Eq. (7.99), using for the perturbation Hamiltonian

$$H' = H - (E_{k_1} + E_{k_2}). \tag{7.106}$$

Therefore,

$$M_{fi} = \langle \psi_{L_1 L_2} | (H - (E_{k_1} + E_{k_2})) | \psi_{k_1 k_2} \rangle = \langle \psi_{L_1 L_2} | x \rangle, \tag{7.107}$$

where we used Eq. (7.95). Using Eqs. (7.96) and (7.88), we see that

$$M_{fi} = \frac{-4A}{N^2} \sum_n \sum_{n_1 n_2} e^{i(k_1 + k_2)n} e^{-i(L_1 n_1 + L_2 n_2)\xi} \langle \varphi_{n_1 n_2} | \varphi_{n, n+1} \rangle$$

$$= \frac{-4A}{N^2} \sum_n e^{i(k_1 + k_2 - L_1 - L2)n} (e^{-iL_2} + e^{-iL_1})\xi \tag{7.108}$$

because

$$\langle \varphi_{n_1 n_2} | \varphi_{n, n+1} \rangle = 1 \text{ when } n_1 = n, \ n_2 = n + 1 \text{ or } n_1 = n + 1, \ n_2 = n$$

$$= 0 \text{ otherwise.}$$

Further, in Eq. (7.108)

$$\sum_n e^{i(k_1 + k_2 - L_1 - L_2)n} = N \text{ when } k_1 + k_2 = L_1 + L_2 + \text{``0''}$$

$$= 0 \text{ otherwise.}$$

Thus, using Eq. (7.97), we have

$$M_{fi} = -\frac{4A}{N}(e^{-iL_1} + e^{-iL_2})(e^{ik_1} + e^{ik_2} - 1 - e^{i(k_1 + k_2)}) \tag{7.109}$$

when $k_1 + k_2 = L_1 + L_2 + \text{``0''}$. Because of this relation between k and L, and also the relation in Eq. (7.100), M_{fi} is symmetric in k and L.

When we expand M_{fi} for small k_i we see that

$$|M|^2 \sim (k_1 \cdot k_2)^2.$$

Therefore, unless k_i is large, the scattering cross section is small.

7.9 OPERATOR METHOD

It is possible to rewrite the Hamiltonian for spin waves in terms of creation and annihilation operators.

Using the spin operators σ_x and σ_y, we define σ_+ and σ_- by

$$\sigma_+ = \frac{\sigma_x + i\sigma_y}{2}; \qquad \sigma_- = \frac{\sigma_x - i\sigma_y}{2}. \tag{7.110}$$

Remembering that

$$\sigma_x = \begin{pmatrix} 0 & 1 \\ 1 & 0 \end{pmatrix} \quad \text{and} \quad \sigma_y = \begin{pmatrix} 0 & -i \\ i & 0 \end{pmatrix}, \tag{7.111}$$

we have

$$\sigma_+ = \begin{pmatrix} 0 & 1 \\ 0 & 0 \end{pmatrix} \quad \text{and} \quad \sigma_- = \begin{pmatrix} 0 & 0 \\ 1 & 0 \end{pmatrix}. \tag{7.112}$$

It is easy to show that

$$\sigma_+ \sigma_- = \tfrac{1}{2} + \tfrac{1}{2}\sigma_z. \tag{7.113}$$

When $|\varphi_0\rangle$ is the "vacuum" state in which all spins are up, we can write the state $|\varphi_N\rangle$ in which the spin is down at N as

$$|\varphi_N\rangle = \sigma_{N-}|\varphi_0\rangle. \tag{7.114}$$

The state $|\varphi_k\rangle$ of Eq. (7.86) (generalized to three dimensions) may be expressed as

$$|\varphi_k\rangle = \left(\frac{1}{\sqrt{N}} \sum_N e^{ik \cdot N} \sigma_{N-}\right)|\varphi_0\rangle = a_k^+|\varphi_0\rangle, \tag{7.115}$$

where we define the creation operator of a magnon to be

$$a_k^+ = \frac{1}{\sqrt{N}} \sum_N e^{ik \cdot N} \sigma_{N-}. \tag{7.116}$$

The destruction operator is then

$$a_k = \frac{1}{\sqrt{N}} \sum_N e^{-ik \cdot N} \sigma_{N+}. \tag{7.117}$$

Further, $|\psi_{k_1, k_2}\rangle$ of Eq. (7.88) is then

$$|\psi_{k_1 k_2}\rangle = a_{k_1}^+ a_{k_2}^+|\varphi_0\rangle. \tag{7.118}$$

To see that this is correct, remember that

$$\sigma_-\sigma_- = 0.$$

Equation (7.117) may be solved for σ_{N+} as

$$\sigma_{N+} = \frac{1}{\sqrt{N}} \sum_k e^{ik \cdot N} a_k. \tag{7.119}$$

The Hamiltonian of Eq. (7.83) includes $p^{N,N+M}$, which contains $\sigma_N \cdot \sigma_{N+M}$ or terms of the form

$$\sigma_x^1\sigma_x^2 + \sigma_y^1\sigma_y^2 + \sigma_z^1\sigma_z^2,$$

in which the first two terms can be rewritten in terms of products of two a's. The last term corresponds to products of four a's because of the result shown in Eq. (7.113).

Now let us find out if the a's satisfy commutation relations such as

$$[a_k, a_L^+] = \delta_{kL}. \tag{7.120}$$

The answer is no, and is related to the fact that the $\psi_{k_1 k_2}$ are not orthogonal to each other. We can disprove Eq. (7.120) using Eqs. (7.116) and (7.117) together with the fact that $[\sigma_+, \sigma_-] = \sigma_z$:

$$[a_k, a_L^+] = \frac{1}{N} \sum_{N_1 N_2} e^{-ik \cdot N_1} e^{iL \cdot N_2} [\sigma_{N_1-}, \sigma_{N_2+}]$$

$$= \frac{1}{N} \sum_N e^{-i(k-L) \cdot N} \sigma_{Nz}$$

$$= \delta_{k,L} + \frac{1}{N} \sum_N e^{-i(k-L) \cdot N}(\sigma_{Nz} - 1), \tag{7.121}$$

because

$$\frac{1}{N} \sum_N e^{-i(k-L) \cdot N} = \delta_{k,L}.$$

On account of the second term in Eq. (7.121), Eq. (7.120) does not hold exactly, although the second term is numerically of the order of $1/N$ when there are only a few particles with spin down.

7.10 SCATTERING OF SPIN WAVES—OSCILLATOR ANALOG

Figure 7.3 shows two waves of momenta K_1 and K_2 interacting and "emerging" with momenta L_1 and L_2. The scattering matrix M_{fi} is given by

$$M_{fi} = -2A \sum_M e^{iQ \cdot M}(1 - e^{-iM \cdot K_2})(1 - e^{iM \cdot K_1}). \tag{7.122}$$

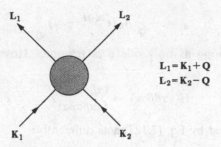

$$L_1 = K_1 + Q$$
$$L_2 = K_2 - Q$$

Fig. 7.3 Scattering of two spin waves.

For small K_1, K_2, $M_{fi} \sim K_1 K_2 \sim K^2$. The scattering cross section σ is given approximately by

$$\sigma \sim |M|^2 \sim K^4. \tag{7.123}$$

At a temperature T, $\hbar K^2/2\mu = kT$ and $K^2 \approx T$. From Eqs. (7.122) and (7.123), $\sigma \sim T^2$. For ordinary particles, $M_{fi} \sim 1$; hence $\sigma \sim 1$, and the virial coefficient $b_2 \sim T^{1/2}$. In the case of spin waves, however, $\sigma \sim T^2$ and thus

$$b_2 \sim T^{5/2}. \tag{7.124}$$

It is plausible to suppose that in a virial expansion we must first consider independent spin waves, for the next term, two-particle collisions, then three, and so on. Such plausibility arguments can be made rigorous by the use of an "harmonic oscillator analog."

In this analog, corresponding to the spin state where the ith spin is down we will take the *oscillator* state where the ith oscillator is excited to the first level and the others are not excited at all. Hence, 00100 corresponds to $\alpha\alpha\beta\alpha\alpha$, and so on. Let α_N^+ be the creation operator for the oscillators. That is, α_N^+ creates a phonon (say) in the Nth oscillator.

$$a_K^+ = \frac{1}{\sqrt{N}} \sum_N \alpha_N^+ e^{-iK \cdot N}. \tag{7.125}$$

Now the real H (for *spin* waves) acts as follows:

$$H(\alpha\alpha\beta\alpha\alpha) \to \frac{(\alpha\beta\alpha\alpha\alpha)}{(\alpha\alpha\alpha\beta\alpha)}. \tag{7.126}$$

We want now to construct an analog H for the oscillators so that

$$H(00100) \to \frac{(01000)}{(00010)}.$$

Now

$$
\begin{aligned}
H &= \sum_i \left[-2A(\alpha_{i+1}^+ \alpha_i - \alpha_i^+ \alpha_i) - 2A(\alpha_{i-1}^+ \alpha_1 - \alpha_i^+ \alpha_i) \right] \\
&= - \sum_N A_M(\alpha_{M+N}^+ - \alpha_N^+)\alpha_M = \sum_K \varepsilon_K a_K^+ a_K,
\end{aligned} \tag{7.127}
$$

where

$$\varepsilon_K = -\sum_M A_M(e^{iK \cdot M} - 1)$$

will do the trick as long as only one β is present. However, consider the following

$$H(\alpha\alpha\beta\beta\alpha\alpha) \rightarrow \frac{(\alpha\beta\alpha\beta\alpha\alpha)}{(\alpha\alpha\beta\alpha\beta\alpha)}. \tag{7.128}$$

However, the H defined by Eq. (7.127) acts differently:

$$\begin{aligned} H(001100) &\rightarrow (010100) \\ &\quad\ (001010) \\ \sqrt{2}\,(000200) &\leftarrow \\ \sqrt{2}\,(002000) &\leftarrow \end{aligned} \tag{7.129}$$

Note that the arrowed functions in Eq. (7.129) have no analog in Eq. (7.128). Therefore we must add terms to H so that these arrowed terms will disappear; that is, we want to define a new $H' = H + \Delta$ such that

$$H'(00100) \rightarrow \frac{(010100)}{(001010)}.$$

Hence the term Δ that is added to H behaves like

$$\Delta(001100) \rightarrow \frac{-\sqrt{2}\,(000200)}{-\sqrt{2}\,(002000)}.$$

Then

$$\Delta = \sum 2A[\alpha_{i+1}^+\alpha_{i+1}^+\alpha_{i+1}\alpha_i - \alpha_{i+1}^+\alpha_i^+\alpha_{i+1}\alpha_i] \tag{7.130}$$

is the new term we need. For example, $\alpha_{i+1}\alpha_i$ annihilates two adjacent phonons at i and $i+1$, and $\alpha_{i+1}^+\alpha_{i+1}^+$ creates two at $i+1$.

$H' = H + \Delta$ defined as above is not Hermitian, and terms such as $\alpha_{i+1}^+\alpha_i^+\alpha_i\alpha_i$ must be added. H is an exact analog for the real H if states with only one β are considered; H' is exact if states with one or two β's are considered; and so forth.

Note that the state (00200) may contribute an energy E_l such that $e^{-\beta E_l}$ is not negligible, in which case the $e^{-\beta E_l}$ term must be subtracted out when forming the partition function, $e^{-\beta E_n}$.

Problem: Using α_N's and α_N^+'s which satisfy anticommutation relations will solve the problem of states with more than one "particle." What are the difficulties?

CHAPTER 8

POLARON PROBLEM

8.1 INTRODUCTION

An electron in an ionic crystal polarizes the lattice in its neighborhood. The interaction changes the energy of the electron, and furthermore, when the electron moves the polarization state must move with it. An electron moving with its accompanying distortion of the lattice has sometimes been called a *polaron*. It has an effective mass higher than that of the electron. We wish to compute the energy and effective mass of such an electron.

Even without any vibration, there is a (periodic) potential acting on the electron. This potential will be approximated by assuming that its presence causes the electron to behave as a free particle but with an effective mass, M. Hence the energy of the electron when there are no lattice vibrations is $P^2/2M$, where P is the electron's momentum.

It will be recalled that the dispersion in a crystal is as shown in Fig. 8.1.

We will be interested in the region described by the almost constant part of the optical branch, that is, the branch for which the positive ions move in a direction opposite to that of the neighboring negative ions.

If we consider the permittivity as a function of frequency (Fig. 8.2), the portion of the curve that will interest us is that part where the lattice vibrations are important.

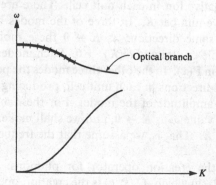

Fig. 8.1 Dispersion in a crystal.

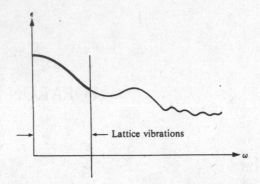

Fig. 8.2 Permittivity as a function of frequency.

Let us take the crystal lattice as being a continuum with a unit cell at each point x. The lattice vibrations are quantized according to Section 6.10, where the transition from a discrete lattice to a continuous one is made according to Eq. (6.201) of that section. Before writing down the Hamiltonian let us consider which vibrational modes actually interact with the electron; the others may be ignored in this problem.

Using the notation of Section 6.10, we assume that the electron's potential due to an undisturbed lattice is either zero or else is taken into account by reducing the mass as already described. If the lattice is displaced, the potential $\Delta V_1(x)$ felt by the electron is due to the charge density $\rho(x)$ resulting from the displacement, and this in turn is due to the polarization $P(x)$:

$$\nabla^2 \, \Delta V_1(x) = e\rho(x) = -e\nabla \cdot P(x) \tag{8.1}$$

(where the electron has charge $-e$). Assume that the crystal lattice has a positive ion and a negative ion in each unit cell. There are then six modes of vibration for each wave number K. In three of the modes both ions move the same distance in the same direction; as $K \to 0$ these modes approach rigid translations of the whole crystal and $\omega(K) \to 0$. Such modes do not contribute much to the polarization $P(x)$. In the other three modes the positive and negative ions move in opposite directions in each unit cell, producing a polarization that is proportional to the amplitude of the mode. For these modes the frequency approaches a nonzero value ω as $K \to 0$, and we shall make the approximation that $\omega(K) = \omega$ for all K. That is, we assume that the frequency is independent of the wave number.

Let $a^+(K, a)$ be the creation operator for phonons. By analogy with Eqs. (6.200) and (6.202) (in which $A^+(k, a)$ is the creation operator for phonons) we see that the polarization (which must be proportional to the displacement) is of the form

$$P(x) = \alpha' \int \frac{d^3K}{(2\pi)^3} \sum_{a=1}^{3} \left[a(K, a)e^{iK \cdot x} e_{K,a} + a^+(K, a)e^{-iK \cdot x} e^*_{K,a} \right]$$

where α' is a real constant. The charge density is then

$$\rho(x) = -\nabla \cdot P(x)$$

$$= i\alpha' \int \frac{d^3K}{(2\pi)^3} \sum_{a=1}^{3} \left[a(K, a)e^{iK \cdot x} K \cdot e_{K,a} - a^+(K, a)e^{-iK \cdot x} K \cdot e^*_{K,a} \right].$$

To a very good approximation we have one e_{Ka} which is in the K direction (the longitudinal optical mode) and therefore two which are perpendicular to it (transversal optical modes). Actually these differ in frequency, and have a frequency ratio of $(\varepsilon_0/\varepsilon_\infty)^{1/2}$, where ε_0 and ε_∞ are the static and high frequency dielectric constants of the crystals. Only the longitudinal mode, the one with $(e_{K,a} \| K)$ contributes to $\rho(x)$; denote the creation operator for that mode simply by a_K^+ and ignore the other modes. Thus

$$\rho(x) = i\alpha' \int \frac{d^3K}{(2\pi)^3} K[a_K e^{iK \cdot x} - a_K^+ e^{-iK \cdot x}], \tag{8.2}$$

so that, from Eq. (8.1), the electron's potential energy due to the lattice vibrations is

$$\Delta V_1(x) = -ie\alpha' \int \frac{d^3K}{(2\pi)^3} \frac{1}{K} [a_K e^{iK \cdot x} - a_K^+ e^{-iK \cdot x}]$$

$$= i(\sqrt{2\pi}\alpha)^{1/2} \left(\frac{\hbar^5 \omega^3}{M} \right)^{1/4} \int \frac{d^3K}{(2\pi)^3} \frac{1}{K} [a_K^+ e^{-iK \cdot x} - a_K e^{iK \cdot x}], \tag{8.3}$$

where α is a dimensionless constant. It can be shown that

$$\alpha = \frac{1}{2} \left(\frac{1}{\varepsilon_\infty} - \frac{1}{\varepsilon_0} \right) \frac{e^2}{\hbar\omega} \left(\frac{2M\omega}{\hbar} \right)^{1/2} \tag{8.4}$$

where ε_0 and ε_∞ are again the static and high-frequency dielectric constants of the crystal. In a typical case such as sodium chloride, α is about 5, and in general it runs from about 1 to 20.

The Hamiltonian for the free electron and lattice is

$$H_{el} + H_{osc} = \frac{P^2}{2M} + \hbar\omega \int \frac{d^3k}{(2\pi)^3} a_k^+ a_k, \tag{8.5}$$

where we are considering only one electron with momentum operator P, and we are ignoring the other lattice modes (which do not interact with the electron according to our assumptions).

To simplify slightly the calculations of the following sections we shall assume the units to be such that $\hbar = M = \omega = 1$. Also we shall sometimes use the notation

$$\sum_{K} = \int \frac{d^3 K}{(2\pi)^3}; \qquad \delta_{KK'} = (2\pi)^3 \delta^3(K - K').$$

With the assumptions made above (due to Fröhlich*), the problem is reduced to that of finding the properties of the following Hamiltonian:

$$H = \tfrac{1}{2}P^2 + \sum_{K} a_K^+ a_K + i(\sqrt{2}\,\alpha\pi)^{1/2} \sum_{K} \frac{1}{K} [a_K^+ e^{-iK \cdot X} - a_K e^{iK \cdot X}]. \qquad (8.6)$$

Here, X is the vector position of the electron, P its conjugate momentum, and a_K^+, a_K are the creation and annihilation operators of a phonon of momentum K.

The polaron problem is not of special importance, but we have now reduced it to mathematics. The method we shall use to solve this mathematical problem will be applicable to different problems of a similar nature. Our method will be valid for arbitrary coupling, but we will first assume small α and will use conventional perturbation theory to get, first, an answer that we can compare with our results for arbitrary coupling, and second, a description of what happens if the electron goes too fast.

8.2 PERTURBATION TREATMENT OF THE POLARON PROBLEM

We wish to find ΔE_0, the perturbation energy when there are originally no lattice vibrations. The Hamiltonian is $H = H_0 + H'$ where

$$H_0 = \frac{P^2}{2} + \sum_{K} a_K^+ a_K$$

$$H' = i(\sqrt{2}\,\pi\alpha)^{1/2} \sum_{K} \frac{1}{K} (a_K^+ e^{-iK \cdot X} - a_K e^{iK \cdot X}).$$

$$\Delta E_0 = H'_{00} + \sum_{n} \frac{H'_{0n} H'_{n0}}{E_0^0 - E_n^0} + \cdots, \qquad (8.7)$$

where $H'_{mn} = \langle m|H'|n \rangle$, $|m\rangle$ and $|n\rangle$ are eigenstates of H_0 and E_n^0 is the eigenvalue of H_0 corresponding to $|n\rangle$.

Because H' acting on a state changes the number of phonons, we have $H'_{00} = 0$. We are interested in the energy of an electron traveling with the momentum P. Therefore, we consider the diagram in Fig. 8.3. Here the initial

---

* H. Fröhlich, *Adv. in Physics*, **3**, 325 (1954).

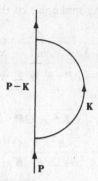

Fig. 8.3 Electron with momentum $(P - K)$ and phonon with momentum K.

state has an electron of momentum P and no phonons. In the intermediate state the electron's momentum is $(P - K)$ and a phonon of momentum K exists. Thus the energies E_0^0 and E_n^0 are

$$E_0^0 = \frac{P^2}{2},$$

$$E_n^0 = \frac{(P - K)^2}{2} + 1. \tag{8.8}$$

Here "1" in E_n^0 is the energy of the phonon (in our present units).

$$H_{n0} = i(\sqrt{2}\,\pi\alpha)^{1/2}\langle n| \sum_K \frac{1}{K} (a_K^+ e^{-iK \cdot X} - a_K e^{iK \cdot X})|P; \text{no phonons}\rangle,$$

which is 0 unless $|n\rangle$ represents a state with one phonon and an electron of momentum P'. If the phonon has momentum K,

$$H_{n0} = i(\sqrt{2}\,\pi\alpha)^{1/2}\langle P', \text{no phonons } |(1/K)e^{-iK \cdot X}|P, \text{no phonons}\rangle. \tag{8.9}$$

Since $e^{-iK \cdot X}|P\rangle = |P - K\rangle$, P' must be $P - K$, as expected, so

$$H_{n0} = i(\sqrt{2}\,\pi\alpha)^{1/2}(1/K)\delta_{P',P-K},$$

and

$$\Delta E_0 = -\sqrt{2}\,\pi\alpha \sum_K \frac{1}{K^2} \frac{1}{(P - K)^2/2 + 1 - P^2/2} \times 2. \tag{8.10}$$

The extra factor of 2 is a consequence of the spin of the electron. There are two intermediate states with momentum $P - K$. Generalizing Eq. (8.10) to three dimensions and writing the sum as an integral, we find

$$\Delta E_0 = -2\sqrt{2}\,\pi\alpha \int \frac{2d^3K/(2\pi)^3}{K^2(K^2 - 2P \cdot K + 2)}. \tag{8.11}$$

In evaluating this integral, we may make use of the identity

$$\frac{1}{ab} = \int_0^1 \frac{dx}{[ax + b(1 - x)]^2}.$$ (8.12)

Regarding $b = K^2$ and $a = K^2 - 2P \cdot K + 2$, we can write Eq. (8.11) as

$$\Delta E_i = -4\sqrt{2}\,\alpha\pi \int \frac{d^3K}{(2\pi)^3} \int_0^1 \frac{dx}{[x(K^2 - 2P \cdot K + 2) + (1 - x)K^2]^2}$$

$$= -4\pi\sqrt{2}\,\alpha \int_0^1 dx \int \frac{d^3K/(2\pi)^3}{(K^2 - 2xP \cdot K + 2x)^2}$$

$$= -4\pi\sqrt{2}\,\alpha \int_0^1 dx \int \frac{d^3K/(2\pi)^3}{[(K - xP)^2 + (2x - x^2P^2)]^2}.$$

Now we use the integral:

$$\int \frac{d^3K/(2\pi)^3}{[K^2 + a]^2} = \frac{1}{8\pi\sqrt{a}}$$

to find

$$\Delta E_i = -4\pi\sqrt{2}\,\alpha \int_0^1 \frac{dx}{8\pi\sqrt{2x - x^2P^2}}$$

$$= -\frac{4\pi\sqrt{2}\,\alpha}{8\pi} \frac{2}{P} \sin^{-1} \frac{P}{\sqrt{2}}.$$

Thus, finally,

$$\Delta E_i = -\alpha \frac{\sqrt{2}}{P} \sin^{-1} \frac{P}{\sqrt{2}}.$$ (8.13)

When $P = 0$, Eq. (8.13) gives

$$\Delta E_i = -\alpha.$$ (8.14)

If the perturbation expansion is carried to higher order,[*] Eq. (8.14) becomes

$$\Delta E = -\alpha + 1.26(\alpha/10)^2.$$

This is the perturbation energy when the electron is at rest. When P is small, we may expand Eq. (8.13) in the form

$$\Delta E_i = -\alpha \frac{\sqrt{2}}{P} \left[\frac{P}{\sqrt{2}} + \frac{1}{6}\left(\frac{P}{\sqrt{2}}\right)^2 + \cdots \right]$$

$$= -\alpha - \frac{P^2}{12}\alpha + \cdots.$$ (8.15)

[*] E. Haga *Progr. Theoret. Physics* (Japan) **11**, 449 (1954).

When we combine this perturbation with the kinetic energy, the total energy if

$$E = \frac{P^2}{2} - \alpha - \frac{P^2}{12}\alpha + \cdots = \frac{P^2}{2(1 + \alpha/6)} - \alpha + \cdots. \qquad (8.16)$$

Equation (8.16) shows that the mass of the electron is increased $(1 + \alpha/6)$ times by the interaction with the phonons. That is,

$$\frac{m_{\text{eff}}}{m} = 1 + \frac{\alpha}{6}. \qquad (8.17)$$

In Eq. (8.13), if $P > \sqrt{2}$, ΔE_i becomes imaginary. This implies that the energy of the electron dissipates by creating a phonon in the same sense as in Čerenkov radiation. Let us evaluate the smallest value of P sufficient for this decay effect. Obviously, the process that could give the smallest possible P is the one in which $\boldsymbol{P}$ is in the direction of $\boldsymbol{K}$. Consider the case shown in Fig. 8.4. By conservation of energy, we know that

$$\frac{(\boldsymbol{P} - \boldsymbol{K})^2}{2} + 1 = \frac{P^2}{2}. \qquad (8.18)$$

Equation 8.18 can be rewritten as

$$\frac{K}{2} + \frac{1}{K} = P. \qquad (8.19)$$

The smallest P that can satisfy Eq. (8.19) corresponds to the minimum value of the left-hand side, which is $\sqrt{2}$, so that

$$P \geq \sqrt{2} \qquad (8.20)$$

is the condition for the "Čerenkov" effect to occur, as we might have expected from Eq. (8.16).

Let us next calculate the rate of this Čerenkov effect. This rate is given as

$$\text{Rate} = \sum_{\substack{\text{Final} \\ \text{states}}} 2\pi |H_{\text{fo}}|^2 \delta(E_{\text{f}} - E_{\text{i}}). \qquad (8.21)$$

The initial state i consists of an electron of momentum $\boldsymbol{P}$, and in the final state we have an electron of momentum $(\boldsymbol{P} - \boldsymbol{K})$ and a phonon of momentum $\boldsymbol{K}$. From Eq. (8.6), we have

$$H_{\text{fi}} = \langle \boldsymbol{P}', 1_K | H_{\text{int}} | \boldsymbol{P} \rangle = \frac{i}{K}(\sqrt{2}\,\pi\alpha)^{1/2}\delta_{\boldsymbol{P}',\boldsymbol{P}-\boldsymbol{K}},$$

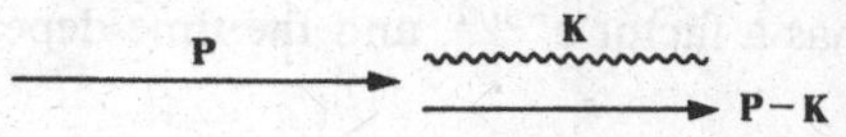

Fig. 8.4 Electron dissipates energy by creating a phonon.

where 1_K means one phonon of momentum K. Thus

$$\text{Rate} = \frac{1}{\tau} = 2\pi \int \frac{\sqrt{2}\,\pi\alpha}{K^2} \, \delta\!\left(\frac{(P-K)^2}{2} + 1 - \frac{P^2}{2}\right) \frac{d^3K}{(2\pi)^3} \, 2. \qquad (8.22)$$

The δ-function insures conservation of energy. We may write

$$d^3K = 4\pi K^2 \, dK (d\Omega/4\pi)$$

so that

$$\text{Rate} = 2\sqrt{2}\,\alpha \int \frac{d\Omega}{4\pi} \int dK \, \delta\!\left(-PK\cos\theta + \frac{K^2}{2} + 1\right), \qquad (8.23)$$

where θ is the angle between P and K.

When K_θ is a solution of

$$-PK_\theta \cos\theta + \frac{K_\theta^2}{2} + 1 = 0, \qquad (8.24)$$

we may transform

$$\int dK \, \delta\!\left(-PK\cos\theta + \frac{K^2}{2} + 1\right) = \int dK \, \delta[\tfrac{1}{2}((K - P\cos\theta)^2 - (K_\theta - P\cos\theta)^2)]$$

$$= \frac{2}{|K_\theta - P\cos\theta|}.$$

Thus Eq. (8.23) becomes

$$\text{Rate} = 4\sqrt{2}\,\alpha \int \frac{d\Omega}{4\pi|K_\theta - P\cos\theta|}. \qquad (8.24)$$

When the momentum of the incoming electron P is given, K_θ of the created phonon is a function of the angle θ, and is given by Eq. (8.24). Solving Eq. (8.24) we obtain

$$\text{Rate} = 2\sqrt{2}\,\alpha \int_0^{P\cos\theta = \sqrt{2}} \frac{\sin\theta\,d\theta}{\sqrt{P^2 \cos^2\theta - 2}} = 2\alpha \frac{\sqrt{2}}{P} \cosh^{-1} \frac{P}{\sqrt{2}}. \qquad (8.25)$$

It should be noticed that since $P/\sqrt{2} > 1$, we may write

$$\sin^{-1} \frac{P}{\sqrt{2}} = \frac{\pi}{2} + i \cosh^{-1} \frac{P}{\sqrt{2}}.$$

Thus, $\cosh^{-1} P/\sqrt{2}$ in Eq. (8.25) is the imaginary part of $\sin^{-1} P/\sqrt{2}$, a factor we obtained in Eq. (8.13). The connection between Eqs. (8.13) and (8.25) can be seen as follows. When the rate of transition is γ, the *amplitude* of remaining in the original state has a factor $e^{-\gamma t/2}$, and the time-dependent wave function has a factor

$$e^{-(\gamma/2)t} e^{-iEt} = e^{-i(E - i\gamma/2)t};$$

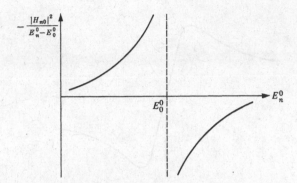

Fig. 8.5 The summand in Eq. (9.7)

thus $-\gamma/2$ is the imaginary part of the energy. Equation (8.25) agrees with Eq. (8.13) because both are for the limit of small α and both are based on a perturbation treatment.

Although we have found Eq. (8.25) to be consistent with Eq. (8.13), you may object; Eq. (8.13) should be valid only when $P < \sqrt{2}$. The situation is clarified as follows. Strictly speaking, Eq. (8.7) does not hold when $E_n^0 = E_0^0$, or when the sum over E_n^0 diverges near E_0^0. Schematically drawn, the summand in Eq. (8.7) has the shape shown in Fig. 8.5. The correct way to take care of the region near E_0^0 is to write the perturbation as

$$\Delta E_0 = \sum_n \frac{H_{0n}H_{n0}}{E_0^0 - E_n^0 + i\varepsilon} \tag{8.7'}$$

and take the limit as $\varepsilon \to 0$.

We see

$$\frac{1}{x + i\varepsilon} = \frac{x}{x^2 + \varepsilon^2} - \frac{i\varepsilon}{x^2 + \varepsilon^2}.$$

The first term has the shape shown in Fig. 8.6, and to use this term is equivalent to taking the principal value when integrating. The imaginary part approaches π times a δ-function as $\varepsilon \to 0$, because

$$\int_{-\infty}^{\infty} \frac{\varepsilon}{x^2 + \varepsilon^2}\, dx = \pi.$$

Thus we may write

$$\lim_{\varepsilon \to 0} \frac{1}{x + i\varepsilon} = \text{principal value} \left(\frac{1}{x}\right) - i\pi\, \delta(x). \tag{8.26}$$

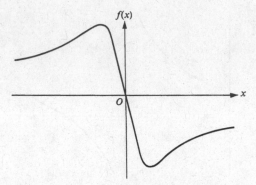

Fig. 8.6 The function $f(x) = \dfrac{-x}{x^2 + \varepsilon^2}$.

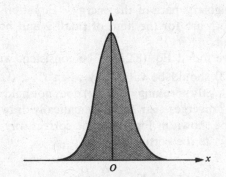

Fig. 8.7 The integral $\displaystyle\int_{-\infty}^{\infty} \dfrac{\varepsilon}{x^2 + \varepsilon^2}\, dx = \pi.$

Thus Eq. (8.7′) may be written as

$$\Delta E_0 = \sum_{\substack{n \\ \text{principal} \\ \text{part}}} \frac{H_{0n}H_{n0}}{E_0^0 - E_n^0} - i\pi \sum_f |H_{f0}|^2\, \delta(E_f - E_0). \tag{8.27}$$

Equation (8.27) shows the consistency of Eqs. (8.7) and (8.21). Although Eq. (8.13) was originally calculated for $P < \sqrt{2}$, the analytic continuation was tacitly made by raising the pole slightly.

Problem: For low-frequency sound waves

$$\omega = K C_s,$$

C_s being the velocity of sound. Calculate the critical value of P for radiation of sound. Calculate the direction of the emerging sound waves as a function of energy. Evaluate the rate of emission.

8.3 FORMULATION FOR THE VARIATIONAL TREATMENT

The partition function may be written as

$$e^{-\beta F} = \text{Tr}\left[e^{-\beta H}\right] = \sum_i e^{-\beta E_i}. \tag{8.28}$$

When $\beta \to \infty$, the leading term is $e^{-\beta E_0}$. Therefore, we claim

$$\lim_{\beta \to \infty}\left[-\frac{1}{\beta}\ln\left(\text{Tr }e^{-\beta H}\right)\right] = E_{\min}. \tag{8.29}$$

Thus, we first calculate the partition function to evaluate $E_{\min}$. We are particularly interested in the case of large β.

Using the path-integral representation, the partition function is written as

$$\text{Tr}\left(e^{-\beta H}\right) = \int e^{-S}\,\mathscr{D}(\text{path}). \tag{8.30}$$

The path-integral representation is used when the Hamiltonian is written in terms of the coordinates and momenta.

Since H is written in terms of the creation and annihilation operators we must first undo it and write it using coordinates and momenta. If we quantize the motion of a crystal, we must choose the creation and annihilation operators so that

$$q_K = \frac{1}{\sqrt{2}}\left(a_K^+ + a_{-K}\right)$$

$$\tag{8.31}$$

$$p_K = \frac{i}{\sqrt{2}}\left(a_{-K}^+ - a_K\right).$$

But then the interaction between the electron and the phonons is of the form

$$H_{\text{int}} = i(\sqrt{2}\,\pi\alpha)^{1/2}\sum_K \frac{1}{|K|}\left[a_K^+ e^{-iK\cdot X} - a_K e^{iK\cdot X}\right]$$

$$= i(\sqrt{2}\,\pi\alpha)^{1/2}\sum_K \frac{1}{|K|}\left(a_{-K}^+ - a_K\right)e^{iK\cdot X}$$

$$= \sqrt{2}\,(\sqrt{2}\,\pi\alpha)^{1/2}\sum_K \frac{1}{|K|}\,p_K e^{iK\cdot X}.$$

We do not find the above form for H_{int} suitable, because in path integrals we want potentials that are functions of position, rather than of momentum. We can exchange the roles of q_K and p_K either before quantization of the crystal motion (by means of a canonical transformation) or afterward, as follows.

Let $a_K' = -ia_{-K}$. Then if we define q_K' and p_K' as in Eq. (8.31) we find

$$q_K' = p_K; \qquad p_K' = -q_K.$$

The resulting Hamiltonian (dropping the primes) is

$$H = \tfrac{1}{2}P^2 + \sum_K (\tfrac{1}{2}p_K^2 + \tfrac{1}{2}q_K^2) + \sqrt{2}\,(\sqrt{2}\,\alpha\pi)^{1/2} \sum_K \frac{q_K e^{iK \cdot X}}{K}. \tag{8.32}$$

Thus, the Hamiltonian is written in terms of the electron coordinate X, momentum P, phonon coordinates q_K and momenta p_K. We write

$$\mathrm{Tr}\,(e^{-\beta H}) = \int_{\substack{X(0)=X(\beta)\\ q_l(0)=q_l(\beta)}} e^{-S}\,\mathcal{D}x(u)\,\mathcal{D}q_1(u)\,\mathcal{D}q_2(u)\cdots \tag{8.33}$$

where the action integral S is

$$S = \int \left[\frac{\dot{X}(u)^2}{2} + \sum_K \tfrac{1}{2}(\dot{q}_K^2(u) + q_K^2(u)) + \sqrt{2}\,(\sqrt{2}\,\alpha\pi)^{1/2} \sum_K \frac{q_K(u)e^{iK \cdot X(u)}}{K} \right] du. \tag{8.34}$$

The advantage of this method is that the path integral over the phonon coordinates can be performed because q_K and $\dot{q}_K$ both appear quadratically (and linearly) in Eq. (8.34). We use the result proved previously*, and find

$$\int \exp\left\{ -\int_0^\beta \left[\tfrac{1}{2}(\dot{q}^2 + \omega^2 q^2) + q(u)\gamma(u) \right] du \right\} \mathcal{D}q(u)$$

$$= \exp\left\{ +\frac{1}{4\omega} \int_0^\beta \int_0^\beta \gamma(t)\gamma(s)e^{-\omega|t-s|}\,dt\,ds \right\}. \tag{8.35}$$

Since γ and q are complex here, we modify Eq. (8.35) to read $\gamma^*(t)\gamma(s)$ instead of $\gamma(t)\gamma(s)$. We will need this modification after Eq. (8.36). Also, Eq. (8.35) is approximate, for $e^{-\beta\omega}$ is neglected.† When we use Eq. (8.35), Eq. (8.33) is written as

$$\mathrm{Tr}\,(e^{-\beta H}) = \int e^{-S}\,\mathcal{D}x(u)$$

$$= \int \exp\left(-\frac{1}{2}\int_0^\beta \dot{X}^2(u)\,du \right) \exp\left(2\sqrt{2}\,\alpha\pi \sum_K \int_0^\beta \int_0^\beta \frac{1}{2K^2} \right)$$

$$\times \exp\left(iK \cdot (X(t) - X(s))e^{-|t-s|}\,dt\,ds \right) \mathcal{D}X(u)$$

---

* Equation (3.39.)

† The complete form is written by replacing $e^{-\omega|t-s|}$ in Eq. (8.35) by

$$\frac{e^{-\omega|t-s|}}{1 - e^{-\omega\beta}} + \frac{e^{\omega|t-s|}e^{-\omega\beta}}{1 - e^{-\omega\beta}}.$$

But remember that β is very large.

or, changing $\sum_K$ into $\int d^3K/(2\pi)^3$,

$$S = \frac{1}{2} \int_0^\beta \dot{X}^2 \, du - \sqrt{2} \, \alpha\pi \int_0^\beta \int_0^\beta \int \frac{d^3K}{(2\pi)^3} \frac{e^{i\mathbf{K} \cdot (\mathbf{X}(t) - \mathbf{X}(s))}}{K^2} e^{-|t-s|} \, dt \, ds.$$

So

$$S = \frac{1}{2} \int \dot{X}^2 \, du - \frac{\alpha}{\sqrt{8}} \int_0^\beta \int_0^\beta \frac{e^{-|t-s|}}{|X(t) - X(s)|} \, dt \, ds. \tag{8.36}$$

Actually, we are in error when we treat all the q_K's as independent real variables as we did in Eq. (8.35) above. The q_K's are complex, with $q_K^* = q_{-K}$. But if we went to the trouble of doing it properly, Eq. (8.36) would still come out as it did above.

Our aim in using this path-integral approach is to combine it with a variational theorem described before. We write

$$\int e^{-S} \mathcal{D}(\text{path}) = \frac{\int e^{-(S-S_0)} e^{-S_0} \mathcal{D}(\text{path})}{\int e^{-S_0} \mathcal{D}(\text{path})} \int e^{-S_0} \mathcal{D}(\text{path})$$

$$= \langle e^{-(S-S_0)} \rangle \int e^{-S_0} \mathcal{D}(\text{path}). \tag{8.37}$$

where S_0 is an appropriately chosen approximation of S. We use the inequality

$$\langle e^{-f} \rangle \geq e^{-\langle f \rangle}$$

to write Eq. (8.37) as

$$\int e^{-S} \mathcal{D}(\text{path}) \geq e^{-\langle S-S_0 \rangle} \int e^{-S_0} \mathcal{D}(\text{path}), \tag{8.38}$$

where

$$\langle S - S_0 \rangle = \frac{\int (S - S_0) e^{-S_0} \mathcal{D}(\text{path})}{\int e^{-S_0} \mathcal{D}(\text{path})}. \tag{8.39}$$

Because for the approximate choice of the Hamiltonian the free energy F_0 is given by

$$e^{-\beta F_0} = \int e^{-S_0} \mathcal{D}(\text{path}),$$

we can write Eq. (8.38) as

$$F \leq F_0 + \frac{1}{\beta} \langle S - S_0 \rangle. \tag{8.40}$$

Before choosing S_0, let us examine the meaning of S in Eq. (8.36). The first term is analogous to a kinetic energy, and the second term represents the potential energy. The special feature of this potential is that the potential at t, where t is thought of as a "time," depends on the past, with weight $e^{-|t-s|}$. This is a type of *retarded potential*, and the significance of the interaction with the past is that the perturbation caused by the moving electron takes "time" to propagate in the crystal.

When α is small, we may take only the first term of Eq. (8.36) as S_0:

$$S_0 = \tfrac{1}{2} \int \dot{X}^2(u) \, du.$$

Then, from, Eqs. (8.40) and (8.29), we obtain

$$E \le E_0 + \lim_{\beta \to \infty} \frac{1}{\beta} \langle S - S_0 \rangle.$$

This gives the old perturbation answer

$$E - E_0 \le -\alpha.$$

8.4 THE VARIATIONAL TREATMENT

We have seen that the use of S_0 for the kinetic energy $\tfrac{1}{2} \int_0^\beta \dot{X}^2 dt$ gives the perturbation result $\Delta E \le -\alpha$ (second-order perturbation) and furthermore this result is shown to be an upper bound (a result proved only by much greater effort with the usual methods).

However, it is possible to imitate the physical situation much better with a more judicious choice of S_0 and thus obtain a much better estimate of E. The next most obvious thing to try for S_0 is the action for an electron bound in a classical potential $V(X)$. This choice can be shown to be equivalent to the use of some trial wave function in the ordinary (Ritz) variational method. In particular, if one chooses $V(X)$ to be a Coulomb potential, one obtains for E (at large α) the same result as that which follows from a trial wave function of the form e^{-Kr}. If one chooses $V(X)$ to be a harmonic potential, then one obtains an improved estimate for E that could also be obtained with a trial wave function of the form e^{-KX^2} (at least at large α).

However, it can be shown that for α less than approximately 6, no $V(X)$ can improve on the result $V = 0$! This indicates that a classical potential is not a very good representation of the physical situation, except possibly at very large binding energies. There are two major reasons for this.

First, the electron is not constrained to any particular part of the crystal but is free to wander. Any potential $V(X)$ obviously tends to keep the electron near its minimum.

Second, we see from the form of the exact action that the potential the electron feels at any "time" depends on its position at previous times, with a weighting factor $e^{-|t-s|}$. That is, the effect the electron has on the crystal propagates at a finite velocity and can make itself felt on the electron at a later

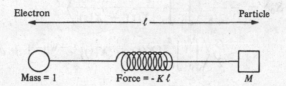

Fig. 8.8 An electron coupled by a "spring" to another particle of mass M.

time. This is less so for tighter binding, where the reaction of the crystal occurs much faster; hence a classical potential might well be expected to be a good approximation only for tight binding.

A model that would not suffer from either of the above objections would be one where, instead of the electron's being coupled to the lattice, it is coupled by some "spring" to another particle and the pair of particles are free to wander. See, for example, Fig. 8.8. The trial action for the system after the coordinates of the mass M have been eliminated is

$$S_0 = \frac{1}{2} \int_0^\beta \dot{X}^2 \, dt + \frac{C}{2} \int_0^\beta \int_0^\beta [X_{(t)} - X_{(s)}]^2 e^{-W|t-s|} \, dt \, ds, \qquad (8.41)$$

where $W \equiv \sqrt{K/M}$ and $C' = MW^3/4$.

Of course, we could have seen mathematically that S_0 would overcome the objections raised above without deriving S_0 from any physical model, because we are free to choose any form for the trial action that we please (if it is not complex). However, the recognition of the physical nature of S_0 greatly helps in avoiding mathematical difficulties when questions of boundary conditions, asymptotic behavior and so on, arise, and also may facilitate computation (see below).

We write according to our variational principle

$$E \leq E_0 + \frac{1}{\beta} \langle S - S_0 \rangle, \qquad (8.42)$$

where E_0 is the binding energy of our model and $\langle \ \rangle$ means "average with weight e^{-S_0}."

Let

$$I(K, t, s) = \langle e^{iK \cdot (X(t) - X(s))} \rangle$$

$$= \frac{\int e^{iK \cdot (X(t) - X(s))} e^{-S_0} \, \mathscr{D}X}{\int e^{-S_0} \, \mathscr{D}X}.$$

If we can evaluate $I(K, t, s)$, then it is easy to find $\langle S - S_0 \rangle$:

$$\langle S - S_0 \rangle = -\frac{\alpha}{\sqrt{8}} \left\langle \int_0^\beta \int_0^\beta \frac{e^{-|t-s|}\, dt\, ds}{|X(t) - X(s)|} \right\rangle$$

$$-\frac{C}{2} \left\langle \int_0^\beta \int_0^\beta \{X(t) - X(s)\}^2 e^{-W|t-s|}\, dt\, ds \right\rangle.$$

For example,

$$\left\langle \int_0^\beta \int_0^\beta |X(t) - X(s)|^2 e^{-W|t-s|}\, dt\, ds \right\rangle = \int_0^\beta \int_0^\beta dt\, ds\, e^{-W|t-s|} \langle |X(t) - X(s)|^2 \rangle$$

$$= \int_0^\beta \int_0^\beta dt\, ds\, e^{-W|t-s|} [-\nabla_K^2 I(K, t, s)]_{K=0}.$$

In general, for any function f of $|X(t) - X(s)|$, we can evaluate $\langle f(|X(t) - X(s)|) \rangle$ by Fourier analyzing f.

$$\left\langle \frac{1}{|X(t) - X(s)|} \right\rangle = \left\langle \int \frac{4\pi e^{iK \cdot [X(t) - X(s)]}}{K^2} \frac{d^3K}{(2\pi)^3} \right\rangle = \int \frac{d^3K}{(2\pi)^3} \frac{4\pi}{K^2} I(K, t, s).$$

To evaluate $I(K, t, s)$ we will work with the more general

$$I(g, t, s) = \left\langle \exp\left[i \int_0^\beta g(u) \cdot X(u)\, du \right] \right\rangle,$$

and then will specialize to the case $g(u) = K[\delta(u - t) - \delta(u - s)]$. We may proceed in several ways to do this general path integral:

a) If we divide "time" into small intervals, we have a problem of the form

$$\int \int \int \cdots dX_1\, dX_2\, dX_3 \cdots \exp\left(-\sum_{ij} A_{ij} X_i X_j + i \sum_i B_i X_i\right). \qquad (8.43)$$

This integral has been developed* for A_{ij} a positive definite and self-adjoint matrix. The result is that the expression 8.43 becomes:

$$(\pi)^{n/2} (\text{Det } A)^{-1/2} \exp\left[-\tfrac{1}{4} \sum_{ij} B_i B_j (A^{-1})_{ij}\right]. \qquad (8.44)$$

Here, n is the number of time intervals in β, and it may tend to infinity after dividing by the proper normalization to make the quantity approach 1 as $B_i \to 0$. With this method we must insert the end-point conditions $X_n = X_0 = 0$ by any one of several artifices (such as an added term in the driving force B).

b) Alternatively we may proceed by recognizing from the physical model that S_0 can be broken up into "normal modes," that is, a free particle with the total mass $M + 1$ plus a harmonic oscillator of frequency $v = \sqrt{W^2 + 4C/W}$

---

* B. Friedman, "*Principles and Techniques of Applied Math*," Wiley, N.Y., 1964.

and reduced mass $M/(1 + M)$. We have already done the path integrals for a driven free particle and a harmonic oscillator, and hence we may write the desired result immediately.

c) Alternatively, we may proceed to perform the integral by a technique* (used in previous lectures for solving the simple harmonic oscillator) that consists of finding the path, $X'(t)$, which makes the driven action an extremum and noticing that (within a normalizing factor) the desired result is

$$ I = \exp \left[\frac{1}{2} \int_0^\beta g(t) \cdot X'(t) \, dt \right]. $$

We will use a fourth method. Let

$$ X(t) = X(0) + \sum_{n=1}^\infty a_n \sin \frac{n\pi t}{\beta}. $$

Note that $X(t)$ satisfies the requirement that $X(0) = X(\beta)$. Integrating over all paths is equivalent to integrating over all $X(0)$ and all a_n. Working in one dimension (for simplicity), we first must find S_0 in terms of $X(0)$ and a_n.

$$ \int_0^\beta \frac{\dot{X}^2}{2} = \frac{1}{2} \int_0^\beta \sum_{n=1}^\infty a_n^2 \frac{n^2\pi^2}{\beta^2} \cos^2 \frac{n\pi t}{\beta} \, dt = \frac{1}{4} \sum_{n=1}^\infty a_n^2 \frac{n^2\pi^2}{\beta}. \tag{8.45} $$

$$ \frac{C}{2} \int_0^\beta \int_0^\beta [X(t) - X(s)]^2 e^{-W|t-s|} \, dt \, ds $$

$$ = \frac{C}{2} \int_0^\beta \int_0^\beta \left[\sum_{n=1}^\infty a_n \left(\sin \frac{n\pi t}{\beta} - \sin \frac{n\pi s}{\beta} \right) \right]^2 e^{-W|t-s|} \, dt \, ds $$

$$ \approx \frac{C}{W} \sum_{n=1}^\infty \left(\frac{n^2\pi^2/\beta}{W^2 + n^2\pi^2/\beta^2} \right) a_n^2. \tag{8.46} $$

Equation (8.46) holds only approximately because we have made certain approximations that are valid in the limit of high β. Then

$$ i \int_0^\beta g(u) X(u) \, du = X(0) b_0 + \sum_{n=1}^\infty a_n b_n = \sum_{n=1}^\infty a_n b_n, \tag{8.47} $$

where

$$ b_n = i \int_0^\beta g(u) \sin \frac{n\pi u}{\beta} \, du = iK \left(\sin \frac{n\pi t}{\beta} - \sin \frac{n\pi s}{\beta} \right). $$

And

$$ I(K, t, s) = \frac{\iiint_{-\infty}^\infty \exp \left[-\sum_{n=1}^\infty (A_n a_n^2 - a_n b_n) \right] \, da_1 \, da_2 \cdots}{\iiint_{-\infty}^\infty \exp - \left(\sum_{n=1}^\infty A_n a_n^2 \right) \, da_1 \, da_2 \cdots}, \tag{8.48} $$

---

* R. P. Feynman, *Phys. Rev.* **97**, 660, (1955).

where

$$A_n = \frac{n^2\pi^2}{4\beta}\left(1 + \frac{4C/W}{W^2 + n^2\pi^2/\beta^2}\right).$$

From Eq. (8.48) it follows that

$$I(K, t, s) = \exp\left(\sum_{n=1}^{\infty}\frac{b_n^2}{4A_n}\right)$$

$$= \exp\left(\sum_{n=1}^{\infty}\frac{-K^2(\sin(n\pi t/\beta) - \sin(n\pi s/\beta))^2}{n^2\pi^2/\beta(1 + (4C/W)/(W^2 + n^2\pi^2/\beta^2))}\right). \quad (8.49)$$

As β approaches ∞, the sum in Eq. (8.49) becomes an integral with $n\pi/\beta \to x$, $\pi/\beta = (\pi/\beta)\,dn \to dx$. This integral can be evaluated to give

$$I(K, t, s) = \exp\left[-\frac{K^2}{2}\left\{\frac{W^2}{V^2}|t - s| + \frac{4C}{WV^3}\left[1 - e^{-|t-s|V}\right.\right.\right.$$

$$\left.\left.\left. + e^{-(t+s)V} - \tfrac{1}{2}e^{-2Vt} - \tfrac{1}{2}e^{-2Vs}\right]\right\}\right], \quad (8.50)$$

where

$$V^2 = W^2 + 4C/W.$$

$I(K, t, s)$ is used only in integrals over s and t with large β. In such integrals there is a contribution only when $K^2W^2/2V^2|t - s|$ is not much greater than unity, that is, in an area of order $2\beta V^2/K^2W^2$ (in the plane formed by the s and t axes). In essentially all of the regions of integration, the fact that β is large implies that $e^{-(t+s)V}$, e^{-2Vt}, and e^{-2Vs} are very small, and we can write

$$I(K, t, s) = \exp\left[-\frac{K^2}{2}\left\{\frac{W^2}{V^2}|t - s| + \frac{4C}{WV^3}\left[1 - e^{-|t-s|V}\right]\right\}\right]. \quad (8.51)$$

From Eq. (8.51) we get for one dimension

$$\left\langle\int_0^\beta\int_0^\beta (X(t) - X(s))^2 e^{-W|t-s|}\, dt\, ds\right\rangle$$

$$= -\int_0^\beta\int_0^\beta dt\, ds\, e^{-W|t-s|}\frac{d^2}{dK^2}I(K, t, s)\bigg|_{K=0} \sim \frac{2\beta}{VW} \text{ as } \beta \to \infty.$$

For three dimensions we must triple the result (since $\nabla \cdot K = 3$ in three dimensions). We find that

$$\frac{C}{2}\left\langle\int_0^\beta\int_0^\beta |X(t) - X(s)|^2 e^{-W|t-s|}\right\rangle = \frac{3\beta C}{VW}. \quad (8.52)$$

We can now find the E_0 we wanted in Eq. (8.42) as follows: $e^{-\beta F_0(C)} = \int e^{-S_0(C)} \mathscr{D}X$ implies that, for large β,

$$E_0'(C) = F_0'(C) = \frac{1}{-\beta e^{-\beta F_0(C)}} \frac{d}{dC} e^{-\beta F_0(C)}$$

$$= -\frac{1}{\beta} \left\langle -\frac{1}{2} \int \int dt\, ds |X(t) - X(s)|^2 e^{-W|t-s|} \right\rangle$$

$$= \frac{3}{VW} = \frac{3}{W\sqrt{W^2 + 4C/W}}.$$

Integrating, and using the fact that $E_0(0) = 0$, we have

$$E_0(C) = \tfrac{3}{2}(V - W). \tag{8.53}$$

To get $\langle S - S_0 \rangle$ we must evaluate

$$\int_0^\beta \int_0^\beta \left\langle \frac{e^{-|t-s|}}{|X(t) - X(s)|} \right\rangle dt\, ds = \int \frac{d^3K}{(2\pi)^3} \frac{4\pi}{K^2} \int_0^\beta \int_0^\beta I(K, t, s) e^{-|t-s|}\, dt\, ds.$$

Let $I(K, t, s)e^{-|t-s|} = g(|t - s|)$.   Here we are assuming Eq. (8.51) to hold because β is large.   Then it is easy to show that

$$\int_0^\beta \int_0^\beta g(|t - s|)\, dt\, ds = 2 \int_0^\beta (\beta - u)g(u)\, du.$$

Since $g(u) \approx 0$ for large u, $\beta g(u) \gg u g(u)$ whenever there is a contribution to the integral; so we can neglect $u g(u)$. We can also let the upper limit of the integral be ∞.   Then

$$\int_0^\beta \int_0^\beta \left\langle \frac{e^{-|t-s|}}{|X(t) - X(s)|} \right\rangle dt\, ds$$

$$= \int \frac{d^3K}{(2\pi)^3} \frac{2\beta(4\pi)}{K^2} \int_0^\beta du \exp\left[-\left\{\frac{K^2}{2}\left[\frac{W^2 u}{V^2} + \frac{4C}{WV^3}(1 - e^{-uV})\right] + u\right\}\right]$$

$$= \frac{4\beta V}{\sqrt{2\pi}} \int_0^\infty \frac{e^{-u}\, du}{\sqrt{W^2 u + [(V^2 - W^2)/V](1 - e^{-uV})}}. \tag{8.54}$$

Using Eqs. (8.52), (8.53), and (8.54), we arrive at the final result:

$$E \le E_0 + \frac{1}{\beta} \langle S - S_0 \rangle$$

$$\underset{\beta \to \infty}{\longrightarrow} \frac{3}{4V}(V - W)^2 - \frac{\alpha V}{\sqrt{\pi}} \int_0^\infty \frac{du\, e^{-u}}{[W^2 u + ((V^2 - W^2)/V)(1 - e^{-uV})]^{1/2}}. \tag{8.55}$$

We may now vary V and W to obtain the lowest upper bound. (These results are the lowest ever obtained as upper bounds for this problem.)

For small α. The best $W = 3$ and $V = 3(1 + 2\alpha(1 - P)/3W)$, where

$$P \equiv \frac{2}{W}[(1 - W)^{1/2} - 1].$$

From this, we obtain

$$E \leq -\alpha - \alpha^2/81 = -\alpha - 1.23(\alpha/10)^2 \qquad (8.56)$$

The correct result from perturbation theory to this order is

$$E = -\alpha - 1.26(\alpha/10)^2.$$

For large α. The best $W = 1$ and $V = (4\alpha^2/9\pi) - (4(\ln 2 + C/2) - 1)$, $C =$ Euler Mascheroni constant $= 0.5772,\ldots$, and

$$E \leq -\alpha^2/2\pi - \tfrac{3}{2}(2\ln 2 + C) - \tfrac{3}{4} + 0(1/\alpha^2)\cdots \qquad (8.57)$$

which, up to $\alpha = 10^3$, is lower than any other estimate, upper bound or otherwise.

For intermediate α. A very great advantage of the method here is that it gives a smooth result for E from weak through intermediate through strong coupling, and it gives the only known reliable results for this problem in the intermediate coupling region. Unfortunately, a little machine calculation is required here to perform the integration in $1/\beta\langle S\rangle$. In case you would like to compare some other calculation, we give in Table 8.1 a few results obtained by Schultz* using Eq. (8.55) and a computer.

Table 8.1

$\alpha =$	3	5	7	9	11
Upper bound of E:	-3.1333	-5.4401	-8.1127	-11.986	-15.710

When using an approximate method, we like to have some idea of the size of the error. Our approximation was to assume that the inequality

$$\langle e^{-f}\rangle \geq e^{-\langle f\rangle}$$

represents an equality. To check the error involved in this assumption, proceed as follows:

$$\langle e^{-f}\rangle = \left\langle 1 - f + \frac{f^2}{2!} + \cdots \right\rangle = 1 - \langle f\rangle + \tfrac{1}{2}\langle f^2\rangle \cdots$$

$$= e^{-\langle f\rangle} + \tfrac{1}{2}(\langle f^2\rangle - \langle f\rangle^2) - \cdots.$$

---

* T. D. Schultz, "Electron-Lattice Interactions in Polar Crystals," Thesis, MIT, 1956.

For small f, a good approximation is to neglect all but the $e^{-\langle f \rangle}$ term on the right. But a better approximation (for small f) is to take

$$\langle e^{-f} \rangle = \exp\left[-\langle f \rangle + \tfrac{1}{2}(\langle f^2 \rangle - \langle f \rangle^2)\right] + \text{higher-order terms.}$$

Taking $e^{-\beta(F - F_0)} = \langle e^{-(S - S_0)} \rangle_0$, we get

$$F \approx F_0 + \frac{1}{\beta}\langle S - S_0 \rangle_0 - \frac{1}{2\beta}\left[\langle |S - S_0|^2 \rangle_0 - \langle S - S_0 \rangle_0^2\right]$$

$$+ \text{higher-order terms in } (S - S_0).$$

If $S - S_0$ is not small, the correction term,

$$-\frac{1}{2\beta}\left[\langle |S - S_0|^2 \rangle - \langle S - S_0 \rangle^2\right],$$

may not be an improvement, but it should be an order-of-magnitude estimate of the size of the error in the uncorrected estimate, $F \approx F_0 + (1/\beta)\langle S - S_0 \rangle$. Such an order-of-magnitude estimate of the error in the polaron energy has been carried out, and it turns out to be small.

8.5 EFFECTIVE MASS

A variational principle along the above lines has not been found for the calculation of the effective mass of the polaron because we need paths $e^{-S'}$ where S' is complex. In fact, for the lowest energy of an electron with velocity U we need to path integrate using

$$S'(X) = \frac{1}{2}\int X^2 \, dt - 2^{-3/2}\alpha \iint |X_t - X_s + 2iU \cdot (t - s)|^{-1} e^{-|t-s|} \, dt \, ds$$

$$= S(X_t + iUt),$$

as we would expect in "imaginary-time" space. However, if U were imaginary, then we would again have a real S' and could use the variational method. Now one can show that $E(|U|)$, considered as function of a complex variable, is analytic near $|U| = 0$. Hence E in the neighborhood of $|U| = 0$ can be obtained by knowing E for imaginary $|U|$. Therefore, one can deduce that for small velocities the effective mass is given by

$$m = 1 + \tfrac{1}{3}\pi^{-1/2}\alpha V^3 \int_0^\infty \frac{d\tau\, \tau^2 e^{-\tau}}{[W^2\tau + ((V^2 - W^2)/V)(1 - e^{-V\tau})]^{3/2}}, \tag{8.58}$$

where V and W are the best parameters found for $E(U = 0)$. It is interesting to note that Eq. (8.58) gives a value for m that is always within a few percent of the total mass, $1 + M = V^2/W^2$, of the trial model.*

---

* For more details, see R. P. Feynman, *Phys. Rev.* **97**, 660 (1955).

CHAPTER 9

ELECTRON GAS IN A METAL

9.1 INTRODUCTION: THE STATE FUNCTION φ

The actual behavior of a metal is very complicated: The metal electrons interact with the lattice, with the lattice vibrations, and with one another. As a first approximation, we will assume the lattice to be rigid and will neglect the mutual electrostatic and magnetic interactions, except insofar as each electron is affected by some average potential due to the other electrons and the periodic lattice.

Hence, the Hamiltonian is given by

$$H = h_1 + h_2 + \cdots; \qquad h_1 = -\frac{\hbar^2 \nabla_1^2}{2m} + V(R_1) \tag{9.1}$$

and the energies given by

$$h_i u_i = \varepsilon_i u_i, \qquad E = \sum_i n_i \varepsilon_i, \qquad \text{and} \qquad \sum_i n_i = N. \tag{9.2}$$

From the Pauli exclusion principle, $n_i = 0$ or 1, and the wave function of the entire electron gas must be antisymmetrical. At moderate temperatures, the electron gas is *almost* in its ground state. That is, the N electrons occupy the lowest N states—*almost*. For a gas with all spins in one direction (for example, all up or all down), the ground-state wave function is

$$\varphi = \frac{1}{\sqrt{N!}} \sum_P (-1)^P u_1(PR_1) u_2(PR_2) \cdots \tag{9.3}$$

$\sum_P$ means a sum over all permutations, an even permutation having weight 1 and an odd permutation, weight -1. PR_1, which might better have been written R_{P1}, means a permutation of the electrons. The normalization factor $1/\sqrt{N!}$ will be derived later.

With these definitions of $\sum_P$ and PR_1, it is seen that Eq. (9.3) can be written as

$$\varphi = \frac{1}{\sqrt{N!}} \times \text{Det}\,[u_i(R_j)] = \frac{1}{\sqrt{N!}} \begin{vmatrix} u_1(R_1) & u_2(R_1) & \cdots \\ u_1(R_2) & u_2(R_2) & \cdots \\ \vdots & & \end{vmatrix}. \tag{9.4}$$

Suppose that $V(R) = 0$. Then (reviewing results obtained in Chapter 1), we have

$$h = -\frac{\hbar^2 \nabla^2}{2m} \quad \text{and} \quad u(R) \equiv \frac{e^{iK \cdot R}}{\sqrt{V}}. \tag{9.5}$$

The energy is*

$$E = \sum_i \varepsilon_i = \sum_K \frac{\hbar^2 K^2}{2m} \to \int \frac{\hbar^2 K^2}{2m} \frac{d^3 K}{(2\pi)^3} VF(K). \tag{9.6}$$

$V =$ volume and $F(K) = 1$ if $|K| \leq K_0$, $F(K) = 0$ if $|K| > K_0$, where $\hbar K_0$ is the maximum momentum, that is, $E_0 = \hbar^2 K_0^2/2m =$ Fermi level.

$$E = V \int_0^{K_0} \frac{\hbar^2 K^2}{2m} \frac{4\pi K^2}{(2\pi)^3} dK = V \frac{1}{5} \frac{\hbar^2}{2m} \frac{1}{2\pi^2} K_0^5,$$
$$N = V \int_0^{K_0} \frac{4\pi K^2}{(2\pi)^3} dK = \tfrac{4}{3}\pi K_0^3 \frac{V}{(2\pi)^3}. \tag{9.7}$$

Thus, if $\rho_0 = N/V$ and $\varepsilon = E/V$,

$$\varepsilon = \frac{3}{5} \rho_0 \frac{\hbar^2 K_0^2}{2m} = \frac{1}{5} \frac{\hbar^2}{4m\pi^2} (6\pi^2)^{5/3} \rho_0^{5/3} = a\rho_0^{5/3}. \tag{9.8}$$

It is sometimes convenient to define r_0 and r_s so that

$$1/\rho_0 = \text{volume per electron} = 4\pi/3 \, r_0^3.$$
$$r_s = r_0/a_0$$

where

$$a_0 = \hbar^2/Zme^2 = \text{Böhr Radius}. \tag{9.9}$$

r_s is of course dimensionless, and for most metals, r_s is between 2 and 6. The energy is often expressed in Rydbergs where 1 Rydberg $= me^4/2\hbar^2$.

The results of Eqs. (9.7) and (9.8), plus, of course, many more, can be obtained from φ of Eq. (9.3). First the normalization factor appearing in Eq. (9.3) will be derived.† We require $\int \varphi^*\varphi \, d^3R_1 \, d^3R_2 \cdots d^3R_n = 1$. Consider a specific term in the summation of Eq. (9.3), without any normalization factor. Such a term is, for example, $u_1(R_2)u_2(R_1)\cdots$. Because the u_i are orthonormal,

$$\int u_1^*(R_1)u_2^*(R_2) \cdots u_1(R_2)u_2(R_1) \cdots d^3R_1 \, d^3R_2 \cdots = 0.$$

Also,

$$\int u_1^*(R_1)u_2^*(R_2) \cdots u_1(R_1)u_2(R_2) \cdots d^3R_1 \, d^3R_2 \cdots = 1.$$

---

* We assume periodic boundary conditions.

† The factor $1/N!$ did not appear in Eq. (6.125) because th enormalization there was $1/N! \int \varphi^*\varphi \, d^3R_1 \cdots d^3R_n = 1$.

In other words,

$$\int \varphi^* u_1(R_2) u_2(R_1) \cdots d^3R_1 \, d^3R_2 \cdots = 1, \quad \text{etc.}$$

There are $N!$ terms of the type $u_1(R_1) u_2(R_2) \cdots$ and thus $\int \varphi^* \varphi \, d^{3N}R = N!$ for an unnormalized φ. When $u_i(R)$ is normalized, the normalization factor is therefore $1/\sqrt{N!}$.

The kinetic energy T can be calculated in the same manner as the normalization integral.

$$T = \left(\varphi, \sum_i - \frac{\hbar^2 \nabla_i^2}{2m} \varphi \right) = - \int \varphi^* \sum_i \frac{\hbar^2 \nabla_i^2}{2m} \varphi \, d^{3N}R. \qquad (9.10)$$

Consider a single term of φ, which is now assumed to be normalized. Such a term is $1/\sqrt{N!} \, u_1(R_2) u_2(R_1) \cdots$ and (dropping the $1/\sqrt{N!}$),

$$-\hbar^2 \nabla_1^2 / 2m u_1(R_2) u_2(R_1) \cdots = u_1(R_2) u_3(R_3) \cdots (-\hbar^2 \nabla_1^2 / 2m u_2(R_1)).$$

Although it is not *a priori* evident that

$$(-\hbar^2 \nabla_1^2 / 2m) u_2(R_1) = \lambda_2 u_2(R_1),$$

where λ_2 is some eigenvalue, this must be so since electron number one has nowhere except u_2 to go. That is, ∇_1^2 operates only on $u_2(R_1)$ and does not bother other electrons. The argument used to evaluate the normalization integral now may be used, and since

$$T = \sum_i - \frac{\hbar^2 \nabla_i^2}{2m}, \qquad T = \sum_i \lambda_i.$$

Since $V = 0$,* the kinetic energy T equals the total energy E and the u_i are plane waves. Hence $\lambda_K = \hbar^2 K^2 / 2m$ and

$$\frac{E}{V} = \sum_K{}' \frac{\hbar^2 K^2}{2m} = \int \frac{\hbar^2 K^2}{2m} F(K) \frac{d^3K}{(2\pi)^3} = \tfrac{3}{5} \rho_0 \frac{\hbar^2 K_0^2}{2m} \qquad (9.11)$$

Equation (9.11) is the same as Eq. (9.8).

The energy per electron $= \tfrac{3}{5}(\hbar^2/2m) K_0^2 = 2.22/r_s^2$ rydbergs, where Eq. (9.9) has been used.

9.2 SOUND WAVES

It is interesting to ask what the velocity of sound c_0 is in an electron gas. It can be found by introducing a perturbation of density $\rho = \rho_{\text{avg}} + (\delta\rho) \cos kx$ and determining how it propagates, or more directly by calculating $c_0^2 = \partial P / \partial \rho$

---

* Try not to confuse V = volume with V = potential.

where P is given in Eq. (1.53) and $\rho = $ the mass density (not the number density ρ_0). Then

$$c_0 = \sqrt{\frac{\partial P}{\partial \rho}} = \frac{\hbar K_0}{\sqrt{3}\, m} = \frac{P_0}{\sqrt{3}\, m}. \tag{9.12}$$

The velocity of an excitation

$$V_g = \frac{1}{\hbar}\frac{\partial \varepsilon}{\partial K_0} = \frac{\hbar K_0}{m} = \frac{P_0}{m}, \qquad V_g > c_0, \tag{9.13}$$

where P_0 is the Fermi momentum (not the pressure).

Equation 9.12 is a general result; if the velocity of a gas particle is a constant, C, the velocity of sound is $C/\sqrt{3}$. For, suppose P is the pressure of the gas, ε its internal energy density and p its internal momentum density. Then, to first order in the sound perturbation, we have, using standard hydrodynamics,

$$\dot{p} = -\nabla P$$

$$\dot{\varepsilon} = -C^2 \nabla \cdot p \quad \text{(energy conservation)}.$$

But for constant velocity we also have $P = \frac{1}{3}\varepsilon$. Combining these equations we obtain

$$\ddot{\varepsilon} - \frac{c^2}{3}\nabla^2 \varepsilon = 0,$$

which represents a wave propagating with velocity $C/\sqrt{3}$.

The perhaps startling conclusion may be drawn that sound propagation in the collisionless electron gas is impossible. In other words, the energy of a density disturbance will be dissipated very quickly. This may be seen as follows: Consider a Fermi sphere (Fig. 9.1).

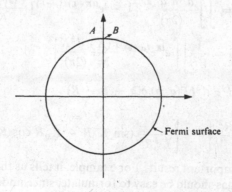

Fig. 9.1 A Fermi sphere.

A given excitation of momentum K can increase the total energy of the gas by as small an amount as we please (for instance when an electron is excited from A to B tangentially to the Fermi surface). Thus a great many energy levels of excitation exist for a given momentum increase. These energy levels are below the energy of a sound wave and hence the sound-wave energy will be quickly dissipated.

9.3 CALCULATION OF $P(R)$

We now remove the restriction $V = 0$. Let

$$V = \sum_{ij} V(R_i, R_j).$$

The reader can easily verify that

$$\left(\varphi, \sum_{ij} V(R_i, R_j)\varphi\right) = \sum_{ij} \int [u_i^*(R_1)u_j^*(R_2) - u_i^*(R_2)u_j^*(R_1)]$$

$$\times V(R_1, R_2)[u_i(R_1)u_j(R_2)]\, d^3R_1\, d^3R_2.$$

$$(9.14)$$

What is the probability, $P(R)$, of two electrons (of same spin) being a given distance R apart?

$$P(a, b) = \left(\varphi, \sum_{ij} \delta(R_i - a)\, \delta(R_j - b)\varphi\right)$$

is the probability that there is a particular at a and one at b. This is of the same form as Eq. (9.14), so

$$P(a, b) = \sum_{ij} \{|u_i(a)|^2|u_j(b)|^2 - [u_i^*(a)u_j(a)][u_j^*(b)u_i(b)]\}. \qquad (9.14)$$

Take $V = 0$. Then $u_K \sim e^{iK \cdot a}/\sqrt{V}$, and $\sum_i \to \int d^3KV/(2\pi)^3$, so that

$$P(a, b) = \iint \frac{d^3K\, d^3L}{(2\pi)^6} [1 - e^{i(K-L)\cdot(a-b)}] \frac{F(K)F(L)}{N^2}$$

$$= \frac{1}{V^2} \left|\int e^{iK\cdot(a-b)} \frac{F(K)}{N} \frac{d^3K}{(2\pi)^3}\right|^2 \qquad (9.15)$$

$$P(R) \equiv \int d^3a\, d^3b P(a, b)\, \delta(a - b - R)$$

$$= \frac{4\pi R^2}{V} \left\{1 - \left[\frac{3}{K_0^3 R^3} (\sin K_0 R - K_0 R \cos K_0 R)\right]^2\right\}. \qquad (9.16)$$

Equation 9.16 is an important result. For example, it tells us that the hard-sphere model of a one-spin gas should be easy to formulate, since under most conditions the electrons are not very close anyway ($P(R) = 0$ at $R = 0$). The short-range

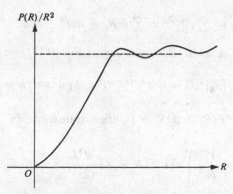

Fig. 9.2 Probability $P(R)$ of two electrons (of same spin) being some distance R apart.

Coulomb interactions are now understandably not important—although the long-range ones are. It is also clear that Eq. (9.14) could have been written as

$$\left(\varphi \sum_{ij} V(R_i, R_j)\varphi\right) = \int V(a, b)P(a, b) \, d^3a \, d^3b. \qquad (9.17)$$

The mean potential energy is the integral over a and b of the product of the potential energy between two particles at a and b and the probability of two particles being at a and b.

Suppose that one wants to know the number n of electrons that are in some region R.

$$n = \sum_i R(R_i) = \sum_i \int \delta(R_i - a)R(a) \, d^3a = \int \rho(a)R(a) \, d^3a$$

where

$$R(a) = \begin{cases} 1 & \text{inside } R \\ 0 & \text{outside } R, \end{cases}$$

and

$$\rho(a) = \sum_i \delta(R_i - a)$$

is the density of electrons at point a.

Fluctuations can also be handled. For example, using the definition of $P(a, b)$, we have

$$(\varphi, n^2\varphi) = \left(\varphi, \sum_{ij} \int \delta(R_i - a) \, \delta(R_j - b) \, R(a)R(b) \, d^3a \, d^3b\varphi\right)$$

$$= \int R(a)R(b)P(a, b) \, d^3a \, d^3b. \qquad (9.18)$$

$$\rho_K = \int \rho(a)e^{+iK \cdot a} \, d^3a$$

or

$$\rho_K = \sum_i e^{iK \cdot R_i} = \sum_i \delta(R_i - a)e^{iK \cdot a} \, d^3a. \qquad (9.19)$$

Clearly $\langle \rho_K \rangle = 0$ for $K \neq 0$. But by squaring Eq. (9.19), we derive

$$\langle |\rho_K^2| \rangle = (\varphi, |\rho_K|^2\varphi) = \int e^{iK \cdot (a-b)}P(a, b) \, d^3a \, d^3b = V S(K),$$

where $S(K) = \int e^{iK \cdot R}P(R, 0) \, d^3R =$ Fourier transform of $P(R, 0)$. From Eq. (9.15) we see

$$S(K) = \frac{V^2}{N^2} \int F(L)F(K - L) \frac{d^3L}{(2\pi)^3}, \qquad K \neq 0. \qquad (9.20)$$

Recalling that

$$F(K - L) = \begin{cases} 1 & \text{if } |K - L| \leq K_0, \\ 0 & \text{if } |K - L| > K_0, \end{cases}$$

we can show that $S(K)$ is the shaded volume in Figure 9.3. Calculating this volume, we obtain

$$S(K) = -\frac{3(2\pi)^3}{4\pi K_0^3}\left[1 - \frac{3}{4}\frac{K}{K_0} + \frac{1}{16}\left(\frac{K}{K_0}\right)^3\right] \quad \text{for } K \leq 2K_0, \quad K \neq 0,$$

$$S(K) = 0, \quad K > 2K_0. \qquad (9.21)$$

Problem: Calculate $\langle |\rho_K|^2 \rangle$ for sound waves. Are the fluctuations here greater or less than the fluctuations of the real electron gas?

9.4 CORRELATION ENERGY

Consider an electron gas with the background of uniform positive charge. The determinantal wave function φ for electrons (Eq. (9.4)) leads to the exclusion-volume radial distribution function $P(R)$ plotted in Fig. 9.2. The Coulomb energy is then calculated as

$$\left(\varphi, \sum_{ij} \frac{e^2}{r_{ij}} \varphi\right) = \int \frac{e^2}{R} P(R) \, d^3R = \int \frac{4\pi e^2}{K^2} S(K) \frac{d^3K}{(2\pi)^3}, \qquad (9.22)$$

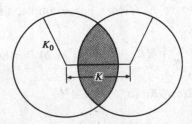

Fig. 9.3 Intersection of two spheres, radius K_0, centers a distance K apart.

where $S(K)$ is the Fourier transform of $P(R)$. Using the form of $S(K)$ in Eq. (9.21), we calculate Eq. (9.22) as

$$\left(\varphi, \sum_{ij} \frac{e^2}{r_{ij}} \varphi\right) = -\frac{0.916}{r_s} \text{ rydbergs.} \tag{9.23}$$

Here r_s is defined in Eq. (9.9) as $r_s = r_0/a_0$ with $a_0 =$ the Böhr radius and $(4\pi/3)r_0^3 =$ the volume per electron.

The total energy of this system is usually written as

$$E = 2.22/r_s^2 - 0.916/r_s + \varepsilon_c \text{ rydbergs.} \tag{9.24}$$

The first term is the kinetic energy calculated in Section 9.1. ε_c is called the *correlation* energy, but actually it is the difference between the true energy E and the first two terms on the right-hand side of Eq. (9.24). ε_c is also sometimes called the *stupidity* energy E_s.

There are attempts to compute ε_c. When the electron density is low and r_s is large, we see from Eq. (9.24) that the kinetic energy is negligible compared to the potential-energy part. When we distribute the electrons in a regular array, the body-centered cubic structure gives the lowest potential energy, which is

$$E_{\text{pot}} = -\frac{1.7}{r_s} \text{ rydbergs for large } r_s. \tag{9.25}$$

Comparing Eqs. (9.24) and (9.25) we see that

$$\varepsilon_c = \frac{-0.88}{r_s} \text{ for large } r_s. \tag{9.26}$$

Wigner examined the case of small r_s also, and his best guess is*

$$\varepsilon_c = \frac{-0.88}{r_s + 7.8} \text{ rydbergs.} \tag{9.27}$$

When Eq. (9.26) was calculated, electrons were put on a body-centered cubic array with the wave function taken to be a δ-function. We may get a better estimate of energy if we use a different wave function. A calculation was done assigning a Gaussian distribution for an electron around a point of the body-centered cubic lattice and varying the shape of the Gaussian distribution. A reasonable numerical result was obtained.

9.5 PLASMA OSCILLATION

In a conductor, electrical resistance increases when impurity ions are introduced, the extra resistivity being due to the scattering of electrons by the extra positive charge. However, the resistivity calculated from the scattering is much larger

---

* See Wigner, E. P., *Phys. Rev.* **46**, 1002 (1934); *Trans. Far. Soc.* **34**, 678 (1938). Also see, for correction, footnote in Pines, D., *Solid State Physics*, **1**, 375 (1955).

than the observed value. This indicates that the extra positive charge is partly neutralized by the higher density of electrons gathered around the positive charge. In the Fermi-Thomas model, for example, the variation of the electron density is related to the (slowly varying) potential φ by

$$\frac{\delta n}{n} = \frac{3}{2}\frac{\delta E_F}{E_F} = \frac{3}{2}\frac{e\varphi}{E_F}$$

where E_F is the Fermi energy. Thus

$$\text{div } \boldsymbol{D} = \text{div } \boldsymbol{E} - 4\pi e\, \delta n = \left[\nabla^2 - \frac{6\pi ne^2}{E_F}\right]\varphi,$$

so that the dielectric constant is

$$\varepsilon(k) = \frac{k^2 + 6\pi ne^2/E_F}{k^2}.$$

Obviously, this means that the field of a point charge decays to zero over a distance of $(E_F/6\pi ne^2)^{1/2}$ (the Fermi-Thomas wavelength), and so does δn in the neighborhood of the point charge. The point charge is therefore screened (if positive) by the electrons, which form a small cluster around it.

The above applies also to each one of the individual electrons in the conductor, δn being now negative, so that each electron creates to himself a "hole" in which it moves. This is beside the effect of Fig. 9.2. The same kind of fluctuation of electron density causes the plasma oscillation. It can be calculated as follows:

Let us consider a one-dimensional sinusoidal fluctuation in the (number) density of electrons:

$$\rho = \rho_{av} + (\delta\rho)\cos kx. \tag{9.28}$$

The potential energy arising from the change of local density is written as

$$E_1 = \frac{1}{2}\int \frac{\partial^2 \varepsilon}{\partial \rho^2}(\delta\rho)^2 \cos^2 kx\, dx, \tag{9.29}$$

where ε is the energy of the Fermi sphere, which depends on the density. From the theory of sound we know that the velocity of a sound wave c_0 is written as

$$\partial^2\varepsilon/\partial\rho^2 = (m/\rho_{av})c_0^2, \tag{9.30}$$

where m is the mass of an electron. For the plasma of charged particles, we have another contribution to the potential energy, namely the electrostatic potential. It is calculated from the Poisson equation

$$-\nabla^2 V = -4\pi e(\delta\rho)\cos kx, \tag{9.31}$$

whose solution is

$$V = -4\pi e(\delta\rho) \frac{\cos kx}{k^2}. \tag{9.32}$$

The electrostatic potential energy is then

$$E_2 = \tfrac{1}{2} \int (-e\rho) \, d(\text{vol})$$

$$= \frac{1}{2} \int \frac{4\pi e^2}{k^2} (\delta\rho)^2 \cos^2 kx \, dx. \tag{9.33}$$

Adding Eqs. (9.29) and (9.33), we find the total potential energy to be

$$E_1 + E_2 = \frac{1}{2} \int \left(\frac{m}{\rho_{av}} c_0^2 + \frac{4\pi e^2}{k^2} \right) (\delta\rho)^2 \cos^2 kx \, dx. \tag{9.34}$$

This quantity is equal to the kinetic energy of the wave in Eq. (9.28):

$$K = \frac{1}{2} \int \frac{m}{\rho_{av}} \frac{\omega_k^2}{k^2} (\delta\rho)^2 \cos^2 kx \, dx, \tag{9.35}$$

where ω is the angular frequency in the wave:

$$u = A e^{i(\omega_k t - k \cdot x)}.$$

Equating Eqs. (9.34) and (9.35) we may write

$$c^2 = c_0^2 + 4\pi e^2 n_0/mk^2, \tag{9.36}$$

where we have put

$$c_k = \omega_k/x,$$

and n_0 is used for ρ_{av}. For the second term of Eq. (9.36) we introduce ω_P, the plasma frequency, as

$$\omega_P = (4\pi e^2 n_0/m)^{1/2}.$$

Then Eq. (9.36) may be written as

$$\omega_k = \sqrt{\omega_P^2 + c_0^2 k^2}. \tag{9.37}$$

The minimum energy required for exciting the plasma oscillation is $\hbar\omega_p$, which is about 10 to 15 eV. It has been observed by sending a beam of electrons through thin films of aluminum or beryllium.

One might wonder how this semi-classical treatment fits into the quantum mechanical picture of the degenerate electron gas. To see this we note that the simplest way of creating an excitation of wave vector q from the Fermi sea is by the operator $a_{k+q}^\dagger a_k$ with a certain allowed k, such that $k + q$ is outside the Fermi sphere whereas k is inside. Since we see no reason to restrict ourselves to a special k, we form

$$\hat{O} = \sum f(k) a_{k+q}^\dagger a_k.$$

The average energy of the state $\hat{O}|0\rangle$ is

$$E = \sum_k \left(\varepsilon_0 + \theta(k, q)\right)|f(k)|^2 + \sum_{k,k'} \left(V(q) - V(k - k')\right) f^*(k) f(k')$$

$$\left(\theta(k, q) = \frac{k \cdot q}{m} + \frac{q^2}{2m}\right)$$

and let us try to minimize it with the normalization condition

$$\sum_{\substack{\text{allowed} \\ k}} |f(k)|^2 = 1.$$

Obviously, this amounts to a diagonalization of a matrix A (allowed k's only),

$$A_{k,k'} = \left(\varepsilon_0 - \lambda + \theta(k, q)\right) \delta_{k,k'} + V(q) - V(k - k').$$

For large q, $\theta(k, q)$ seems to control the result so that the favorable excitations are electron-hole pairs, $f(k) = \delta_{k,k_0}$. But when q is small enough, the long range of the Coulomb potential suggests that $V(q)$ controls the result, so that the minimizing $f(k)$ is $f(k) \sim$ const. Then the favorable $\hat{O}$ is

$$\hat{O} \sim \sum_k a_{k+q}^\dagger a_k,$$

which is exactly the operator for a density fluctuation with wave number q.

We therefore conclude, by this qualitative argument, that for small q, density fluctuations should play an important role.

Let us now see how we can make the microscopic picture more exact.

9.6 RANDOM PHASE APPROXIMATION

Bohm and Pines discussed the plasma oscillation in the following way. The Hamiltonian is

$$H = \sum_i \frac{-\hbar^2}{2m} \nabla_i^2 + \frac{1}{2} \sum_{i \neq j} \frac{e^2}{r_{ij}}. \tag{9.38}$$

The potential energy is Fourier transformed and written as

$$H = \sum_i \frac{-\hbar^2}{2m} \nabla_i^2 + \frac{1}{2V} \sum_{K'} \frac{4\pi e^2}{K'^2} \left[\sum_i e^{iK' \cdot R_i} \sum_j e^{-iK' \cdot R_j} - N\right]. \tag{9.39}$$

(Subtracting N eliminates the self-energy term with $i = j$.)

We will proceed to write the equation of motion of density fluctuations. The Fourier components of the density operator

$$\rho(x) = \sum_i \delta(R_i - x) \tag{9.40}$$

are

$$\rho_K = \sum_i e^{iK \cdot R_i}. \tag{9.41}$$

Let us examine the Heisenberg equation of motion of ρ_K.

The equation of motion of ρ_K is calculated as follows

$$\dot{\rho}_K = i(H\rho_K - \rho_K H) = i \sum_i e^{iK \cdot R_i} \frac{K \cdot (P_i + K/2)}{m} \tag{9.42}$$

where P_i is the momentum operator of the lth electron. In deriving Eq. (9.42), we used the fact that ρ_K of Eq. (9.41) commutes with the potential energy of Eq. (9.39), and also we used the relation

$$P_l e^{iK \cdot R_l} - e^{iK \cdot R_l} P_l = K e^{iK \cdot R_l} \tag{9.43}$$

(remember that in our units $\hbar = 1$). In evaluating $\ddot{\rho}_K$, we note that Eq. (9.42) no longer commutes with the potential part of Eq. (9.39), so that we obtain

$$\ddot{\rho}_K = -\sum_i e^{iK \cdot R_i} \frac{[K \cdot (P_i + K/2)]^2}{m} - \frac{1}{V} \sum_{\substack{K' \\ l,j}} \frac{4\pi e^2}{mK'^2} e^{iK \cdot R_l} e^{-iK' \cdot R_l} e^{iK' \cdot R_j} K \cdot K'. \tag{9.44}$$

Now we make a crude approximation:

$$\sum_l e^{iK \cdot R_l} \frac{[K \cdot (P_l + K/2)]^2}{m} \approx \frac{K^2 p_f^2}{3m} \sum_l e^{iK \cdot R_l} = \frac{K^2 p_f^2}{3m} \rho_K, \tag{9.45}$$

where p_f is the momentum at the Fermi surface. The "3" in the denominator comes from averaging over three directions.

For the second term of Eq. (9.44), we separate the $K' = K$ term from $K' \neq K$. The former gives

$$\frac{4\pi e^2}{m} \frac{N}{V} \sum_j e^{iK \cdot R_j} = \frac{4\pi e^2 n}{m} \rho_K, \tag{9.46}$$

where $n = N/V$ is the number of electrons per unit volume. For $K \neq K'$, we may write

$$\sum_{\substack{K',j \\ K' \neq K}} \frac{4\pi e^2}{mK'^2} K \cdot K' e^{iK' \cdot R_j} \sum_l e^{i(K-K') \cdot R_l} \tag{9.47}$$

Bohm and Pines showed that this term is small for the relatively high-density case. When we neglect this term we call the result the *random phase approximation*. This approximation is interpreted physically as the fact that the sum $\sum_l$ is small if R_l are distributed over a wide variety of positions, so that the various components making up the term tend to cancel. It should be emphasized that Bohm and Pines proved this approximation rather than simply assuming it.

After these considerations, Eq. (9.44) simplifies to

$$\ddot{\rho}_K = -\omega_K^2 \rho_K \tag{9.48}$$

with

$$\omega_K^2 = p_f^2 K^2 / 3m^2 + 4\pi e^2 n/m. \tag{9.49}$$

The last term gives the plasma frequency

$$\omega_P^2 = 4\pi e^2 n/m. \tag{9.50}$$

9.7 VARIATIONAL APPROACH

The plasma state can be investigated using a variational method. It can be regarded as the state of free electrons modified by plasma modes. Thus, we may use as a trial function

$$\varphi_{\text{Trial}} = \exp\left[-\sum_K \alpha_K |\rho_K|^2 \right] \varphi_{\text{Free}}, \tag{9.51}$$

where ρ_K is the Fourier component of the density operator

$$\rho_K = \sum_i e^{iK \cdot R_i}.$$

In Eq. (9.51) α_K is the variational parameter. When we minimize the energy we find that Bohm and Pines's treatment corresponds to the following choice of α_K:

$$\alpha_K = \begin{cases} \dfrac{2\pi e^2}{\hbar \omega_P K^2} & \text{for} \quad K < K_{\text{critical}} \\ 0 & \text{for} \quad K > K_{\text{critical}} \end{cases} \tag{9.52}$$

Thus, the ground state for the plasma oscillation is

$$\varphi = \exp\left[-\sum_{|k| < K_{\text{cr}}} \frac{2\pi e^2}{\hbar \omega_P K^2} |\rho_K|^2 \right] \varphi_{\text{Free}}. \tag{9.53}$$

Variational calculations done on functions of the above type* give correlation energies which were slightly larger for small r_s, but coincide with Wigner's values for $r_s \gtrsim 4$. These correlation energy values are also larger than values obtained by modifications of the RPA theory for larger (and more realistic) densities.†

---

* T. Gaskell, *Proc. Phys. Soc.* (London), **72**, 685 (1958).

† J. Hubbard, *Proc. Roy. Soc.* (London), **A243**, 336 (1957); P. Nozières and D. Pines *Phys. Rev.* **111**, 442 (1958).

9.8 CORRELATION ENERGY AND FEYNMAN DIAGRAMS

We will be concerned next with finding the correlation energy of an electron gas. Although there is an easier way to do this (to be shown later) the present method, based on a paper by Brueckner and Gell-Mann* has the advantage of exemplifying the Feynman-diagram procedure. The diagram method has proved very valuable and has wide application in quantum electrodynamics. Indeed it was for this latter field that the diagrams were invented (see Chapter 6, Section 12).

The problem can be stated as follows: In a metal electron gas, the electrons interact via the Coulomb potential. The Hamiltonian is given by

$$H = - \frac{\hbar^2}{2m} \sum_i \nabla_i^2 + \frac{1}{2} \sum_{i \neq j} \frac{e^2}{r_{ij}}. \tag{9.54}$$

We wish to find the correlation energy ΔE due to the Coulomb interaction. Equation (9.54) can be converted into creation and annihilation operator language. Let a_P^+ be an electron-creation operator, that is, a_P^+ creates an electron of momentum $P = hK$ with wave function $e^{iK \cdot R}$. Then a_P annihilates an electron, and as usual $a_{P'} a_P^+ + a_P^+ a_{P'} = \delta_{P,P'}$. With the methods of Chapter 6, Section 8, Eq. (9.54) can now be written as

$$H = \sum_P \frac{P^2}{2m} a_P^+ a_P + \frac{2\pi e^2 \hbar^2}{V} \sum_Q \frac{1}{Q^2} \sum_{P_1} \sum_{P_2} a_{P_1 - Q}^+ a_{P_2 + Q}^+ a_{P_2} a_{P_1}$$

$$= H_0 + H_1, \tag{9.55}$$

where the Q-dependence may be obtained formally by Fourier analyzing Eq. (9.54). However, Q is nothing more than the momentum transferred by a Coulomb interaction; $4\pi e^2/Q^2$ is the amplitude of such a momentum transfer. Several simplifications are possible. Chief among these is the introduction of b and b^+, the annihilation and creation operators of holes. At $T = 0$, all states less than the Fermi level $\varepsilon_0 = P_0^2/2m$ are filled. We then define b^+ and b as follows:

$$\begin{aligned}
a_P &= a_P & \text{if} \quad |P| > P_0 \\
a_P &= b_{-P}^+ & \text{if} \quad |P| < P_0 \\
a_P^+ &= a_P^+ & \text{if} \quad |P| > P_0 \\
a_P^+ &= b_{-P} & \text{if} \quad |P| < P_0.
\end{aligned} \tag{9.56}$$

In other words, if an electron below the Fermi level jumps above the Fermi level we say that a hole and an electron have been created, and so on. We also have

---

* Some references related to the correlation energy are: D. Pines, Solid State Physics I, 367 (1955). D. Bohm and D. Pines, *Phys. Rev.* **92**, 609 (1953). M. Gell-Mann and K. A. Brueckner, *Phys. Rev.* **106**, 2, 364 (1957). D. Pines and P. Nozières, *The Theory of Quantum Liquids*, W. A. Benjamin, Inc., Menlo Park, California, 1966.

$P_0 = \hbar/\alpha r_0$, where $V = N(4\pi/3)r_0^3$ defines r_0 (N is the number of electrons, V is volume of the gas) and $\alpha = (4/9\pi)^{1/3}$. The momentum, P, will be expressed in Fermi momentum units ($P_0 = 1$) and the energy in rydbergs.

From ordinary perturbation theory

$$\Delta E = \langle 0|H_1|0 \rangle + \sum_{n \neq 0} \frac{(H_1)_{on}(H_1)_{no}}{E_0 - E_n} + \sum_{n \neq 0} \sum_{m \neq 0} \frac{(H_1)_{om}(H_1)_{mn}(H_1)_{no}}{(E_0 - E_m)(E_0 - E_n)} + \cdots.$$
(9.57)

Equation (9.57) may be written as

$$\Delta E \text{ "equals" } \langle 0|H_1|0 \rangle + \langle 0|H_1 \frac{1}{E_0 - H_0} H_1|0 \rangle$$

$$+ \langle 0|H_1 \frac{1}{E_0 - H_0} H_1 \frac{1}{E_0 - H_0} H_1|0 \rangle + \cdots. \qquad (9.58)$$

The quotation marks around "equals" in Eq. (9.58) are to indicate that the equation is valid only if a comment is added to it. This comment will turn out to be surprisingly simple when stated in Feynman-diagram language. We will come to this shortly. The notation of Eqs. (9.57) and (9.58) is familiar. That is, $\langle 0|H_1|0 \rangle = \int \varphi_0^* H_1 \varphi_0 \, dx$ and

$$(H_1)_{mn} = \int \varphi_m^* H_1 \varphi_n \, dx, \text{ etc.,} \quad \text{where} \quad H_0 \varphi_n = E_n \varphi_n. \qquad (9.59)$$

Before continuing, let us try to see the "sense" of Eq. (9.58). First, it is easy to demonstrate the (near) equivalence of Eqs. (9.57) and (9.58). By the ordinary rules of matrix multiplication,

$$\langle 0|H_1 \frac{1}{E_0 - H_0} H_1|0 \rangle \equiv \left(H_1 \frac{1}{E_0 - H_0} H_1 \right)_{00}$$

$$= \sum_{n,m} (H_1)_{om} \left(\frac{1}{E_0 - H_0} \right)_{mn} (H_1)_{no}. \qquad (9.60)$$

But, excluding pathological cases,

$$\int \varphi_m^* f(H_0) \varphi_n \, dx = f(E_n) \int \varphi_m^* \varphi_n \, dx = \delta_{mn} f(E_n).$$

Hence

$$\left(\frac{1}{E_0 - H_0} \right)_{mn} = \delta_{mn} \frac{1}{E_0 - E_n},$$

and Eq. (9.60) equals

$$\sum_n \frac{(H_1)_{on}(H_1)_{no}}{E_0 - E_n}.$$

The comment to be added to Eq. (9.58) is that when we multiply out all matrices we exclude terms with $n = 0$. One could express this more precisely by writing $(1 - |0\rangle\langle 0|)/(E_0 - H_0)$ instead of $1/(E_0 - H_0)$.

Second, it is possible to understand more directly why Eq. (9.58) represents a perturbation expansion. This will not be demonstrated here except to say that the key to the demonstration is the matrix identity convergence*

$$\frac{1}{A + B} = \frac{1}{A} - \frac{1}{A} B \frac{1}{A} + \frac{1}{A} B \frac{1}{A} B \frac{1}{A} - \cdots. \qquad (9.61)$$

We return now to the mainstream of our problem. Using Eqs. (9.55) and (9.56), we can illustrate the possible fundamental processes allowed by H_1 by the set of diagrams in Fig. 9.4. A downward arrow represents a hole and an upward arrow an electron. The space below a dotted line stands for "before the interaction" and above, "after the interaction."

Some comments on Fig. 9.4 may prove helpful. Process (a) is the only one that can start from vacuum or the ground state. The interpretation of (a) is that an electron and hole are created, the difference in momentum Q is carried along the dotted line, and another electron-hole pair created, the momenta being such that the total momentum is conserved. In general, each diagram junction represents a two-particle interaction with conservation of momentum. As another example, (c) describes the annihilation of one electron and the creation of two electrons (of different momenta) and a hole.

To demonstrate how Fig. 9.4 can be used to calculate the desired matrix elements in Eq. (9.58), we first focus on the second-order term,

$$\langle 0|H_1 \frac{1}{E_0 - H_0} H_1|0\rangle = \varepsilon^{(2)}.$$

The first operator, H_1, operates on the ground-state wave function and thus the first operation must be (a) in Fig. 9.4.

The next H_1 operation must get us back to the ground state, because $\langle 0|A|0\rangle = 0$ unless $A\varphi_0 = a\varphi_0$. (The $(E_0 - H_0)^{-1}$ operation does not change the state.) All the possible operations are given in Fig. 9.4, and it is quickly

---

* Equation (9.61) can be proved as follows. Write

$$S = \frac{1}{A} - \frac{1}{A} B \frac{1}{A} + \frac{1}{A} B \frac{1}{A} B \frac{1}{A} - \cdots.$$

Multiply this with A from the left:

$$AS = 1 - B\frac{1}{A} + B\frac{1}{A} B\frac{1}{A} - \cdots = 1 - BS,$$

$$(A + B)S = 1.$$

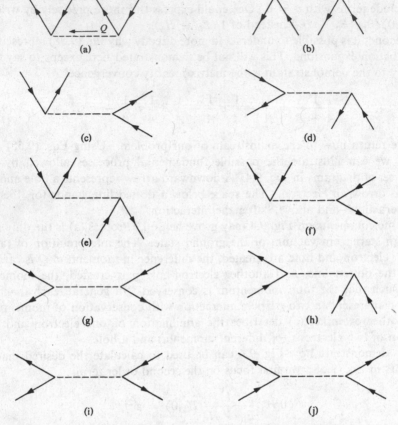

Fig. 9.4 The fundamental processes allowed by H_1.

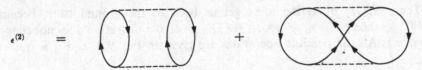

Fig. 9.5 The operations of figure 9.4a and b added together.

seen that only (b) added onto (a) can return us to the ground state or to vacuum conditions. See Fig. 9.5.

Before discussing the amplitudes or numbers associated with each diagram of Fig. 9.5, let us pause to look at some third and fourth-order diagrams—Figs. 9.6 and 9.7. Figures 9.6(a) and (b) are both composed of the fundamental diagrams shown in Figs. 9.4(a), (b), and (e).

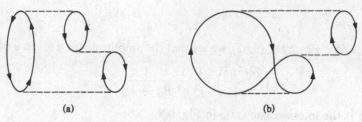

Fig. 9.6 Two third-order diagrams.

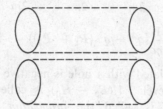

Fig. 9.7 A fourth-order diagram.

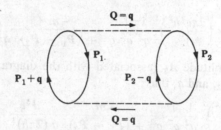

Fig. 9.8 The first diagram of figure 9.5.

When we reach fourth order a new ingredient appears. The diagram in Fig. 9.7 illustrates an evidently valid process. Note, however, that the diagram in this case consists of two independent parts—or unlinked clusters. It can be shown that unlinked clusters must be forgotten and not counted (because they involve intermediate states $|n\rangle = |0\rangle$ in the expansion of Eq. (9.58)). In other words, the process illustrated in Fig. 9.7 is not valid and will not occur. The statement that unlinked cluster diagrams must not be counted is the afore-mentioned comment that must be added to Eq. (9.58). With the addition of this statement the quotation marks can be removed from that equation.

Returning to the calculation of $\varepsilon^{(2)}$ we ask, what is the number associated with each diagram of Fig. 9.5? First, all fundamental diagrams (Fig. 9.4) have amplitude $2\pi e^2 \hbar^2 / VQ^2$ where Q is the Coulomb interaction momentum (see Eq. (9.55)). Now consider the first diagram of Fig. 9.5, shown again in Fig. 9.8. q, P_1, and P_2 are arbitrary but then all other quantities are fixed.

$$\varepsilon^{(2)} = \langle 0|H_1 \frac{1}{E_0 - H_0} H_1|0\rangle \tag{9.62}$$

from Eq. (9.58). From Eq. (9.62), we see that the amplitude for a given q must be

$$\left(\frac{2\pi e^2 \hbar^2}{\text{vol}}\right)^2 \frac{1}{q^2} \frac{1}{q^2} \frac{1}{E_0 - H_n}, \tag{9.63}$$

where $|n\rangle$ is the intermediate state in Fig. 9.8. Now

$$E_n - E_0 = \frac{(P_1 + q)^2}{2m} - \frac{P_1^2}{2m} + \frac{(P_2 - q)^2}{2m} - \frac{P_2^2}{2m}$$

$$= \frac{1}{m}\left(q^2 + q \cdot (P_1 - P_2)\right). \tag{9.64}$$

Note that the energy associated with a hole is negative since we are measuring energy from the Fermi level. $1/(E_0 - H_0)$ is called the *propagator*, and $-m/[q^2 + (P_1 - P_2) \cdot q]$ is called the *propagator factor*. $(2\pi e^2 h^2/\text{vol})^2$ $(1/q^2)(1/q^2)$ is the coupling. Thus for a given P_1, P_2, q the amplitude associated with Fig. 9.8 is

$$A = \left(\frac{2\pi e^2 \hbar^2}{V}\right)^2 \frac{1}{q^2} \frac{1}{q^2} \frac{-m}{q^2 + (P_1 - P_2) \cdot q}. \tag{9.65}$$

To find the total amplitude A_T (associated with the diagram of Fig. 9.8) we must sum over P_1, P_2, and q, thus:

$$A_T = \iiint \left(\frac{2\pi e^2 \hbar^2}{V}\right)^2 \frac{1}{q^2} \frac{1}{q^2} \frac{-m}{q^2 + (P_1 - P_2) \cdot q} \frac{d^3q}{(2\pi h)^3} V \frac{d^3P_1}{(2\pi h)^3} \frac{d^3P_2}{(2\pi h)^3}. \tag{9.66}$$

The integral is such that if R stands for the Fermi sphere, then P_2 is in R, P_1 in R, and $P_1 + q$ and $P_2 + q$ are outside R. Summing over electron spins gives us a factor of 4, two for each loop.

Consider next the second diagram of Fig. 9.5, shown again as Fig. 9.9.

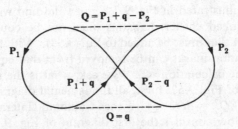

Fig. 9.9 The second diagram of figure 9.5.

An examination of Fig. 9.9. enables us to write down the amplitude immediately:

$$A_T = (-1) \iiint \left(\frac{2\pi e^2 \hbar^2}{VV}\right) \frac{1}{q^2} \left(\frac{2\pi e^2 \hbar^2}{V}\right) \frac{1}{(P_1 + q - P_2)^2} \frac{-m}{q^2 + (P_1 - P_2) \cdot q}$$

$$\times V^3 \frac{d^3q \, d^3P_1 \, d^3P_2}{(2\pi\hbar)^9} f(P_1)f(P_2)[1 - f(P_1 + q)][1 - f(P_2 - q)].$$

$$(9.67)$$

Here

$$f(P) = \begin{cases} 1 & \text{if } |P| < P_0 \\ 0 & \text{if } |P| > P_0. \end{cases}$$

The introduction of the last factor (involving the $f(P)$) enables us to dispense with the comments after Eq. (9.66) explaining the limits of integration. Now the integrations can go from $-\infty$ to ∞. The sign of each amplitude is determined as follows. For Fermi-Dirac statistics (the present case), the sign is -1 for every closed "matter" loop. For Bose-Einstein statistics the sign is $+1$ for every closed matter loop. Figures 9.10(a) and (b) are each considered as one closed matter loop. Figure 9.11 is considered as two closed matter loops. Hence the sign of the first term in Fig. 9.5 is $(+)$ and the sign of the second term is $(-)$. To sum over spins for the second diagram of Fig. 9.5 gives another factor of 2 in Eq. (9.67).

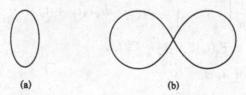

(a)   (b)

Fig. 9.10 Each of these is considered as one closed matter loop.

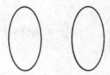

Fig. 9.11 Two closed matter loops.

9.9 HIGHER-ORDER PERTURBATION

Consider the diagram of Fig. 9.12. It should be noticed that the interaction is always wave vector q. The perturbing energy contributed from this diagram is calculated (including the sum over spins) as

$$(-1)^3 2^3 \int \left(\frac{2\pi e^2 \hbar^2}{Vq^2}\right)^3 \frac{d^3q}{(2\pi)^3} \int \frac{d^3P_1}{(2\pi)^3} \int \frac{d^3P_2}{(2\pi)^3} \int \frac{d^3P_3}{(2\pi)^3} V^4$$

$$\times \frac{-m}{(\frac{1}{2}q^2 + q \cdot P_1) + (\frac{1}{2}q^2 + q \cdot P_2)} \frac{-m}{(\frac{1}{2}q^2 + q \cdot P_1) + (\frac{1}{2}q^2 + \dot{q} \cdot P_3)}, \quad (9.68)$$

where the integration is over the appropriate range.

Because of the existence of the powers of q in the denominator, diagrams of this type diverge when integrated over q, even with the restriction on the range of integration. However, the final quantity of interest is the sum of contributions from these diagrams, and the final result is expected not to diverge. In order to arrive at the final result without getting involved in the divergence, $1/q^2$ is first modified into $1/(q^2 + \varepsilon^2)$ using a small ε, the integrations are carried out including ε, and at the last stage ε is brought to zero.

The worst divergences comes from diagrams such as those of Fig. 9.13. These are called the *sausage diagrams*, and the q's are the same for all interactions shown by dashed lines.

Consider the integrals

$$F_q(t) = \int \frac{d^3p}{(2\pi)^3} F(p)(1 - F(p + q))e^{-|t|(q^2/2 + q \cdot p)} \quad (9.69)$$

and

$$A_n = \frac{1}{n} \int_{-\infty}^{\infty} \cdots \int_{-\infty}^{\infty} dt_1 \cdots dt_n \, \delta(t_1 + t_3 + \cdots + t_n)$$

$$\times F_q(t_1) F_q(t_2) \cdots F_q(t_n). \quad (9.70)$$

The simplest case of A_n is

$$A_2 = \frac{1}{2} \int_{-\infty}^{\infty} dt_1 \int_{-\infty}^{\infty} dt_2 \int \delta(t_1 + t_2)e^{-|t_1|(q^2/2 + q \cdot p_1)} e^{-|t_2|(q^2/2 + q \cdot p_2)}$$

$$\times \frac{d^3p_1}{(2\pi)^3} \frac{d^3p_2}{(2\pi)^3}$$

$$= \int \frac{d^3p_1}{(2\pi)^3} \int \frac{d^3p_2}{(2\pi)^3} \frac{1}{2} \int_{-\infty}^{\infty} dt_1 e^{-|t_1|(q^2 + q \cdot (p_1 + p_2))}$$

$$= \int \frac{d^3p_1}{(2\pi)^3} \int \frac{d^3p_2}{(2\pi)^3} \frac{1}{q^2 + q \cdot (p_1 + p_2)}, \quad (9.71)$$

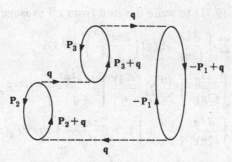

Fig. 9.12 Diagrams of this type diverge when integrated over q.

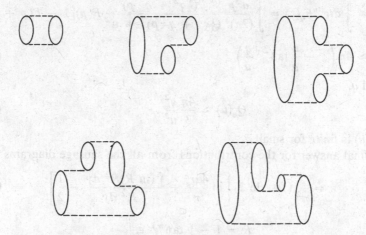

Fig. 9.13 Sausage diagrams.

where all integrals are over the appropriate ranges.

Thus we see that the contribution from an n-loop diagram is

$$C_n = \int \frac{d^3q}{(2\pi)^3} \left(\frac{2\pi e^2 \hbar^2}{Vq^2}\right)^n A_n(q)(-1)^n. \qquad (9.72)$$

In order to simplify $A_n(q)$, use the integral expression of the δ-function

$$\delta(t) = \int_{-\infty}^{\infty} e^{iut} \frac{du}{2\pi}$$

so that Eq. (9.70) becomes

$$A_n = \frac{1}{n} \int dt_1 \cdots dt_n e^{i(t_1 - t_2 + \cdots + t_n)u} F_q(t_1) F_q(t_2) \cdots F_q(t_n) \frac{du}{2\pi}$$

$$= \frac{1}{n} \int \frac{du}{2\pi} \left[\int dt e^{itu} F_q(t)\right]^n. \qquad (9.73)$$

Use Eq. (9.72) in Eq. (9.71) to write the sum from all sausage diagrams as

$$\sum_n C_n = \sum_n \int \frac{d^3q}{(2\pi)^3} A_n(q) \left(\frac{2\pi e^2 \hbar^2}{Vq^2}\right)^n (-1)^n$$

$$= \int \frac{d^3q}{(2\pi)^3} \int \frac{du}{2\pi} \sum_n \frac{(-1)^n}{n} \left[\frac{2\pi e^2 \hbar^2}{Vq^2} \int dt e^{itu} F_q(t)\right]^n$$

$$= \int \frac{d^3q}{(2\pi)^3} \int \frac{du}{2\pi} \ln\left[1 + \frac{2\pi e^2 \hbar^2}{Vq^2} \int dt e^{itu} F_q(t)\right]. \tag{9.74}$$

The integral over $\int dt \cdots$ has been worked out for small q and is

$$Q_q(u) = \int dt e^{itu} F_q(t) = \int \frac{d^3p}{(2\pi)^3} \frac{2(\tfrac{1}{2}q^2 + q \cdot p)}{(\tfrac{1}{2}q^2 + q \cdot p)^2 + u^2} F(p)[1 - F(p + q)]$$

$$\approx 4\pi \left(1 - \frac{u}{q} \tan^{-1} \frac{q}{u}\right). \tag{9.75}$$

For small q,

$$Q_q(u) \approx \frac{4\pi}{3} \frac{q^2}{u^2}. \tag{9.76}$$

Thus $Q_q(u)$ is finite for small q.

The final answer for the contributions from all the sausage diagrams is

$$\varepsilon = \frac{2}{\pi^2} (1 - \ln 2) \left[\ln \frac{4\alpha r_s}{\pi} + \frac{\int (\ln R) R^2 \, dy}{\int R^2 \, dy} - \frac{1}{2}\right] \tag{9.77}$$

where

$$R = 1 - y \tan^{-1} y,$$
$$\alpha = (4/9\pi)^{1/3}. \tag{9.78}$$

r_s is the average spacing of electrons measured in units of the Bohr radius. It should be noticed that a $\ln r_s$ term appears in Eq. (9.76). The final result for the total energy is

$$E = \frac{2.22}{r_s^2} - \frac{0.916}{r_s} + 0.0622 \ln r_s - 0.096 + 0(r_s). \tag{9.79}$$

Here, the last constant term -0.096 is the exchange term calculated apart from the diagram sum.

It should be remembered that the mathematics in this section summed only the sausage diagrams as shown in Fig. 9.13. Although it is true that each of these diagrams causes the worst divergence, the theory would not be complete if the rest of the diagrams like the one in Fig. 9.9 were neglected. Gell-Mann and Brueckner examined the contribution from these diagrams also and -0.096 is the result; as was expected, it is smaller (for small r_s) than the contribution from the sausage diagrams, but it should not be neglected.

CHAPTER 10

SUPERCONDUCTIVITY

10.1 EXPERIMENTAL RESULTS AND EARLY THEORY

Just about any material can be brought into a superconducting state, in which there is no measurable electrical resistance, by cooling it below a certain critical temperature. That temperature depends on the material; it gets as high as $\sim 21°K$ for $Nb_{12}Al_3Ge$. Although the vanishing of the resistance is perhaps the most spectacular effect in superconductors, we can perhaps find clues to the cause of superconductivity by examining other effects. One such effect is the discontinuity in the specific heat at the critical temperature.

The specific heat less the aT^3 contribution of the lattice gives the specific heat of the electrons, C_e. Comparison of normal elements with superconducting ones gives the two curves shown in Fig. 10.1. For the normal element C_e varies as T, because the thermal energy is kT and only those electrons within an energy range kT near the Fermi surface can be excited. Thus, the thermal energy varies as T^2.

Integrating the specific-heat curve shows that the superconductor is lower in energy than the normal element.

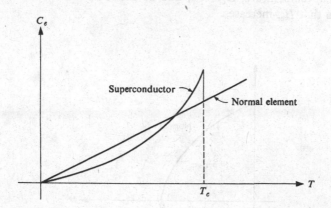

Fig. 10.1 Specific heat of electrons as a function of temperature for a superconductor and a normal element.

Meissner Effect. A changing magnetic field produces an electric field, so something with zero resistance cannot contain a changing magnetic field. If a superconducting sphere is placed in a magnetic field, the field lines will be forced away from the sphere. If a ring is cooled in a magnetic field until it is a superconductor, the flux through the ring will remain after the external source of the field is turned off. Apparently the resistance is exactly zero, for experimentally the flux remains constant indefinitely (provided the ring is not allowed to warm up).

More surprising is the Meissner effect. If a solid (simply connected) piece of superconducting material is placed in a magnetic field and then cooled below the critical temperature, the magnetic field is pushed out of the superconductor. Technically, some lines might be trapped in the object, because some parts reach the superconducting state before others. Furthermore, if the magnetic field is strong enough, it might not be pushed out at all. In such a case, the material does not become superconducting. Its resistance and specific heat are normal. Because of its magnetic domains, iron cannot be cooled into superconductivity.

The Gibbs function is defined as $G = F + PV$ where P is the pressure and V is the volume. Here the pressure can be taken to be the energy per unit volume required to push the field out. From classical thermodynamics, the Gibbs function has the property that in a reversible change of phase at constant temperature and pressure, G does not change. So the critical field is characterized by

$$F_{\text{supercond}} + \frac{\mu H_{cr}^2}{8\pi} V = G_{\text{supercond}} = G_{\text{normal}} = F_{\text{normal}}. \tag{10.1}$$

At the critical temperature, $H_{cr} = 0$, and as the temperature decreases F_{normal} increases, so that H_{cr} increases.

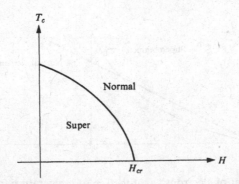

Fig. 10.2 Variation of critical field with temperature.

London observed, on the Meissner effect, that if n = density of electrons, mass = m, charge = $-e$ and electric field = E, then acceleration is given by

$$-eE = m\ddot{x},$$

and the current density j is

$$-ne\dot{x} = j.$$

In a magnetic field of vector potential A (if we use cgs units),

$$E = -\frac{1}{c}\frac{\partial A}{\partial t}$$

gives a rate of change of current density of

$$\frac{dj}{dt} = -\Lambda\frac{\partial A}{\partial t}.$$

Hence,

$$-\Lambda(A - A_{const}) = j,$$

where $\Lambda = ne^2/mc$ = constant. A_{const} is constant in time, but may vary with position. It is fixed by the Meissner effect, so that no arbitrary magnetic fields can be put into the superconducting region.

London proposed that A_{const} be taken equal to zero for superconductors. We then must satisfy the boundary condition that $j_{normal} = 0$ at the surface of the superconductor, so that $A_{normal} = 0$. We do this by an appropriate choice of gauge, called "transverse gauge."

We conclude then that

$$j = -\Lambda A. \tag{10.2}$$

Equation (10.2) implies a modification of the statement that there is no magnetic field in a superconductor. Since

$$\nabla^2 A = -\frac{4\pi}{c}j = \frac{4\pi}{c}\Lambda A, \tag{10.3}$$

there is no sudden drop in the magnetic field to zero when we enter the superconductor. In one dimension, for example,

$$A \propto \exp\left(\pm\sqrt{4\pi\Lambda/c}\,x\right),$$

where the sign in the exponential is chosen so that A decreases as the distance into the solid increases. The magnetic field penetrates to a depth of order 700 Å ($\sqrt{c/4\pi\Lambda}$). The existence of a finite penetration depth can be determined experimentally by measuring the diamagnetic susceptibility of small drops of superconducting material, or by working with thin films.

The above theory suggests where the Λ might come from and approximately how big it should be. But remember that electrons in a metal are not free, so Λ

is not exactly ne^2/mc. The theory can be made to correspond with reality better by taking $j = -\Lambda A'$, where A' is A averaged over the position using an appropriate function. For example, in one dimension the average might be taken as follows:

$$A' = \text{normalization constant} \times \int A(y)e^{-|x-y|/\xi}\,dy.$$

For "hard" or "type II" superconductors, ξ is much smaller than the penetration depth. For "soft" or "type I" superconductors we have large ξ. Impurities make a superconductor hard.

A further contribution by London came from consideration of the quantum-mechanical electric current, which is $-e$ times the probability current:

$$j = \frac{-\hbar e}{2im}(\psi^*\nabla\psi - (\nabla\psi)^*\psi).$$

In a magnetic field the momentum operator becomes $p + e/cA$, so that the current is then

$$j = \frac{-e}{2m}\left\{\psi^*\left(\frac{\hbar}{i}\nabla + \frac{eA}{c}\right)\psi + \left[\left(\frac{\hbar}{i}\nabla + \frac{eA}{c}\right)\psi\right]^*\psi\right\}$$

$$= \frac{-\hbar e}{2im}[\psi^*\nabla\psi - (\nabla\psi)^*\psi] - \frac{e^2 A}{mc}\psi^*\psi. \tag{10.4}$$

With $A = 0$ we get

$$j = \frac{-\hbar e}{2im}[\psi_0^*\nabla\psi_0 - (\nabla\psi_0)^*\psi_0] = 0.$$

When $A \neq 0$, ψ_0 changes into ψ. If $\psi \approx \psi_0$, then Eq. (10.4) gives

$$j = -\frac{e^2 A}{mc}\psi^*\psi = -\frac{ne^2}{mc}A = -\Lambda A. \tag{10.2}$$

We see then that if the wave function is "rigid" (that is, if it does not change when A is introduced) then Eq. (10.4) implies Eq. (10.2). What could cause this rigidity?

From perturbation theory, we have

$$\psi = \psi_0 + \sum_{n \neq 0}\frac{\langle n|H_{\text{pert}}|0\rangle}{E_n - E_0}|n\rangle,$$

where E_0 is the ground-state energy and E_n is the energy of an excited state. If there is a gap between the ground-state energy and the energy of the first excited state, then $E_n - E_0$ is large and $\psi \approx \psi_0$.

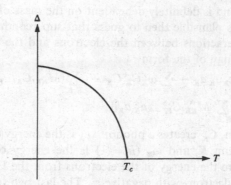

Fig. 10.3 Variation of energy gap Δ with temperature.

The assumption of an energy gap can also explain the anomaly in the specific heat. Instead of the energy varying at T^2, a gap would cause it to vary as $e^{-\Delta/kT}$, where Δ is the size of the gap. More direct confirmation of the existence of a gap has been afforded by experiments involving microwaves. The energy needed to excite the material across the gap could even be measured as a function of temperature, and Δ decreases as the temperature increases.

It took almost fifty years for the problem of superconductivity to be reduced to that of explaining the gap. In what follows we will explain the gap following the theory of Bardeen, Cooper, and Schreiffer. This theory is essentially correct, but I believe it needs to be made more obviously correct. As it stands now there are a few seemingly loose ends to be cleared up.

10.2 SETTING UP THE HAMILTONIAN

The energy gap is $\Delta \simeq kT_\alpha \simeq 10^{-3}$ eV/electron, which is a small quantitative effect to produce a large qualitative effect. Thus the Coulomb correlation energy is too big, and can therefore be neglected.

If this reason for neglecting Coulomb effects seems odd, remember that we are not trying to explain and predict everything about the solid. We are just trying to understand superconductivity. If some effect is associated with an energy that is too large, then we know *that* effect is not the cause of superconductivity. Similarly, as we know that the energy associated with superconductivity is small, we can predict that certain phenomena (such as $e^+ e^-$ annihilation) are not going to be affected much by superconductivity.

If we change the isotope out of which the superconductor is made, the critical temperature changes. Spin–spin and spin–orbit interactions do not change with changes in the isotope, so they should be neglected.

The speed of sound is definitely dependent on the mass of the atoms of the superconductor. It is plausible then to guess that superconductivity has something to do with interactions between the electrons and the phonons. Let us try, then, a Hamiltonian of the form

$$H = \sum_K \varepsilon_K a_K^+ a_K + \sum_K \omega_K C_K^+ C_K + \sum_{K,K'} M_{KK'} C_{K'-K} a_{K'}^+ a_K$$
$$+ \sum_{K,K'} M_{KK'}^* C_{K'-K}^+ a_K^+ a_{K'}. \tag{10.5}$$

a_K^+ creates an electron, C_K^+ creates a phonon. ε_K is the energy of an independent electron of momentum K, and ω_K ($\hbar = 1$) is the energy of an independent phonon. We measure the energy of the electrons from the Fermi surface, and holes are treated as electrons with negative ε. The last two terms of Eq. (10.5) represent the interaction between phonons and electrons. Usually this interaction is taken to be responsible for resistance. High resistance at normal temperatures implies high M, which in turn implies a special propensity towards superconductivity.

There are many effects that cause large energy changes, but which are easily understood and have nothing to do with superconductivity. For example, the effect represented by Fig. 10.4 changes the energy of the electrons by an amount larger than the gap, but does not produce a gap.

The diagram in Fig. 10.5 just corrects the properties of the phonon.

Fig. 10.4 An effect that causes large energy changes.

Fig. 10.5 An effect that corrects the properties of the phonon.

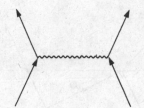

Fig. 10.6 A two-electron process.

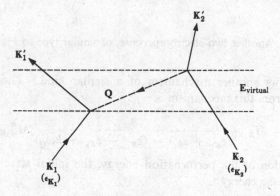

Fig. 10.7 One electron distorts the lattice, the other is affected by the distortion.

We see, finally, that the superconductivity must involve more than only one electron or one phonon. We must consider diagrams such as that in Fig. 10.6. Physically, this diagram can be interpreted as a consequence of one electron distorting the lattice, which in turn affects another electron. Figure 10.7 shows that two electrons K_1 and K_2 interact via a phonon, causing K_1' and K_2' to come out. For the momenta the relations are

$$K_1' - K_1 = Q,$$
$$K_2 - K_2' = Q.$$

The initial energy is $E_{\text{initial}} = \varepsilon_{K_1} + \varepsilon_{K_2}$ in the intermediate region, and the virtual energy is

$$E_{\text{virtual}} = \varepsilon_{K_1'} + \varepsilon_{K_2} + \hbar\omega_Q.$$

The perturbation energy due to the mechanism of Fig. 10.7 is

$$V_1 = M_{K_2 K_2'} \frac{1}{E_{\text{initial}} - E_{\text{virtual}}} M_{K_1' K_1}^*$$

$$= M_{K_2 K_2'} \frac{1}{(\varepsilon_{K_1} + \varepsilon_{K_2}) - (\varepsilon_{K_1'} + \varepsilon_{K_2} + \hbar\omega_Q)} M_{K_1' K_1}^*. \qquad (10.6)$$

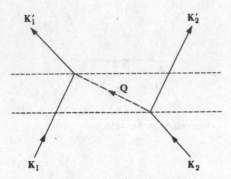

Fig. 10.8 Another two-electron process, of similar type to Fig. 10.7.

Figure 10.8 shows another mechanism of a similar kind. The perturbation energy coming from this mechanism is

$$V_2 = M_{K_1 K_1'} \frac{1}{(\varepsilon_{K_1} + \varepsilon_{K_2}) - (\varepsilon_{K_1} + \varepsilon_{K_2'} + \hbar\omega_Q)} M_{K_2' K_2}^*. \qquad (10.7)$$

For the calculation of the perturbation energy, the initial state and the final state have the same energy:

$$\varepsilon_{K_1} + \varepsilon_{K_2} = \varepsilon_{K_1'} + \varepsilon_{K_2'} \qquad \text{or} \qquad \varepsilon_{K_2} - \varepsilon_{K_2'} = -(\varepsilon_{K_1} - \varepsilon_{K_1'}).$$

The perturbation energy is the sum $V_{K_1' K_2'; K_1 K_2}$. We are mainly concerned with electrons near the Fermi surface, so we can take all ε's to be approximately equal. Then

$$V_{K_1' K_2'; K_1 K_2} \approx -\frac{1}{\hbar\omega_Q}(M_{K_2 K_2'} M_{K_1' K_1}^* + M_{K_1 K_1'} M_{K_2' K_2}^*). \qquad (10.8)$$

If the M's are approximately equal, the perturbation energy is negative; so the electrons near the Fermi surface attract each other.

In metals, the Fermi surface is curved. If we disregard the curvature of the surface, the Hamiltonian can be treated exactly. However, when we do that we find no superconductivity. Therefore the curvature of the Fermi surface is essential for superconductivity.

The Coulomb interaction among electrons gives the interaction energy:

$$V_{\text{coulomb}} = \frac{4\pi e^2}{|K_1 - K_1'|^2 + (\text{const})^2} \qquad (10.9)$$

which is always positive, so the interaction is repulsive. The "(const)^2" is a consequence of shielding. In order to achieve overall attractive interaction between electrons we need

$$V + V_{\text{coulomb}} < 0.$$

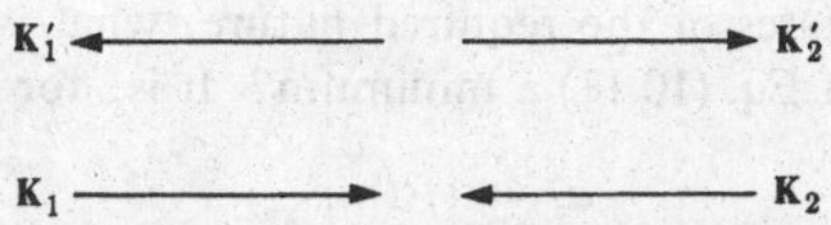

$$K_1' \longleftarrow \qquad \longrightarrow K_2'$$

$$K_1 \longrightarrow \qquad \longleftarrow K_2$$

Fig. 10.9 A case where $|K_1 - K_1'|$ is large.

Looking at Eqs. (10.8) and (10.9) we see that for overall attraction when $\varepsilon_{K_1} \approx \varepsilon_{K_1'}$, larger values of $|K_1 - K_1'|$ are more favorable. For example, look at the case shown in Fig. 10.9.

We can summarize these results by writing a new Hamiltonian for the electrons alone. The phonons have the effect of modifying the ε_K for the electrons and of modifying the interaction between the electrons so that the interaction is, under certain circumstances, attractive. We will be working with the Hamiltonian

$$H = \sum_K \varepsilon_K a_K^+ a_K + \sum_{K_1' K_2'; K_1 K_2} V_{K_1' K_2'; K_1 K_2} a_{K_1'}^+ a_{K_2'}^+ a_{K_1} a_{K_2}. \qquad (10.10)$$

Our problem will be to discover how Eq. (10.10) can lead to a ground state with especially low energy.

10.3 A HELPFUL THEOREM

Consider a Hamiltonian

$$H = H_0 + U.$$

The eigenvalues and the eigenfunctions of H_0 are written as E_i and φ_i.

$$U_{ij} = (\varphi_i, U\varphi_j).$$

If all the E_i's are nearly equal, and also all the U_{ij}'s are nearly equal, then a large amount of lowering of energy can be achieved. This is shown as follows: Try a wave function

$$\psi = \sum_i a_i \varphi_i$$

and evaluate the energy expectation value for this state:

$$\xi = \sum_i E_i |a_i|^2 + \sum_{ij} U_{ij} a_i^* a_j.$$

Following the assumption let us put

$$E_i \approx E_0, \qquad U_{ij} \approx -V.$$

The normalization of the a_i's is

$$\sum_i |a_i|^2 = 1.$$

Then

$$\xi = E_0 - V \sum_{ij} a_i^* a_j. \qquad (10.11)$$

Suppose there are m states of the required nature; what is the best choice of the a_i's that makes ξ of Eq. (10.11) a minimum? It is* for

$$a_i = 1/\sqrt{m},$$

when Eq. (10.11) becomes

$$\xi = E_0 - mV.$$

Thus, when V is positive, or U_{ij} is negative, the stabilization is m-fold intensified.

10.4 GROUND STATE OF A SUPERCONDUCTOR

With the theorem of the previous section as a guide, let us proceed to find a set of such φ_i's for which the U_{ij}'s are nearly equal and negative, and the E_i's are all nearly equal.

In the k-space, a wave function is defined by the configuration of occupancy and vacancy of states. For U_{ij}, consider

$$\langle k_\alpha, k_\beta, \ldots, k'_1, k'_2 | V | k_\alpha, k_\beta, \ldots, k_1, k_2 \rangle,$$

where

$$V = \sum_{k'_1 k'_2; k_1 k_2} a^+_{k'_1} a^+_{k'_2} a_{k_1} a_{k_2}.$$

We can consider this sum to be restricted to, say, $k'_1 > k'_2$ and $k_1 > k_2$. Remember that $a_{k_1} a_{k_2} = -a_{k_2} a_{k_1}$ and $a^+_{k'_1} a^+_{k'_2} = -a^+_{k'_2} a^+_{k'_1}$. The two states considered differ only in k'_1, k'_2 and k_1, k_2. This matrix element is

$$\langle k_\alpha, k_\beta, \ldots, k'_1, k'_2 | V | k_\alpha, k_\beta, \ldots, k_1, k_2 \rangle = \pm V_{k'_1 k'_2; k_1 k_2}. \tag{10.12}$$

The $\pm$ sign comes from the ordering of k's, and the rule is easy to obtain from Fermi statistics. The sign is $(-1)^{N+N'}$ where N = number of states between k_1 and k_2 in the function describing the initial configuration, and N' = number of states between k'_1 and k'_2 in the function describing the final configuration. For example,

$$\langle k_\alpha, k'_1, k_\beta, k'_2 | V | k_1, k_\alpha, k_\beta, k_2 \rangle = (-1)^{1+2} V_{k'_1 k'_2; k_1 k_2}.$$

Bardeen, Cooper and Schrieffer proceed to show that if a particular set of configurations is chosen, we can make $V_{k'_1 k'_2, k_1 k_2}$ in Eq. (10.12) all real and negative, and the sign in front of it always positive. The idea is to take the states in k-space always in pairs. This means that the pair of states k_1 and k_2 are either

---

* To prove this, use the Schwarz inequality:

$$\sum_{ij} a^*_i a_j = \left| \sum_i 1 \cdot a_i \right|^2 \leq \sum_i 1^2 \sum_j |a_j|^2 = m$$

with equality if $(a_1, a_2, \ldots) \propto (1, 1 \cdots)$, that is, if the a_i's are equal (and may be chosen real.)

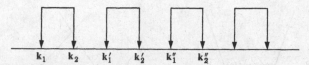

Fig. 10.10 k-states arranged with pairing states side-by-side.

both occupied or both empty, and the configuration in which one of them is occupied and the other empty is not allowed. When we make this requirement on the configurations we can show that the sign in Eq. (10.12) is always positive. Suppose all the k-states are arranged in such an order that the pairing states sit side by side as shown in Fig. 10.10. Using the rule $(-1)^{N+N'}$ and the fact that $N = N' = 0$, we see that the sign is always positive.

Because $V_{k_1'k_2', k_1 k_2}$ should not vanish, we require for every pair

$$k_1' + k_2' = k_1 + k_2 = p = \text{const.}$$

This constant p can be chosen as any vector, but for the ground state we choose

$$p = 0.$$

Thus, the pairs we consider are

$$k_2 = -k_1.$$

How about spins? In computing the potential, we included the diagram in Fig. 10.12a but neglected Fig. 10.12b, plus and minus. The sign is $+$ for states that are symmetric in space, that is, for states that are antisymmetric in spin. To make the potential as large and negative as possible, we take the states with antisymmetric total spin, that is with total spin equal to zero. We choose

$$\sigma_2 = -\sigma_1.$$

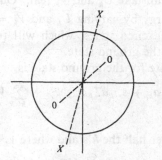

Fig. 10.11 Pairs in the k-space.

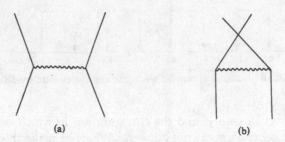

Fig. 10.12 Included and excluded states in the potential calculation.

To simplify the notation, the pair $(k\uparrow; -k\downarrow)$ is denoted by "k" in this section. Consider all the pairs in the k-space and arrange them in a certain order

$$k_1, k_2, k_3, \ldots$$

A pair k_i is either occupied or unoccupied.

Let us define:

$|\varphi_k(0)\rangle$ = the wave function for $k\uparrow$ and $-k\downarrow$ both unoccupied, and $|\varphi_K(1)\rangle$ = the wave function for $k\uparrow$ and $-k\downarrow$ both occupied.

Then a wave function that is a possible candidate for the ground state is

$$|\psi\rangle = \prod_i |\psi_{k_i}\rangle, \tag{10.13}$$

where

$$|\psi_k\rangle = U_k|\varphi_k(1)\rangle + V_k|\varphi_k(0)\rangle$$

and

$$|U_k|^2 + |V_k|^2 = 1.$$

The actual ground state is some linear combination of states of the form given in Eq. (10.13), but we will make the simplifying assumption that with appropriate choice of U_k and V_k, Eq. (10.13) is the ground state. By adjusting the phase of $|\varphi_k(1)\rangle$ and $|\varphi_k(0)\rangle$, we can take U_k and V_k real. Our program will first be to find the ground-state energy by varying U_k and $V_k = \sqrt{1 - U_k^2}$. Then we will find the energy of the excited states, which will turn out to be a finite, macroscopic amount above the ground state.

The energy of a candidate for the ground state is

$$E = \langle\psi|H|\psi\rangle = \sum_k \langle\psi|\varepsilon_k a_k^+ a_k + \varepsilon_{-k} a_{-k}^+ a_{-k}|\psi\rangle + \sum_{k',k} V_{k'k}\langle\psi|a_k^+ a_{-k'}^+ a_k a_{-k}|\psi\rangle \tag{10.14}$$

where all sums are taken over half the k's and where $V_{k'k} = V_{k',-k';k,-k}$. Then

$$E = \sum_k \varepsilon_k\langle\psi_k|a_k^+ a_k + a_{-k}^+ a_{-k}|\psi_k\rangle + \sum_{k',k} V_{k'k}\langle\psi_{k'}|\langle\psi_k|a_k^+ a_{-k'}^+ a_k a_{-k}|\psi_k\rangle|\psi_{k'}\rangle.$$

Let

$$s_k = \langle \psi_k | a_k^+ a_k + a_{-k}^+ a_{-k} | \psi_k \rangle$$

and   (10.15)

$$t_k = \langle \psi_k | a_k^+ a_{-k}^+ | \psi_k \rangle.$$

Then

$$E = \sum_k \varepsilon_k s_k + \sum_{k',k} V_{k'k} t_{k'} t_k^*.$$ (10.16)

If

$$|\psi_k\rangle = U_k |\varphi_k(1)\rangle + V_k |\varphi_k(0)\rangle,$$

then

$$s_k = 2U_k^2, \qquad t_k = U_k V_k = t_k^*.$$

$$E = \sum_k 2\varepsilon_k U_k^2 + \sum_{k'k} V_{k'k} U_{k'} V_{k'} U_k V_k.$$ (10.17)

The U_k's for the unperturbed state are

$$U_k = 1 \qquad \text{for} \quad k < k_{\text{Fermi}}.$$

$$U_k = 0 \qquad \text{for} \quad k > k_{\text{Fermi}}.$$

The best choice of the ground state wave function is obtained by minimizing Eq. (10.17) with respect to the U_k's, V_k being a function of U_k. In the minimization it is not necessary to fix the total number of electrons explicitly. By varying U_k and V_k, we vary the number of electrons. If we choose an appropriate zero of the energy ε_k, we wind up with the correct number density of electrons after minimizing the total energy. We will take the zero of the energy to be at the Fermi surface, so that $\varepsilon_{\text{Fermi}} = 0$.

10.5 GROUND STATE OF SUPERCONDUCTOR (CONTINUED)

Following our program we must next minimize Eq. (10.17) with respect to the U_k's. A set of equations results:

$$4\varepsilon_k U_k + 2\sum_{k'} V_{kk'} U_{k'} V_{k'} \left(V_k - \frac{U_k^2}{V_k} \right) = 0.$$ (10.18)

Because $V_{kk'}$ is real and negative, we can define

$$\Delta_k = -\sum_{k'} V_{kk'} U_{k'} V_{k'} > 0$$ (10.19)

and we write Eq. (10.18) as

$$2\varepsilon_k U_k = \Delta_k \frac{1 - 2U_k^2}{\sqrt{1 - U_k^2}}.$$ (10.20)

Introduce x as

$$U_k^2 = \tfrac{1}{2}(1 + x)$$ (10.21)

so that

$$1 - U_k^2 = \tfrac{1}{2}(1 - x)$$

and   (10.22)

$$1 - 2U_k^2 = -x.$$

Square Eq. (10.20) and use Eqs. (10.21) and (10.22) to solve for x^2:

$$x^2 = \frac{\varepsilon_k^2}{E_k^2},$$ (10.23)

where we define

$$E_k \equiv \sqrt{\varepsilon_k^2 + \Delta_k^2}.$$ (10.24)

Taking the right sign we obtain from Eq. (10.23):

$$-(1 - 2U_k^2) = x = -\frac{\varepsilon_k}{E_k}$$ (10.25)

so that for $\varepsilon_k < 0$ (below Fermi level) $x > 0$ and $U_k > V_k$, as it should be.
Putting this back into Eq. (10.21) we have

$$\left. \begin{aligned} U_k^2 &= \frac{1}{2}\left(1 - \frac{\varepsilon_k}{E_k}\right), \\ V_k^2 &= \frac{1}{2}\left(1 + \frac{\varepsilon_k}{E_k}\right). \end{aligned} \right\}$$ (10.26)

U_k^2 is the probability that the pair k is occupied. The form of U_k^2 in
Eq. (10.26) shows qualitatively that the distribution U_k^2 is a rounded Fermi
distribution. Assuming Δ_k is small, we see

$$U_k^2 \to 1 \text{ deep inside the Fermi sphere } (\varepsilon_k < 0)$$

and   (10.27)

$$U_k^2 \to 0 \text{ far outside the Fermi sphere } (\varepsilon_k > 0).$$

These limits suggest the distribution shown in Fig. 10.13.

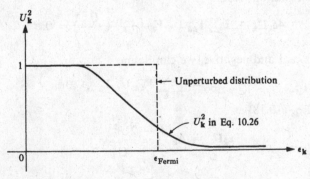

Fig. 10.13 The rounded Fermi distribution.

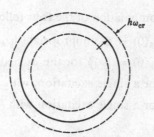

Fig. 10.14 The range of nonzero $V_{kk'}$, as assumed by Bardeen, Cooper, and Schrieffer.

To complete the calculation we have to solve for Δ_k. If we are to use Eq. (10.19), we shall find $U_k V_k$. Because Eq. (10.26) gives

$$U_k V_k = \frac{1}{2} \sqrt{1 - \frac{\varepsilon_k^2}{E_k^2}} = \frac{\Delta_k}{2E_k}, \qquad (10.28)$$

Eq. (10.19) can be written as

$$\Delta_k = - \sum_k V_{kk'} \frac{\Delta_{k'}}{2\sqrt{\varepsilon_{k'}^2 + \Delta_{k'}^2}}. \qquad (10.29)$$

This is the equation to be solved for Δ_k.

In order to obtain some insight, Bardeen, Cooper, and Schrieffer assumed $V_{kk'}$ is a constant for k and k' within the range $\hbar\omega_{cr}$ above and below the Fermi surface, and zero otherwise. See Fig. 10.14. When we make this assumption, Δ_k of Eq. (10.28) becomes a constant independent of k, and the equation for it is

$$1 = -V \sum_k \frac{1}{2\sqrt{\varepsilon_k^2 + \Delta^2}}.$$

Changing $\sum_k$ into an integral over ε, we have

$$1 = \left| \frac{V}{2} \right| \int_{-\hbar\omega_{cr}}^{\hbar\omega_{cr}} \frac{M(\varepsilon)\, d\varepsilon}{\sqrt{\varepsilon^2 + \Delta^2}} \approx |V| M(0) \int_0^{\hbar\omega_{cr}} \frac{d\varepsilon}{\sqrt{\varepsilon^2 + \Delta^2}}.$$

$M(\varepsilon)$ is the density of states for energy ε; it is close to a constant $M(0)$ near the Fermi surface. Solving for Δ, we obtain

$$\Delta = \frac{\hbar\omega_{cr}}{\sinh(1/|V|M(0))}. \qquad (10.30)$$

10.6 EXCITATIONS

Now we are ready to find the excitation energy. To describe excited states we will have to consider the possibility of a given pair of momenta being half-occupied. Define

$$\varphi_k(2) = a_{k\uparrow}^+ \varphi_k(0), \qquad \varphi_k(3) = a_{-k\downarrow}^+ \varphi_k(0). \qquad (10.31)$$

Then $|\psi_n\rangle$ is some linear combination of the four following orthonormal states:

$|\psi_k(0)\rangle = U_k\varphi_k(1) + V_k\varphi_k(0) = $ the appropriate $|\psi_k\rangle$ for the ground state;

$|\psi_k(1)\rangle = V_k\varphi_k(1) - U_k\varphi_k(0) = |\psi_k\rangle$ for the excited state of a pair;

$|\psi_k(2)\rangle = \varphi_k(2) = |\psi_k\rangle$ for a single excitation, with $-k\downarrow$ unoccupied;

$|\psi_k(3)\rangle = \varphi_k(3) = |\psi_k\rangle$ for a single excitation with $k\uparrow$ unoccupied.

$$(10.32)$$

If $\psi_{k_1} = \varphi_{k_1}(2)$ and $\psi_k = U_k\varphi_k(1) + V_k\varphi_k(0)$ for $k \neq k_1$, then the energy that we will call E', is given by Eq. (10.16). Take E_0 as the ground-state energy. Then $s_{k_1} = 1$ and $t_{k_1} = 0$, so

$$E' - E_0 = \varepsilon_{k_1} - 2\varepsilon_{k_1}U_{k_1}^2 - \sum_k V_{kk_1}(U_kV_kU_{k_1}V_{k_1}) - \sum_k V_{k_1k}(U_{k_1}V_{k_1}U_kV_k)$$

$$= \varepsilon_{k_1}(1 - 2U_{k_1}^2) - 2\left(\sum_k V_{k_1k}U_kV_k\right)U_{k_1}V_{k_1}.$$

We have used the fact that $V_{kk_1} = V_{k_1k}^* = V_{k_1k}$. Using Eqs. (10.19), (10.25), and (10.28) we see that

$$E' - E_0 = \varepsilon_{k_1}\left(\frac{\varepsilon_{k_1}}{E_{k_1}}\right) - 2(-\Delta_{k_1})\left(\frac{\Delta_{k_1}}{2E_{k_1}}\right) = \frac{\varepsilon_{k_1}^2 + \Delta_{k_1}^2}{E_{k_1}} = E_{k_1}. \qquad (10.33)$$

If $\psi_{k_1} = \varphi_{k_1}(3)$, $s_{k_1} = 1$ and $t_{k_1} = 0$, so Eq. (10.33) still holds. If $\psi_{k_1} = V_{k_1}\varphi_{k_1}(1) - U_{k_1}\varphi_{k_1}(0)$, the energy becomes E''. It is easy to show that $s_{k_1} = 2V_{k_1}^2$, $t_{k_1} = -U_{k_1}V_{k_1}$, and

$$E'' - E_0 = 2E_{k_1}. \qquad (10.34)$$

Note that $E_{k_1} > \Delta_{k_1}$.

Suppose $\psi_{k_1} = \varphi_{k_1}(2)$ and $\psi_{k_2} = \varphi_{k_2}(2)$. Then, if we are careful not to count $V_{k_1k_2}$ terms too often, we get

$$E - E_0 = E_{k_1} + E_{k_2} + 2V_{k_1k_2}U_{k_1}V_{k_1}U_{k_2}V_{k_2} \approx E_{k_1} + E_{k_2},$$

because $V_{k_1k_2}U_{k_1}V_{k_1}U_{k_2}V_{k_1}$ is an infinitesimal quantity. Thus,

$$E_{\text{two excitations}} - E_0 = E_{k_1} + E_{k_2}. \qquad (10.35)$$

If $\psi_{k_1} = V_{k_1}\varphi_{k_1}(1) - U_{k_1}\varphi_{k_1}(0)$, it is plausible to consider the excitation to be a double excitation, with both k_1 and $-k_1$ excited.

We conclude that there is a gap between the ground state and the excited states, and that excitations consist of the breaking of pairs. The excitation energy is the sum of the E_k for each excited electron in each pair.

10.7 FINITE TEMPERATURES

To find the energy of a system at a finite temperature, we will use:

Expectation value of the energy $= E$

$$= \sum_i (\text{Probability of state } i) \times (\text{energy of state } i).$$

Describe a state by the product $\prod_j |\psi_{k_j}(n_j)\rangle$, or by the sequence $n_1, n_2, \ldots,$ where $n_j = 0, 1, 2,$ or 3.

If $P_k(n)$ is the probability that $|\psi_k\rangle = |\psi_k(n)\rangle$, then the probability of a state is $\prod_j P_{k_j}(n_j)$. Of course, $\sum_n P_k(n) = 1$. Then

$$E = \sum_{n_1, n_2, \ldots} \prod_j P_{k_j}(n_j) \left[\sum_l \varepsilon_{k_l} s_{k_l}(n_l) + \sum_{lm} V_{k_l k_m} t_{k_l}(n_l) t^*_{k_m}(n_m) \right]$$

$$= \sum_l \left(\sum_{\substack{n_1, \ldots \\ \text{not } n_l}} \prod_{j \neq l} P_{k_j}(n_j) \right) \sum_{n_l} P_{k_l}(n_l) \varepsilon_{k_l} s_{k_l}(n_l)$$

$$+ \sum_{l,m} \left(\sum_{\substack{n_1, \ldots \\ \text{not } n_l, n_m}} \prod_{j \neq l, m} P_{k_j}(n_j) \right) \sum_{n_l, n_m} P_{k_l}(n_l) P_{k_m}(n_m) V_{k_l k_m} t_{k_l}(n_l) t^*_{k_m}(n_m).$$

But

$$\sum_{\substack{n_1, \ldots \\ \text{not } n_l}} \prod_{j \neq l} P_{k_j}(n_j) = \prod_{j \neq l} \sum_{n_j} P_{k_j}(n_j) = \prod_{j \neq l} 1 = 1,$$

and similarly,

$$\sum_{\substack{n_1, \ldots \\ \text{not } n_l, n_m}} \prod_{j \neq l, m} P_{k_j}(n_j) = 1.$$

Thus,

$$E = \sum_l \sum_n P_{k_l}(n) \varepsilon_{k_l} s_{k_l}(n) + \sum_{l,m} \sum_{n, n'} P_{k_l}(n) P_{k_m}(n') V_{k_l k_m} t_{k_l}(n) t^*_{k_m}(n'). \tag{10.36}$$

Suppose we call f_k the probability that a k state is excited. Then the probability that a given pair of k states is unexcited is

$$P_k(0) = (1 - f_k)^2. \tag{10.37a}$$

The probability that one of the states (say k) is excited, and the other is not, is

$$P_k(2) = P_k(3) = f_k(1 - f_k). \tag{10.37b}$$

The probability that both states of a pair are excited is

$$P_k(1) = f_k^2. \tag{10.37c}$$

From Eqs. (10.37) and (10.15),

$$S_k(0) = 2U_k^2, \qquad\qquad t_k(0) = U_k V_k,$$
$$S_k(1) = 2V_k^2, \qquad\qquad t_k(1) = -U_k V_k,$$
$$S_k(2) = S_k(3) = 1, \qquad t_k(2) = t_k(3) = 0.$$

Thus, from Eqs. (10.36) and (10.37),

$$E = \sum_k 2\varepsilon_k[U_k^2(1 - 2f_k) + f_k] + \sum_{kk'} V_{kk'} U_{k'} V_{k'} U_k V_k (1 - 2f_{k'})(1 - 2f_k). \tag{10.38}$$

Here we assume that $U_k^2 + V_k^2 = 1$, but we do not assume that U_k is such that E is a minimum for zero temperature. U_k and f_k are functions of temperature.

For a given set of orthonormal states U_k and f_k, define a set of $P_i = $ probability of state i. Then

$$E = \sum_i E_i P_i$$

$$S = -k \sum_i P_i \ln P_i \quad (k = \text{Boltzmann constant})$$

$$F(P_i) = E - TS. \tag{10.39}$$

If the P_i are proportional to $e^{-E_i/kT}$, then $F(P_i) = $ free energy. It is easy to prove that if $F(P_i)$ is minimized with respect to P_i (subject to the condition that $\sum P_i = 1$), then the P_i have the correct values for temperature T.

Using the method of Lagrange multipliers, we find that

$$0 = \frac{\partial F}{\partial P_i} - \lambda \frac{\partial \sum P_i}{\partial P_i} = E_i + kT(\ln P_i + 1) - \lambda.$$

Then $P_i = e^{(\lambda/kT - 1)}e^{-E_i/kT}$.

λ must be chosen so that $\sum P_i = 1$, in which case the P_i are correctly given for equilibrium.

For the case of superconductors at a finite temperature, minimizing F with respect to P_i is equivalent to minimizing with respect to U_k and f_k.

Before we do this, note that we are really considering an F that has an undetermined number of electrons. We then minimize subject to the condition that the expectation value of the number of particles is a given number. So, for example, P_i should be written $P(n_1, n_2, \ldots)$ where $n_1, n_2, \ldots$ are the number of particles in state $1, 2, \ldots$ respectively. In addition to the condition that

$$\sum_{n_1, n_2, \ldots} P(n_1, n_2, \ldots) = 1$$

we have

$$\sum_{n_1, n_2, \ldots} (n_1 + n_2 + n_3 + \cdots)P(n_1, n_2, \ldots) = \text{fixed number}.$$

The method of Lagrange multipliers then gives what amounts to the chemical potential of Chapter 1, Section 6. When working with superconductivity, we do not explicitly write the chemical potential because it is taken care of by an appropriate choice of the zero of the energy.

$$S = \text{entropy} = -k \sum_i \text{probability of state } i \times \ln \left[\text{probabilty of state } i\right]$$

$$= -k \sum_k \sum_n P_k(n) \ln P_k(n).$$

It follows that

$$TS = -2\beta^{-1} \sum_k \left[f_k \ln f_k + (1 - f_k) \ln (1 - f_k)\right]$$

$$0 = \frac{\delta F}{\delta f_k} = \frac{\delta E}{\delta f_k} - \frac{\delta TS}{\delta f_k}$$

$$= 2\varepsilon_k(1 - 2U_k^2) - 4 \sum_{k'} V_{kk'} U_{k'} V_{k'} U_k V_k(1 - 2f_{k'}) + 2\beta^{-1} \ln \frac{f_k}{1 - f_k}.$$

$$(10.40)$$

Set

$$\mathscr{E}_k = \varepsilon_k(1 - 2U_k^2) - 2 \sum_{k'} V_{kk'} U_{k'} V_{k'} U_k V_k(1 - 2f_{k'}) \qquad (10.41)$$

Then

$$f_k = 1/(e^{\beta \mathscr{E}_k} + 1). \qquad (10.42)$$

If we now take

$$0 = \frac{\delta F}{\delta U_k} = \frac{\delta E}{\delta U_k} = \left[4\varepsilon_k U_k + 2 \sum_{k'} V_{kk'} U_{k'} V_{k'} \left(V_k - \frac{U_k^2}{V_k}\right)(1 - 2f_{k'})\right](1 - 2f_k),$$

$$(10.43)$$

we have (in principle) equations that can be solved for both U_k and f_k. Now, define

$$\Delta_k = - \sum_{k'} V_{kk'} U_{k'} V_{k'}(1 - 2f_{k'}). \qquad (10.44)$$

Then Eq. (10.43) becomes identical to Eq. (10.20). All our results from Eq. (10.20) through Eq. (10.28) are true at finite temperatures. Equation (10.41) becomes

$$\mathscr{E}_k = \varepsilon_k(1 - 2U_k^2) + 2U_k V_k \Delta_k = \varepsilon_k \left(\frac{\varepsilon_k}{E_k}\right) + 2 \frac{\Delta_k}{2E_k} \Delta_k = E_k, \qquad (10.45)$$

where we have used Eq. (10.44), then Eqs. (10.25) and (10.28), and finally Eq. (10.24).

The final result is

$$\Delta_k = - \sum_{k'} V_{kk'} \frac{\Delta_{k'}}{2E_{k'}} \tanh \frac{E_{k'}}{2kT},$$

$$E_k^2 = \varepsilon_k^2 + \Delta_k^2. \tag{10.46}$$

To simplify this equation, assume

$$V_{kk'} = -V < 0 \quad \text{for} \quad |\varepsilon_k| < \hbar\omega_{cr}.$$

Then Δ_k becomes independent of k, and Eq. (10.46) reduces to

$$\frac{1}{M(0)V} = \int_0^{\hbar\omega_{cr}} \frac{d\varepsilon}{E} \tanh \frac{E}{2kT}, \tag{10.47}$$

where $M(0)$ is the number of states per unit energy at the Fermi surface. The "2" in the denominator of Eq. (10.46) drops out because $\int_{-\hbar\omega_{cr}}^0 d\varepsilon$ and $\int_0^{\hbar\omega_{cr}} d\varepsilon$ are equal. When $T \to 0$, Eq. (10.47) gives

$$\frac{1}{M(0)V} \approx \sinh^{-1} \frac{\hbar\omega_{cr}}{\Delta} - 2 \int_0^{\hbar\omega_{cr}} \frac{d\varepsilon \exp\left[-\sqrt{\varepsilon^2 + \Delta^2}/kT\right]}{\sqrt{\varepsilon^2 + \Delta^2}}. \tag{10.48}$$

Assuming $M(0)V$ is small, we find

$$\frac{1}{M(0)V} \approx \ln \frac{2\hbar\omega_{cr}}{\Delta(0)}; \qquad \Delta(0) = \Delta \text{ at } T = 0. \tag{10.48'}$$

It is easy to see that as T increases, Δ decreases. There cannot be a value for Δ at all temperatures, because in the limit of high temperatures we get an equation independent of Δ:

$$\frac{1}{M(0)V} \approx \int_0^{\hbar\omega_{cr}} \frac{d\varepsilon}{2kT} = \frac{\hbar\omega_{cr}}{2kT},$$

and this equation may simply not hold.

In Eq. (10.47), Δ appears only as a square (in $E = \sqrt{\varepsilon^2 + \Delta^2}$). The effect of decreasing Δ is to decrease $E(\varepsilon)$. But $E(\varepsilon)$ cannot be lowered below the value it has for $\Delta = 0$. The critical temperature, then, is at $\Delta = 0$, above which there is no solution for Δ. Putting $\Delta = 0$ in Eq. (10.47) we have

$$\frac{1}{M(0)V} = \int_0^{\hbar\omega_{cr}} \frac{d\varepsilon}{\varepsilon} \tanh\left(\frac{\varepsilon}{2kT_c}\right) \approx \ln\left(1.14 \frac{\hbar\omega_{cr}}{kT_c}\right) \tag{10.49}$$

Equating Eqs. (10.48') and (10.49), we get,

$$2\Delta(0) = 3.52 kT_c. \tag{10.50}$$

We write "$2\Delta(0)$" because in experiments in which a superconductor is excited, we create both a hole and an electron. Equation (10.50) is experimentally

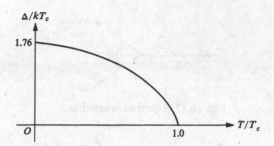

Fig. 10.15 $\Delta(T)$ as a function of T/T_c.

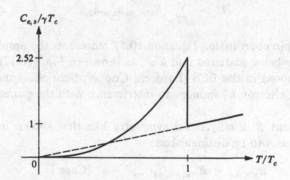

Fig. 10.16 Specific heat of a superconductor, $C_{e,s}$, as a function of T/T_c.

fairly good. Let $T = T_c - \tau$. As τ increases from zero it is clear that if Eq. (10.47) is to remain true, E must change from ε by an amount proportional to τ. But $E = \sqrt{\varepsilon^2 + \Delta^2} \approx \varepsilon + \Delta^2/2\varepsilon$, so Δ^2 is proportional to τ. That is, if $T \approx T_c$,

$$\Delta^2 = A(T_c - T)$$

for some positive constant, A. We conclude that the general shape of $\Delta(T)$ is as in Fig. 10.15.

Another check with experiment is the specific heat whose electronic part is of concern to us. The normal specific heat $C_{e,n}$ is proportional to T:

$$C_{e,n} = \gamma T,$$

γ being a constant. For the superconductor, $C_{e,s}$ is plotted in units of γT_c. It is as shown in Fig. 10.16, which is in agreement with experiment.

10.8 REAL TEST OF EXISTENCE OF PAIR STATES AND ENERGY GAP

Any phenomenon in which scattering of electrons is involved will serve as a test for the existence of the pair states. Attenuation of phonons and paramagnetic relaxation are examples.

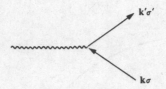

Fig. 10.17 Electron scattering.

In such phenomena the interaction part of the Hamiltonian includes the form

$$H' = \sum_{k\sigma, k'\sigma'} B_{k'\sigma', k\sigma} a^{+}_{k'\sigma'} a_{k\sigma}, \tag{10.51}$$

where σ is the spin coordinate. Equation 10.51 represents the amplitude for an electron with $k\sigma$ being scattered into $k'\sigma'$, as shown in Fig. (10.17). When the pair states proposed in the BCS (Bardeen, Cooper, Schrieffer) theory exist, a scattering of an electron $k\uparrow$ induces an interference with the paired electron at $-k\downarrow$.

The coefficient B usually has a symmetry like that shown in Fig. 10.18. There are two cases to be distinguished:

$$
\begin{aligned}
B_{k'\sigma';k\sigma} &= B_{-k;-\sigma,-k',-\sigma'} \qquad &\text{(Case I),} \\
B_{k'\sigma';k\sigma} &= -B_{-k,-\sigma;-k',-\sigma'} \qquad &\text{(Case II).}
\end{aligned}
\tag{10.52}
$$

In the ordinary scattering case we write a table:

	$k\uparrow$	$k'\uparrow$
Initial	*occupied*	empty
final	empty	*occupied*

That is, we consider the scattering of an electron from $k\uparrow$ to $k'\uparrow$.

For the BCS case, the table should include $k\uparrow$ and $-k\downarrow$ together. When we speak of the scattering of a $k\uparrow$ electron into $k'\uparrow$, we mean that we take the initial wave function with $\psi_k = \varphi_k(2)$ and $\psi_{k'} = U_{k'}\varphi_{k'}(1) + V_{k'}\varphi_{k'}(0)$. Then final

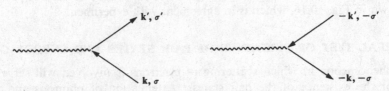

Fig. 10.18 Symmetry of the coefficient B.

wave function is $\psi_{k'} = \varphi_{k'}(2)$ and $\psi_k = U_k\varphi_k(1) + V_k\varphi_k(0)$. Then the matrix element for the transition is

$$\langle\psi_2|H'|\psi_1\rangle = \langle\varphi_{k'}(2)|\langle U_k\varphi_k(1) + V_k\varphi_k(0)|(B_{k'\uparrow k\uparrow}a^+_{k'\uparrow}a_{k\uparrow}$$
$$+ B_{-k\downarrow -k'\downarrow}a^+_{-k\downarrow}a_{-k'\downarrow})$$
$$\times |\varphi_k(2)\rangle|U_{k'}\varphi_{k'}(1) + V_{k'}\varphi_{k'}(0)\rangle$$
$$= B_{k'\uparrow k\uparrow}V_kV_{k'} + B_{-k\downarrow -k'\downarrow}U_kU_{k'}$$
$$= B_{k'\uparrow k\uparrow}(V_kV_{k'} \pm U_kU_{k'}). \tag{10.53}$$

The $\pm$ sign gives the two alternative cases (I) and (II) in Eq. (10.52). Changing the sign from plus to minus causes a big difference in the scattering. Qualitatively we can understand the change when we consider the case $k' \approx k$. Then we see

$$V_k^2 + U_k^2 = 1; \qquad V_k^2 - U_k^2 = 1 - 2U_k^2 = \varepsilon_k/E_k = \varepsilon_k/\sqrt{\varepsilon_k^2 + \Delta_k^2}. \tag{10.54}$$

When the sign is positive, the scattering is enhanced, whereas when it is negative the effect is reduced.

Phenomena such as attentuation of sound, thermal conductivity, and paramagnetic loss have been examined and the theory can explain the experiments well.

Let us now discuss some gap experiments. Suppose we form a junction by placing a superconductor with a thin insulating layer (such as an oxide of the superconducting material) against another metal, which can be either a normal metal or a superconductor. If a voltage is placed across the junction, the amount of current will be affected by the presence of the gap. Semiquantitatively, we can set up the problem as follows:

Label the metals 1 and 2. Electrons in a state described by "k" in metal 1 will have an amplitude to go to state "q" of metal 2. Let that amplitude be a constant, M, in the energy range that interests us (near the Fermi levels). The probability of transition from state k of metal 1 to some state of metal 2 is

$$\frac{2\pi}{\hbar} \sum_q |M|^2 \text{ (probability of state } k \text{ being occupied)}$$

$$\times (1 - \text{probability of state } q \text{ being occupied)} \times \delta(E_k - E_q),$$

where E_k and E_q are the energies of the respective states. The probability of state k being occupied is

$$f(E_k) = \frac{1}{\exp[E_k/kT] + 1},$$

and similarly for state q. If voltage $-V$ is applied to metal 2, the energy of state q becomes $E'_q = E_q - V$, and the probability of q being occupied is the same as before, namely $f(E_q) = f(E'_q + V)$.

The current from metal 1 to metal 2 is

$$\frac{2\pi e}{\hbar} \sum_{kq} |M|^2 f(E_k)[1 - f(E'_q + V)]\, \delta(E_k - E'_q).$$

If

I = the total current = current from 1 to 2 − current from 2 to 1,

we find that

$$I = \frac{2\pi e}{\hbar} \sum_{k,q} |M|^2 \{f(E_k)[1 - f(E'_q + V)] - f(E'_q + V)[1 - f(E_k)]\}\, \delta(E_k - E'_q)$$

$$= \frac{2\pi e}{\hbar} |M|^2 \sum_{k,q} [f(E_k) - f(E'_q + V)]\, \delta(E_k - E'_q).$$

Now $\sum_{k,q}$ can be replaced by $\int dE_k\, dE'_q \rho_1(E_k)\rho_2(E_q)$, where $\rho_1(E_k)$ and $\rho_2(E_q)$ are the density of state functions in the respective metals. Then

$$I(T) = \frac{2\pi e}{\hbar} |M|^2 \int dE[f(E) - f(E + V)]\rho_1(E)\rho_2(E + V).$$

At absolute zero,

$$I(0) = \frac{2\pi e}{\hbar} |M|^2 \int_{-V}^{0} dE \rho_1(E)\rho_2(E + V). \tag{10.55}$$

Thus the current depends on the density of states near the Fermi level, E_F, for the two metals. For normal metals we can consider $\rho(E)$ to be roughly constant, but for superconductors the interactions between electrons cause the spectrum to be distorted. The state that would have had energy ε_k without the interaction actually has energy $|E_k| = \sqrt{\varepsilon_k^2 + \Delta_k^2}$. For simplicity, take $\Delta_k = \Delta$ independent of k. Thus there is a gap around $E_k = 0$. While $dN/d\varepsilon$ (which is the density of states if there were no interactions) is roughly constant,

$$\rho(E_k) = \left|\frac{dN}{dE_k}\right| = \frac{dN}{d\varepsilon_k} \frac{1}{|dE_k/d\varepsilon_k|} = \frac{dN}{d\varepsilon} \frac{|E_k|}{\sqrt{E^2 - \Delta^2}}$$

rises to infinity as the gap is approached, then drops to zero within the gap.

If metals 1 and 2 are normal, $\rho_1(E)$ and $\rho_2(E + V)$ are constants, and I is proportional to V. But if metal 1 is a superconductor, $\rho_1(E)$ is zero for all E in the integral of Eq. (10.55), unless $V > \Delta$. Thus for $V < \Delta$ there is no current. For $V > \Delta$, it is easy to show with our simplifications that $I \propto \sqrt{V^2 - \Delta_1^2}$, and looks like the function in Fig. 10.19. At finite temperatures, the curve looks like the function in Fig. 10.20.

This situation, with metal 1 a superconductor and metal 2 normal, can be visualized as shown in Fig. 10.21. At $V = 0$, E_{F_1} is the same as E_{F_2}. When V lowers E_{F_2} more than Δ_1 below E_{F_1}, current can flow from 1 to 2.

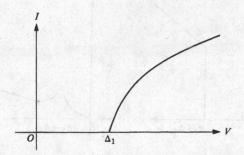

Fig. 10.19 Current across a junction with one superconducting metal at zero temperature.

Those are both metals 1 and 3 are superconductors. The current as a function of voltage rises then curves to Fig. 10.20. From the second curve in Fig. 10.22, both Δ_1 and Δ_3 can be obtained. For more details concerning the energy gap, see *Progress in Low Temperature Physics*, Vol. IV, and 1967 and no especially page 76–90.

10.9 SUPERCONDUCTION WITH CURRENT

In Section 10.4 we assumed that the electrons form pairs that have total momentum zero; that is $\mathbf{k}_1 + \mathbf{k}_2 = 0$. The more general situation is $\mathbf{k}_1 + \mathbf{k}_2 = \mathbf{Q} \neq 0$, a current, in which case there exists a current. As before, we can describe the pair by total momentum vector $\hbar \mathbf{Q}$. But in this case,

$$\mathbf{k}_1 + \mathbf{k}_2 = \mathbf{Q}$$

Formerly the energy of a pair was $\varepsilon = \varepsilon_{-k} + \varepsilon_k$, but

$$W = \frac{\hbar^2 (\frac{1}{2}Q - k)^2}{2m} + \frac{\hbar^2 (\frac{1}{2}Q + k)^2}{2m}$$

is analogue to Fig. 10.9 would appear as defined in Section. This is the difference, however, that the values at temperatures are the new case; the differ ent, or order Ω to become, and to form they were for some superconducting entity, from atomic objects. Doppler the magnetic field by the. 10.31 Xexult is increased by ΩV away.

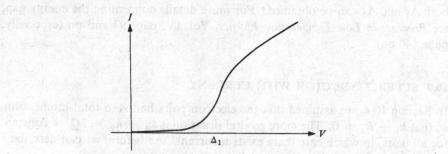

Fig. 10.20 Current in the same junction at finite temperatures.

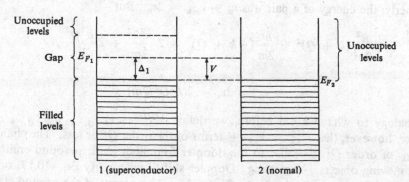

Fig. 10.21 Levels in a junction of two metals, separated by an insulating layer. Metal 1 is a superconductor and metal 2 is normal.

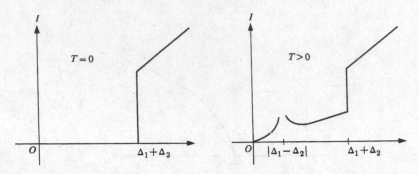

Fig. 10.22 Current in a junction of the superconductors, at zero and finite temperatures.

Suppose both metals 1 and 2 are superconductors. The current as a function of voltage looks like the curves in Fig. 10.22. From the second curve in Fig. 10.22, both Δ_1 and Δ_2 can be obtained. For more details concerning the energy gap, see *Progress in Low Temperature Physics*, Vol. IV, page 97 and on (especially, page 140 on).

10.9 SUPERCONDUCTOR WITH CURRENT

In Section 10.4, we assumed that the electron pairs had zero total momentum, so that $k_1 + k_2 = 0$. The more general situation is $k_1 + k_2 = 2Q =$ constant for all pairs, in which case there exists a current. As before, we can describe a pair by a single momentum vector, k, but in this case,

$$k_1 = k + Q,$$
$$k_2 = -k + Q.$$

Formerly, the energy of a pair was $\varepsilon_k + \varepsilon_{-k} = 2\varepsilon_k$. But

$$\frac{\hbar^2}{2m}(k + Q)^2 + \frac{\hbar^2}{2m}(-k + Q)^2 = 2\frac{\hbar^2 k^2}{2m} + 2\frac{\hbar^2 Q^2}{2m},$$

so

$$\varepsilon_{k_1} + \varepsilon_{k_2} = 2\varepsilon_k + 2(\hbar^2 Q^2/2m).$$

In analogy to what we did before, we define $V_{kk'}^{(Q)}$ as $V_{k+Q,-k+Q;k'+Q,-k'+Q}$. Notice, however, that $V_{kk'}^{(Q)} \approx V_{kk'} +$ terms of the order Q^2 or less. The change in $V_{kk'}$ of order Q^2 is similar to the doppler frequency shift in sound emitted from moving objects. Neglecting "Doppler shift'" we modify Eq. (10.17) only by the replacement of $2\varepsilon_k$ by $2\varepsilon_k + 2\hbar^2 Q^2/2m$. The energy of the ground state is increased by $(\hbar^2 Q^2/2m) \sum_k 2U_k^2$.

The number density of electrons is $\mathcal{N} = \int 2U_k^2$, and we can define $v = \hbar Q/m$. Then

$$\text{excess energy} = \mathcal{N}(mv^2/2). \qquad (10.56)$$

The current is

$$J = \left\langle \psi \left| \frac{-e}{m} p \right| \psi \right\rangle,$$

where p is the momentum operator. It follows that

$$J = \frac{-e\hbar}{m} \sum_k [(k + Q) + (-k + Q)]U_k^2 = -e \frac{\hbar Q}{m} \sum_k 2U_k^2 = -\mathcal{N}ev. \qquad (10.57)$$

How is the wave function changed by replacing pairs $\pm k$ with pairs $\pm k + Q$? The equations for U_k can be brought up to date by replacing ε_k with $\varepsilon_k + \hbar^2 Q^2/2m$. As well as $\hbar^2 Q^2/2m$, we must add another constant to make the zero of the energy appropriate for the correct number density of electrons. The net result is that the zero of energy is again at the Fermi surface, and the U_k come out the same as before. But now the U_k describe the amplitude for $\pm k + Q$ to be occupied.

Suppose we describe the ground-state wave function by $\psi_0(x_1, x_2, \dots)$. We want to modify this wave function to become $\psi_Q(x_1, x_2, \dots)$. This modification is effected by replacing each component of momentum p by a component of the same amplitude, but of momentum $p + \hbar Q$. That is, $e^{ip_e \cdot x_e/\hbar}$ is multiplied by $e^{iQ \cdot x_e}$. Then

$$\psi_Q = \exp\left(iQ \cdot \sum_e x_e\right)\psi_0 = \exp\left[i\frac{mv}{\hbar} \cdot \sum_e x_e\right]\psi_0. \qquad (10.58)$$

Equations (10.56) and (10.58), it should be noted, represent simply the consequences of a Galilean transformation of Schrödinger's equation. According to our approximations, a superconductor with a current is the same as a super-conductor in its ground state as seen by a moving observer.

In a real metal the current can vary from place to place. Because k is large near the Fermi surface, the lowest-energy wave functions are not affected much in energy if the variation in current is slow.

To see how to deal with variation of current with respect to position, assume a metal is divided into regions. Each electron is constrained to move within a given region, and for each region there is a different current. Then the wave function becomes

$$\psi = \exp\left(i\frac{m}{\hbar} v_1 \cdot \sum_{\substack{\text{electrons} \\ \text{in region 1}}} x_e\right) \exp\left(i\frac{m}{\hbar} v_2 \cdot \sum_{\text{region 2}} x_e\right) \cdots \psi_0$$

$$= \exp\left(\frac{im}{h} \sum_e v(x_e) \cdot x_e\right) \psi_0. \qquad (10.59)$$

We can now let $v(x_e)$ vary gradually with x_e. But this is not quite right. To see why this modification of ψ is inaccurate, remember that the equation for current density involves derivatives. The fact that $v(x_e)$ is not constant modifies the current.

To improve our description of a current in a superconductor, let us next try

$$\psi = \exp\left(i \sum_e \theta(x_e)\right)\psi_0.$$

The number density of electrons is given by

$$\mathcal{N}(R) = \sum_e \int \psi^*\psi \, \delta(x_e - R) \, dx_1 \, dx_2 \cdots,$$

and the current density is given by

$$j(R) = \sum_e \frac{-\hbar e}{2im} \int [\psi^* \nabla_e \psi - (\nabla_e \psi)^* \psi] \, \delta(x_e - R) \, dx_1 \, dx_2 \cdots.$$

If we remember that the current for $\psi = \psi_0$ is zero, we find

$$j(R) = -\mathcal{N}e \frac{\hbar}{m} \nabla\theta. \tag{10.60}$$

For example, if $\theta = mv \cdot x/\hbar$, $j(R)$ is the same as that for Eq. (10.57).

Note that we have only described currents for which $\nabla \times j = 0$. For the steady state, $\nabla \cdot j = 0$.

In a ring, the above conditions can be satisfied with nonzero current. In this case, $\oint j \cdot ds = -\mathcal{N}e(\hbar/m) \Delta\theta$, where $(\Delta\theta)$ is the difference in θ as we go around the ring. $\Delta\theta$ need not be zero, but it must be a multiple of 2π, for we require that the wave function be single valued. If we define $v = -j/\mathcal{N}e = (\hbar/m)\nabla\theta$, then

$$\oint v \cdot ds = 2\pi n \frac{\hbar}{m}, \tag{10.61}$$

where n is an integer. If the ring's central hole shrinks to a point, the only way n can be kept greater than zero is to have lines along which the superconductor breaks down.

By calling $v = -j/\mathcal{N}e$, we have implied that this quantity can be thought of as a velocity. This interpretation is confirmed by computation of the expected value of the energy.

$$E = \int \psi^* H \psi = \int \psi^* \left[\sum_e \frac{-\hbar^2}{2m} \nabla_e^2 + V \right] \psi$$

$$= \frac{\hbar^2}{2m} \sum_e \int \nabla_e \psi^* \cdot \nabla_e \psi + \int \psi^* V \psi$$

$$= \frac{\hbar^2}{2m} \sum_e \int d^3R \int (\nabla_e \psi^* \cdot \nabla_e \psi) \, \delta(x_e - R) + \int \psi^* V \psi$$

$$\psi^* V \psi = \psi_0^* V \psi_0. \tag{10.62}$$

Also,

$$\nabla \psi^* \cdot \nabla \psi = \nabla \psi_0^* \cdot \nabla \psi_0 + i \nabla \theta \cdot [\psi_0 \nabla \psi_0^* - \psi_0^* \nabla \psi_0] + (\nabla \theta)^2 \psi_0^* \psi_0.$$

Equation (10.62) becomes (if E_0 is the energy for $\psi = \psi_0$):

$$E = E_0 + \frac{\hbar^2}{2m} \int d^3R \, i \nabla \theta(R) \cdot \int \sum_e [\psi_0 \nabla_e \psi_0^* - \psi_0^* \nabla_e \psi_0] \, \delta(x_e - R)$$

$$+ \frac{\hbar^2}{2m} \int d^3R |\nabla \theta(R)|^2 \sum_e \psi_0^* \psi_0 \delta(x_e - R)$$

$$= E_0 + \frac{\hbar}{e} \int d^3R \nabla \theta(R) \cdot j_0(R) + \frac{\hbar^2}{2m} \int d^3R |\nabla \theta(R)|^2 \, \mathcal{N}(R). \tag{10.63}$$

But $j_0(R)$ = current for ground state = 0. Then

$$E = E_0 + \int d^3R (\tfrac{1}{2} m v^2) \, \mathcal{N}(R), \tag{10.64}$$

as we would expect from the interpretation of $v(R)$ as a velocity.

10.10 CURRENT VERSUS FIELD

In the equation $j(r) = -\Lambda A'(r)$, A' is averaged over some region about the point r. In other words, we must use a sort of "spread-out" $A(r)$. We can interpret this spreading as an effect of the size of the electron pairs, and we can calculate the spreading by means of the BCS theory.

Empirically we have

$$A'(r') = C \int \frac{[A(r)e^{-|r-r'|/\xi_0} \cdot (r - r')](r - r')}{|r - r'|^4} \, d^3r, \tag{10.65}$$

where C is an appropriate normalization factor (so that constant A gives $A' = A$). Note that Eq. (10.65) has the form of a convolution. If we write

$$j(q) = \int j(r)e^{-iq \cdot r}\, d^3r, \qquad A(q) = \int A(r)e^{-iq \cdot r}\, d^3r,$$

$$\tag{10.66}$$

$$K(q) = \int K(r)e^{-iq \cdot r}\, d^3r, \qquad K(r) = \frac{Ce^{-r/\xi_0}rr}{r^4},$$

then

$$j(q) = A(q) \cdot K(q). \tag{10.67}$$

In what follows we will find the component of $j(q)$ that is in the direction of $A(q)$.

Suppose that $A(r)$ is in the z direction and has magnitude $A_0 e^{-iq \cdot x}$, where q is perpendicular to A (we are using a gauge in which $\nabla \cdot A = 0$). Then

$$A_z(q') = \int A_z(x)e^{-iq' \cdot x}\, d^3x = A_0\, \delta_{q,q'},$$

where everything is normalized within a unit volume.

In Section 10.1, we showed that if the wave function is "rigid," $j_D = -\Lambda A$. In so far as the wave function is not rigid, we get a contribution from the paramagnetic current of

$$j = \frac{-\hbar e}{2im} \left[\psi^* \nabla \psi - (\nabla \psi)^* \psi \right] \cdot j_{\text{total}} = j + j_D.$$

Set $\hbar = 1$. By considering ψ as an operator (as in second quantization) we could get the operator for $j(r)$. But we will use a slightly different approach.

For a single particle, the operator for $j_1(r)$ is

$$j_1(r) = \frac{-e}{2m} \left[P_1\, \delta(r - r_1) + \delta(r - r_1)P_1 \right],$$

where r_1 is the operator for position and r is a vector (a set of three numbers). Then

$$j_1(q) = \int \langle j_1(r) \rangle e^{-iq \cdot r} = \int \langle \psi | j_1(r) | \psi \rangle e^{-iq \cdot r}\, d^3r$$

$$= \frac{-e}{2m} \int \int d^3r_1\, d^3r \psi^*(r_1) \left[\frac{1}{i} \nabla_1\, \delta(r - r_1) + \delta(r - r_1) \frac{1}{i} \nabla_1 \right] \psi(r_1)e^{-iq \cdot r}.$$

After some manipulation we find

$$j_1(q) = \frac{-e}{2m} \langle \psi | P_1 e^{-iq \cdot r} + e^{-iq \cdot r_1} P_1 | \psi \rangle.$$

Summing over all particles, we get

$$j(q) = \sum_l j_l(q) = \frac{-e}{2m} \langle \psi | \sum_l [P_l e^{-iq \cdot r_l} + e^{-iq \cdot r_l} P_l] | \psi \rangle. \qquad (10.68a)$$

By, say, $\sum_l e^{-iq \cdot r_l} P_l$, we are representing the operator, acting on all the particles, that on one particle is $e^{-iq \cdot r} P$ (r and P are operators), Using

$$e^{-iq \cdot r} = \int d^3x e^{-iq \cdot x} |x\rangle\langle x|$$

and

$$P = \int \frac{d^3P}{(2\pi)^3} P |P\rangle\langle P|$$

it is easy to show that

$$e^{-iq \cdot r} P = \sum_k k |k - q\rangle\langle k|.$$

In Chapter 6, Section 8, we learned how to find the corresponding operator on arbitrary numbers of particles. The result is

$$\sum_l e^{-iq \cdot r_l} P_l = \sum_k k a^+(k - q) a(k).$$

Similarly,

$$\sum_l P_l e^{-iq \cdot r_l} = \sum_k (k - q) a^+(k - q) a(k).$$

Equation (10.68a) becomes

$$j(q) = \frac{-e}{2m} \langle \psi | \sum_k (2k - q) a_{k-q}^+ a_k | \psi \rangle. \qquad (10.68b)$$

Because q is perpendicular to the z axis (i.e., to A),

$$j_z(q) = \frac{-e}{m} \langle \psi | \sum_k k_z a_{k-q}^+ a_k | \psi \rangle. \qquad (10.69)$$

The wave function $|\psi\rangle$ can be easily found by means of perturbation theory.
The Hamiltonian for the superconductor with no field has a term

$$\sum_l \frac{P_l^2}{2m}.$$

This is modified to become

$$\frac{1}{2m} \sum_l \left(P_l + \frac{eA(x_l)}{c} \right)^2.$$

Using the fact that $\nabla \cdot A = 0$ and neglecting terms of second order in A, we see that the Hamiltonian is modified by

$$H' = \frac{e}{mc} \sum_l A(x_l) \cdot P_l = \frac{e}{mc} \sum_l A_0 e^{iq \cdot x_l} P_{l_z} \to \frac{eA_0}{mc} \sum_k k_z a_{k+q}^+ a_k. \qquad (10.70)$$

The wave function is

$$|\psi\rangle = |\psi_0\rangle + \sum_n \lambda_n |\psi_n\rangle$$

where

$$\lambda_n = \frac{\langle n|H'|0\rangle}{E_0 - E_n} = \frac{-eA_0}{mc} \frac{\langle n| \sum_k k_z a_{k+q}^+ a_k|0\rangle}{E_n - E_0}. \qquad (10.71)$$

Here we are using $|n\rangle$ and $|\psi_n\rangle$ interchangeably. $|\psi_0\rangle$ is the ground-state wave function, described by Sections 10.4 and 10.5. Putting together Eqs. (10.71) and (10.69), we obtain

$$j_z(q) \approx \frac{-e}{m} \left[\langle 0| \sum_k k_z a_{k-q}^+ a_k|0\rangle + \sum_n \lambda_n^* \langle n| \sum_k k_z a_{k-q}^+ a_k|0\rangle \right.$$
$$\left. + \sum_n \lambda_n \langle 0| \sum_k k_z a_{k-q}^+ a_k|n\rangle \right].$$

We will calculate the third terms of the above result, and it should then be easy to see that the first two give zero.

$$\langle 0| \sum_n k_z a_{k-q}^+ a_k|n\rangle = \langle n| \left(\sum_n k_z a_{k-q}^+ a_k \right)^+ |0\rangle^*.$$

From

$$\left(\sum_k k_z a_{k-q}^+ a_k \right)^+ = \sum_k k_z a_k^+ a_{k-q} = \sum_k k_z a_{k+q}^+ a_k \qquad \text{(because } k_z + q_z = k_z)$$

and setting $k + q = k'$, we find

$$j_z(q) = \frac{e^2 A_0}{m^2 c} \sum_n \frac{|\langle n| \sum_k k_z a_k^+ a_k|0\rangle|^2}{E_n - E_0}. \qquad (10.72)$$

The ground-state wave function is a product of terms such as

$$\psi_k(0) = U_k|k\uparrow, -k\downarrow\rangle + V_k|0, 0\rangle.$$

$k_z a^+_{k'\uparrow} a_{k\uparrow}$ operating on $\psi_k(0)\psi_{k'}(0)$ gives $k_z U_k V_{k'} |0, -k\downarrow\rangle |k'\uparrow, 0\rangle$. In other words, $k_z a^+_{k'\uparrow} a_{k\uparrow}$ takes the state

$$
\begin{array}{cc}
k\uparrow, \quad -k\downarrow & k'\uparrow, \quad -k'\downarrow \\
\text{occupied} & \text{empty}
\end{array}
$$

into the state

$$
\begin{array}{ccc}
(-k)\downarrow, \quad (k'\uparrow) & k\uparrow, \quad -k'\downarrow \\
\text{occupied} & \text{empty}
\end{array}
$$

with amplitude $k_z U_k V_{k'}$. We can get a similar final state by using the operator $-k_z a^+_{-k\uparrow} a_{-k'\downarrow}$. This operator takes

$$
\begin{array}{cc}
k\uparrow, \quad -k\downarrow & k'\uparrow, \quad -k'\downarrow \\
\text{empty} & \text{occupied}
\end{array}
$$

into the same state as that found by using $k_z a^+_{k'\uparrow} a_{k\uparrow}$, but with amplitude $-k_z U_{k'} V_k$. Then

$$
j_z(q) = \frac{e^2 A_0}{m^2 c} \sum_k \frac{|\langle k \text{ and } k' \text{ excited}|(k_z a^+_{k'\uparrow} a_{k\uparrow} - k_z a^+_{-k\downarrow} a_{-k'\downarrow})|0\rangle|^2}{E_k + E_{k'}}
$$

$$
= \frac{e^2 A_0}{m^2 c} \sum_k \frac{k_z^2 (U_k V_{k'} - V_k U_{k'})^2}{E_k + E_{k'}}. \tag{10.73}
$$

The algebra in the above derivation is not rigorous. For example, the signs have not been unequivocally proved. The reader should do the algebra himself to check such things as the correctness of the signs. It is important to be consistent; decide, for example, whether $|k\uparrow, -k\downarrow\rangle$ is $a^+_{k\uparrow} a^+_{-k\downarrow} |0, 0\rangle$ or $a^+_{-k\downarrow} a^+_{k\uparrow} |0, 0\rangle$. The following steps are straightforward:

$$
(U_k V_{k'} - V_k U_{k'})^2 = U_k^2 V_{k'}^2 + V_k^2 U_{k'}^2 - 2U_k V_k U_{k'} V_{k'}.
$$

From Eqs. (10.26) and (10.28), the right-hand side becomes

$$
\frac{1}{2}\left(1 - \frac{\Delta_k \Delta_{k'} \varepsilon_k \varepsilon_{k'}}{E_k E_{k'}}\right),
$$

and Eq. (10.73) becomes (because $j_z(q) = K_{zz}(q) A_z(q)$)

$$
K_{zz}(q) = \frac{e^2}{2m^2 c} \sum_k \frac{k_z^2}{E_k + E_{k'}}\left(1 - \frac{\Delta_k \Delta_{k'} \varepsilon_k \varepsilon_{k'}}{E_k E_{k'}}\right)
$$

with $E_K = \sqrt{\varepsilon_k^2 + \Delta_k^2}$ (Eq. (10.24)).

Bardeen, Cooper, and Schrieffer find what amounts to $K(q)$ for arbitrary components and nonzero temperature. They also show that at zero temperature, the result approximates the empirical formula Eq. (10.65), with $\zeta_0 = 0.15$ to 0.27 in place of the factor 0.18. Considering the approximations we (and Bardeen, Cooper, and Schrieffer) made, the theoretical results are very good.

10.11 CURRENT AT A FINITE TEMPERATURE

To compute the energy of a superconductor with a current and at a nonzero temperature, we must modify Eqs. (10.36) and (10.37) as shown in Table 10.1.

Table 10.1

What was formerly	Becomes
$\varepsilon_k s_k(0) = 2\varepsilon_k U_k^2$	$(\varepsilon_{k+Q} + \varepsilon_{-k+Q})U_k^2$
$\varepsilon_k s_k(1) = 2\varepsilon_k V_k^2$	$(\varepsilon_{k+Q} + \varepsilon_{-k+Q})V_k^2$
$\varepsilon_k s_k(2) = \varepsilon_k$	ε_{k+Q}
$\varepsilon_k s_k(3) = \varepsilon_k$	ε_{-k+Q}
$P_k(0) = (1 - f_k)(1 - f_{-k}) = (1 - f_k)^2$	$(1 - f_{k+Q})(1 - f_{-k+Q})$ = Probability that neither $(k + Q)$ nor $(-k + Q)$ are excited.
$P_k(1) = f_k^2$	$f_{k+Q}f_{-k+Q}$
$P_k(2) = f_k(1 - f_k)$	$f_{k+Q}(1 - f_{-k+Q})$ = Probability that $(k + Q)$ is occupied (excited) and $(-k + Q)$ is not.
$P_k(3) = f_k(1 - f_k)$	$f_{-k+Q}(1 - f_{k+Q})$

The energy is

$$E = \sum_k (1 - f_{k+Q})(1 - f_{-k+Q})(\varepsilon_{k+Q} + \varepsilon_{-k+Q})U_k^2$$

$$+ \sum_k f_{k+Q}f_{-k+Q}(\varepsilon_{k+Q} + \varepsilon_{-k+Q})V_k^2$$

$$+ \sum_k f_{k+Q}(1 - f_{-k+Q})\varepsilon_{k+Q} + \sum_k f_{-k+Q}(1 - f_{k+Q})\varepsilon_{-k+Q}$$

$$+ \sum_{k,k'} V_{kk'}^{(Q)} U_k V_k U_{k'} V_{k'}[(1 - f_{k+Q})(1 - f_{-k+Q}) - f_{k+Q}f_{-k+Q}]$$

$$\times [(1 - f_{k'+Q})(1 - f_{-k'+Q}) - f_{k'+Q}f_{-k'+Q}]$$

$$= \sum_k [(\varepsilon_{k+Q} + \varepsilon_{-k+Q})U_k^2(1 - 2f_{k+Q}) + 2\varepsilon_{k+Q}f_{k+Q}]$$

$$+ \sum_{k,k'} V_{kk'}^{(Q)} U_k V_k U_{k'} V_{k'}(1 - 2f_{k+Q})(1 - 2f_{k'+Q}). \qquad (10.74)$$

Here we have used $U_k = U_{-k}$, $V_k = V_{-k}$ and $V_{kk'} = V_{-kk'} = V_{k-k'}$.

So far, we know neither what f_k is nor what U_k is. As before, we must minimize $F = E - TS$, where

$$
\begin{aligned}
TS &= -\beta^{-1} \sum_k \sum_k P_k(n) \log P_k(n) \\
&= -\beta^{-1} \sum_k \big[f_{k+Q} \log f_{k+Q} + f_{-k+Q} \log f_{-k+Q} \\
&\quad + (1 - f_{k+Q}) \log (1 - f_{k+Q}) + (1 - f_{-k+Q}) \log (1 - f_{-k+Q}) \big] \\
&= -2\beta^{-1} \sum_k \big[f_{k+Q} \log f_{k+Q} + (1 - f_{k+Q}) \log (1 - f_{k+Q}) \big]. \quad (10.75)
\end{aligned}
$$

If we vary with respect to U_k and let

$$
\Delta_k = -\sum_{k'} V_{kk'}^{(Q)} U_{k'} V_{k'} (1 - 2f_{k'+Q}), \quad (10.76)
$$

then we get an equation similar to Eq. (10.20). In fact, if we neglect errors of order Q^2, we get Eq. (10.20) exactly, and everything from Eq. (10.20) to Eq. (10.28) is again true. More generally, we can define $\varepsilon(k) = \frac{1}{2}(\varepsilon_{k+Q} + \varepsilon_{-k+Q})$, and Eqs. (10.20) to (10.28) hold with $\varepsilon(k)$ in place of ε_k.

Next, vary with respect to f_{k+Q}. The result is

$$
\begin{aligned}
0 &= -2\varepsilon(k)U_k^2 + \varepsilon_{k+Q} + 2\Delta_x U_x V_x + \beta^{-1} \log \frac{f_{k+Q}}{1 - f_{k+Q}} \\
&= \varepsilon_{k+Q} - \varepsilon(k) + E_k + \beta^{-1} \log \frac{f_{k+Q}}{1 - f_{k+Q}} \\
&= \frac{\hbar^2 k \cdot Q}{m} + E_k + \beta^{-1} \log \frac{f_{k+Q}}{1 - f_{k+Q}}, \quad (10.77)
\end{aligned}
$$

from which we find

$$
f_{k+Q} = \frac{1}{\exp\{[E_k + \hbar^2(k \cdot Q/m)]\beta\} + 1}. \quad (10.78)
$$

Equation (10.78) is what we would expect from a Galilean transformation of velocity $v = \hbar Q/m$.

We might also expect that the current should be the velocity times the total density. In fact, this is not the case. The actual current is less than this, for the energy of excitations causes them to prefer to move in such a way as to oppose the flow.

The effect described above should not be confused with "back flow," which is instead an effect that comes about as follows. Consider the problem of a wave packet of excited electrons in a superconductor. Such a wave packet can be described by an approximate wave function, which, in turn, can be used to find energies and localized currents. But it turns out that the approximate wave function first used did not conserve current. An improved trial wave function (which does not improve the energy much) does conserve current. "Back flow" is the additional current that arises from the correction to the wave function.

Now, let us see how much current we get from the wave function we have chosen for a uniform current in a superconductor.

Electric current $\times$ $m(-e)$ = momentum

$$= \sum_i \text{Probability of state } i \times \text{momentum of } i$$

$$= \sum_k \sum_n P_k(n) \times \text{contribution of the } \pm k + Q \text{ states to the expectation value of the momentum}$$

$$= \sum_k (1 - f_{k+Q})(1 - f_{-k+Q})U_k^2[(k + Q) + (-k + Q)]$$

$$+ \sum_k f_{k+Q}(1 - f_{-k+Q})(k + Q) + \sum_k f_{-k+Q}(1 - f_{k+Q})(-k + Q)$$

$$+ \sum_k f_{k+Q}f_{-k+Q}V_k^2[(k + Q) + (-k + Q)]$$

$$= Q \sum_k [2U_k^2(1 - 2f_{k+Q}) + 2f_{k+Q}] + \sum_k k(f_{k+Q} - f_{-k+Q}). \tag{10.79}$$

The term multiplying Q can be written as

$$Q \sum_k 2U_k^2 + Q \sum_k (1 - 2U_k^2)(2f_{k+Q}) = \mathcal{N}Q + 2Q \sum_k f_{k+Q}\frac{\varepsilon(k)}{E_k}$$

(see Eq. (10.25)).

$\sum_k f_{k+Q}\varepsilon(k)/E_k$ is roughly zero. To see why, consider f_{k+Q} and $\varepsilon(k)/E_k$ as functions of energy, ε. (See Fig. 10.23.) f_{k+Q} is roughly symmetric about the Fermi energy, and $\varepsilon(k)/E_k$ is roughly antisymmetric. Their product is roughly antisymmetric; so it integrates out to zero.

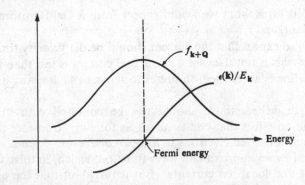

Fig. 10.23 The functions f_{k+2} and $\varepsilon(k)/E_k$.

The term in Q is therefore, $\mathcal{N} Q$. The other term in the current is

$$\sum_k k(f_{k+Q} - f_{-k+Q}) = \sum_k 2kf_{k+Q} = 2 \int \frac{d^3k}{(2\pi)^3} \frac{k}{\exp\{[E_k + (k \cdot Q/m)]\beta\} + 1}.$$

Up to first order in Q, the integrand of this term is

$$\frac{k}{\exp\{[E_k + (k \cdot Q/m)]\beta\} + 1} \approx \frac{k}{e^{E_k\beta} + 1} - \frac{k(k \cdot Q/m)\beta e^{E_k\beta}}{(e^{E_k\beta} + 1)^2}.$$

The only contribution after integration comes from the second term in the integrand, which by symmetry gives

$$-2 \int \frac{d^3k}{(2\pi)^3} \frac{k(k \cdot Q/m)\beta e^{E_k\beta}}{(e^{E_k\beta} + 1)^2} = -\frac{2\beta Q}{3m} \int \frac{d^3k}{(2\pi)^3} \frac{k^2 e^{E_k\beta}}{(e^{E_k\beta} + 1)^2}.$$

Summarizing, we see that Eq. (10.79) leads to:

$$j = -\mathcal{N}_s ev,$$

where $\mathcal{N}_s$ = effective density of superconducting electrons = $\mathcal{N} - \mathcal{N}_n$;

$$\mathcal{N} = 2 \int \frac{d^3k}{(2\pi)^3} U_k^2 = \text{numerical density of electrons},$$

$$\mathcal{N}_n = \frac{2\beta}{3m} \int \frac{d^3k}{(2\pi)^3} \frac{k^2 e^{E_k\beta}}{(e^{E_k\beta} + 1)^2},$$

$$v = \frac{Q}{m}. \tag{10.80}$$

As the temperature tends to zero, β approaches infinity, $\mathcal{N}_n$ approaches zero, and $\mathcal{N}_s$ tends toward $\mathcal{N}$. $\mathcal{N}_n$ comes from the excitations, and can be considered as a number density of normal (nonsuperconducting) electrons.

To help us see how $\mathcal{N}_n$ behaves, let us make some approximations. Suppose that the integrand of $\mathcal{N}_n$ is appreciable only near the Fermi surface in k space. This assumption is a consequence of the assumption that f_{k+Q} contributes only near the Fermi surface, as in Fig. 10.23. Then we can replace k^2 with k_F^2, and

$$\int \frac{d^3k}{(2\pi)^3} \to \int_{-\infty}^{\infty} M(0) \, d\varepsilon,$$

where $M(0)$ is the density of states at $\varepsilon = 0$. We have

$$M(\varepsilon) \, d\varepsilon = \frac{4\pi k^2 \, dk}{(2\pi)^3} = \frac{4\pi km}{(2\pi)^3} \frac{k \, dk}{m} = \frac{4\pi km}{(2\pi)^3} \, d\varepsilon.$$

Therefore

$$M(0) = \frac{4\pi k_F m}{(2\pi)^3}.$$

Also

$$\mathcal{N} = 2 \times \frac{4}{3} \frac{\pi k_F^3}{(2\pi)^3}.$$

The factor of 2 comes from spin.

Putting all this together yields

$$\mathcal{N}_n \approx \beta \mathcal{N} \int_{-\infty}^{\infty} d\varepsilon \frac{e^{E_k\beta}}{(e^{E_k\beta} + 1)^2} = \mathcal{N} \int_{-\infty}^{\infty} dx \beta\Delta \frac{e^{\beta\Delta\sqrt{1+x^2}}}{(e^{\beta\Delta\sqrt{1+x^2}} + 1)^2} = \mathcal{N} y(\beta\Delta).$$

(10.81)

Here y is a function defined as

$$y(z) \equiv \int_{-\infty}^{\infty} \frac{e^{z\sqrt{1+x^2}}}{(e^{z\sqrt{1+x^2}} + 1)^2} \, dx.$$

We already know that at absolute zero, $y = 0$. At the critical temperature, $\Delta = 0$, $E_k = \varepsilon$, and $y = 1$. In other words, $\mathcal{N}_s = 0$ at the critical temperature, and there is no superconducting current. $\mathcal{N}_s$ is given qualitatively as a function of temperature by Fig. 10.24.

Without much effort we can generalize this derivation. Suppose that instead of simply minimizing F, we minimize F subject to the condition that there is a fixed expectation value of the momentum. We then do not get a stable state, but instead a state that will decay into a stable one. An outline of the procedure follows.

Where before we minimized F, we now minimize $F + V_n \cdot \langle P \rangle$. The components of V_n are Lagrange multipliers with the dimensions of velocity.

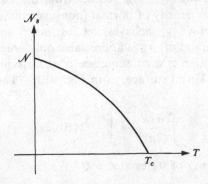

Fig. 10.24 Number density of superconducting electrons N_s, as a function of temperature.

We fix the value of $\langle P \rangle$ at the beginning of the problem, and at the end of the problem we adjust V_n so that $\langle P \rangle$ has the right value. The result is

$$j = - \mathcal{N} e V \frac{2e}{m} \int \frac{d^3 k}{(2\pi)^3} \frac{k}{\exp \{[E_k + k \cdot (V - V_n)]\beta\} + 1}$$

where $V = Q/m$. We have already worked with the integral in this equation, and it is easy to see that

$$j = - \mathcal{N} e V + \mathcal{N}_n e (V - V_n) = - \mathcal{N}_s e V - \mathcal{N}_n e V_n. \qquad (10.82)$$

Another result obtained using the same ideas is

$$\text{Energy (at fixed entropy)} = \tfrac{1}{2} \mathcal{N}_s m V^2 + \tfrac{1}{2} \mathcal{N}_n m V_n^2. \qquad (10.83)$$

It can be shown (with some difficulty) that V_n is the velocity of wave packets in the superconductor.

Equations (10.82) and (10.83) can be summarized by the following slightly nonrigorous statement:

In a superconductor there are two types of electrons—normal electrons, which carry currents that decay due to resistance, and superconducting electrons, which "short circuit" the current carried by normal electrons.

This statement is misleading because $\mathcal{N}_s$ and $\mathcal{N}_n$ do not really count electrons. $\mathcal{N}_n V$ is defined in terms of a certain integral, and $\mathcal{N}_s$ is given by $\mathcal{N} - \mathcal{N}_n$. In a similar manner, liquid helium can also be thought of in terms of a two-fluid model, and here, too, the model is satisfactory if used carefully, although it should not be taken too seriously.

10.12 ANOTHER POINT OF VIEW

Schrödinger's equation can be found from a variational principle of the form

$$\delta \int \mathcal{L} \, d^3 x \, dt = 0, \qquad (10.84)$$

where the Lagrangian density is

$$\mathcal{L}(\psi, \psi^*, \nabla\psi, \nabla\psi^*, \psi, \psi^*)$$

$$= \frac{1}{2m} \left[\left(\frac{\hbar}{i} \nabla - \frac{qA}{c} \right) \psi \right]^* \cdot \left[\left(\frac{\hbar}{i} \nabla - \frac{qA}{c} \right) \psi \right] + q V \psi^* \psi + \frac{\hbar}{i} \psi^* \dot{\psi}. \qquad (10.85)$$

Finding ψ and ψ^* such that $\int \mathcal{L} \, d^3 x \, dt$ is an extremum is equivalent to solving the equations

$$\left. \begin{array}{l} \delta \mathcal{L} / \delta \psi = 0 \\ \delta \mathcal{L} / \delta \psi^* = 0 \end{array} \right\} \qquad (10.86)$$

where, for example,

$$\frac{\delta \mathscr{L}}{\delta \psi} = \frac{\partial \mathscr{L}}{\partial \psi} - \frac{d}{dt} \frac{\partial \mathscr{L}}{\partial \dot{\psi}} - \sum_{j=1}^{3} \frac{d}{dx_j} \frac{\partial \mathscr{L}}{\partial(\partial \psi / \partial x_j)}. \tag{10.87}$$

Using Eq. (10.87) with Eq. (10.86), we get Schrödinger's equation and its complex conjugate.

If we write $\psi = \sqrt{\rho}\, e^{i\theta}$, we can consider $\mathscr{L}$ as a function of ρ and θ and their derivatives. Then

$$\frac{\delta \mathscr{L}}{\delta \rho} = \frac{\delta \mathscr{L}}{\delta \psi} \frac{\partial \psi}{\partial \rho} + \frac{\delta \mathscr{L}}{\delta \psi^*} \frac{\partial \psi^*}{\partial \rho} = 0 \tag{10.88}$$

and

$$\delta \mathscr{L} / \delta \theta = 0.$$

In terms of ρ and θ,

$$\mathscr{L} = \frac{1}{2m} \rho \left(\hbar \nabla \theta - \frac{qA}{c} \right)^2 + \frac{\hbar^2}{2m} (\nabla \sqrt{\rho}) \cdot (\nabla \sqrt{\rho}) + qV\rho + \hbar \rho \theta + \frac{\hbar}{2i} \dot{\rho}.$$

It follows that

$$\tfrac{1}{2} m v^2 + qV + U = -\hbar \dot{\theta},$$
$$\nabla \cdot \rho v + \dot{\rho} = 0, \tag{10.89}$$

where

$$v = (1/m)(\hbar \nabla \theta - qA/c) \quad \text{and} \quad U = -(\hbar^2/2m) 1/\sqrt{\rho}\, \nabla^2 \sqrt{\rho}. \tag{10.90}$$

Incidentally,

$$j = \frac{1}{2m} \left\{ \psi^* \left(\frac{\hbar}{i} \nabla - \frac{qA}{c} \right) \psi + \left[\left(\frac{\hbar}{i} \nabla - \frac{qA}{c} \right) \psi \right]^* \psi \right\} = \rho v$$

is the probability current, not the electric current.

Now, how does all this apply to superconductivity? In superconductors, the electrons in the ground state come in pairs with opposite spin. The linear combination of such pairs with the lowest energy can be considered a particle. Since the spin is zero, the "particle" obeys Bose, statistics, and the ground state of a superconductor has huge numbers of "particles" in the same state. If ρ is such that ψ is normalized, Eq. (10.89) describes the motion of a single "particle," but because the "particles" are bosons, any number can be in the same state and ρ can be the number density.

Note that $q = 2e$ and $m =$ twice the effective mass of the electrons. Note also that if ρ is the number density, we are obliged to modify $\mathscr{L}$ in some way so that ρ does not deviate too much from some value characteristic of the material, say ρ_s. We must add to $\mathscr{L}$ a term

$$\frac{\alpha}{2} (\rho - \rho_s)^2,$$

α does not come from electrostatics, for the electrostatics specifies only that $\mathscr{N}_s + \mathscr{N}_n = $ constant, where $\mathscr{N}_s = \rho$. We will not discuss α any further.

Equation (10.89) can be replaced by the equations

$$\nabla \cdot j + \frac{\partial \rho}{\partial t} = 0,$$

$$m\left[\frac{\partial v}{\partial t} + (v \cdot \nabla)v\right] = q(E + v \times B) + \nabla U, \qquad (10.91)$$

where

$$E = -\nabla V - \frac{1}{c}\frac{\partial A}{\partial t}, \qquad B = \nabla \times A.$$

Equations (10.91) are merely the equations of hydrodynamics. The first equation is the equation of conservation of the fluid, and the second is $F = ma$, with U interpreted as a pressure. Because of the $\alpha/2(\rho - \rho_s)^2$ term in $\mathscr{L}$, U is no longer given by Eq. (10.90). Just as Maxwell's equations describe the motion of many photons in the same state, Eq. (10.91) are macroscopic equations for many superconducting "particles" in the same state.

From Eq. (10.90), $mv + qA/c = \hbar\nabla\theta$ is a gradient; therefore its curl is zero. If $B = \nabla \times A$ is zero, $\nabla \times v = 0$ and the fluid is irrotational. From Eq. (10.90),

$$\hbar(\theta_2 - \theta_1) = \int_1^2 \left(mv + \frac{qA}{c}\right) \cdot ds.$$

If point 1 = point 2, we get

$$\hbar(2\pi n) = \oint\left(mv + \frac{qA}{c}\right) \cdot ds = \frac{q}{c}\oint\left(\frac{mcv}{q} + A\right) \cdot ds$$

$$= \frac{q}{c}\oint(\Lambda' j + A) \cdot ds,$$

where n is an integer and $\Lambda' = mc/q\rho_s = -2m_e c/2e\rho_s$.

$$\oint(\Lambda' j + A) = -\frac{2\pi\hbar c}{2e}n \approx -(2.09 \times 10^{-7} \text{ gauss cm}^2)n. \qquad (10.92)$$

Hereafter, replace n by $-n$.

For a simply connected superconductor (i.e., one without holes) V and A are continuous everywhere; so $n = 0$. It follows that $A = -\Lambda' j$.* But suppose

---

* This equation is the same as Eq. (10.2). Remember that the j of Eq. (10.2) is the charge current, not the probability current as above.

the superconductor is in the form of a ring. The field does not penetrate very deeply into the ring (see Section 10.1), so there is no current deep in the ring. For a path of integration near the middle of a reasonably thick ring,

$$2.09 \times 10^{-7} n = \oint (\Lambda' \boldsymbol{j} + \boldsymbol{A}) \cdot d\boldsymbol{s} = \oint \boldsymbol{A} \cdot d\boldsymbol{s} = \iint \boldsymbol{B} \cdot d\boldsymbol{S}. \qquad (10.93)$$

Equation (10.93) predicts that the flux through a ring is quantized, and this prediction is confirmed by experiment. Furthermore, if q were e instead of $2e$, the coefficient of n would be twice as large as experiment gives.

There are some other, very interesting, applications of the point of view introduced in this section. Consider an insulator between two identical pieces of superconductor (Fig. 10.25).

If Δx is sufficiently large, the two superconductors do not affect each other, and

$$-\frac{\hbar}{i} \psi_1 = E_1 \psi_1,$$

$$-\frac{\hbar}{i} \psi_2 = E_2 \psi_2.$$

E_1 can differ from E_2 if there is a voltage across the insulator. If Δx is very small, the phenomenon of barrier penetration allows some of ψ_2 to leak into the region of ψ_1, and vice versa. Since Schrödinger's equation is linear and homogeneous, the rate of leakage into region 1 is proportional to ψ_2 and vice versa. It can be shown that

$$-\frac{\hbar}{i} \psi_1 = E_1 \psi_1 + a \psi_2,$$

$$-\frac{\hbar}{i} \psi_2 = E_2 \psi_2 + a^* \psi_1 \qquad (10.94)$$

where a is real if there is no vector potential. If there is a vector potential,

$$a_A = a_0 \, e^{(iqA/\hbar c) \cdot \Delta x}.$$

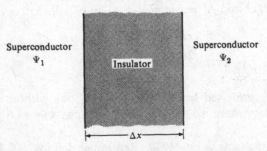

Fig. 10.25 An insulator, width Δx, between two identical superconductors.

We will take a real, and leave the more general case for the reader to work out. If we write $\psi = \sqrt{\rho}\, e^{i\theta}$, Eq. (10.94) yields

$$-\frac{\hbar}{i}\frac{1}{2\sqrt{\rho_1}}\,\dot{\rho}_1 + \hbar\dot{\theta}_1\sqrt{\rho_1} = E_1\sqrt{\rho_1} + a\sqrt{\rho_2}\exp\left[i(\theta_2 - \theta_1)\right] \qquad (10.95)$$

and a similar equation with "1" and "2" exchanged. Equating real parts of Eq. (10.95) we find that

$$\hbar\dot{\theta}_1 = a\,\sqrt{\frac{\rho_2}{\rho_1}}\cos(\theta_2 - \theta_1) + E_1$$

and exchanging "1" and "2" gives

$$\hbar\dot{\theta}_2 = a\,\sqrt{\frac{\rho_1}{\rho_2}}\cos(\theta_2 - \theta_1) + E_2.$$

Similarly,

$$\dot{\rho}_1 = -a\,\frac{2a}{\hbar}\sqrt{\rho_1\rho_2}\sin(\theta_2 - \theta_1) = -\dot{\rho}_2.$$

$j = -\dot{\rho}_1 = $ the current that flows through the junction. Define

$$\delta \equiv (\theta_2 - \theta_1), \qquad j_0 \equiv \frac{2a}{\hbar}\sqrt{\rho_1\rho_2} \approx \frac{2a\rho_s}{\hbar}.$$

If $\rho_1 \approx \rho_2$, we get

$$\left.\begin{array}{l} j = j_0\sin\delta, \\[4pt] \dfrac{d\delta}{dt} = \dfrac{E_2 - E_1}{\hbar} = \dfrac{qV}{\hbar} = \dfrac{2eV}{\hbar}, \end{array}\right\} \qquad (10.96)$$

where V is the voltage across the junction.

Now, suppose we construct the device shown in Fig. 10.26. If $\delta = \theta_2 - \theta_1$

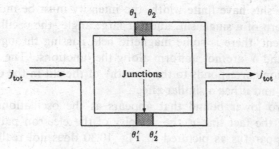

Fig. 10.26 A two-junction device.

and $\delta' = \theta'_2 - \theta'_1$, then

j_{tot} = current that flows freely through the device = $j_0(\sin \delta + \sin \delta')$.   (10.97)

Inside the superconducting part of the loop $\nabla\theta = (1\ \hbar)(m\boldsymbol{v} + (q\boldsymbol{A}/c))$, so

$2\pi n$ = change in θ as we pass around the loop = $\delta' - \delta + \oint \frac{1}{\hbar}\left(m\boldsymbol{v} + \frac{q\boldsymbol{A}}{c}\right)\cdot d\boldsymbol{s}$.

But if the wires and junctions are thick enough, $\boldsymbol{v}$ is small, and

$$2\pi n = \delta' - \delta + \frac{q}{\hbar c}\oint \boldsymbol{A}\cdot d\boldsymbol{s} = \delta' - \delta + \frac{q\varphi}{\hbar c},\qquad (10.98)$$

where $\varphi = \int \boldsymbol{B}\cdot d\boldsymbol{S}$ is the flux through the loop.

Equation (10.97) can be graphically represented if we notice that

$$j_{tot} = \text{the imaginary part of } j_0(e^{i\delta} + e^{i\delta'}).$$

In the complex plane, j behaves as shown in Fig. 10.27. A small voltage can cause δ and δ' to change with time, but $\delta' - \delta$ is a function only of the flux through the loop. With a given flux through the loop, the maximum value j_{tot} can have is by simple geometry

$$j_{max} = 2j_0 \cos\frac{(\delta' - \delta)}{2} = 2j_0 \cos\frac{e\varphi}{\hbar c}.\qquad (10.99)$$

We predict J_{max} to vary periodically with φ as shown in Fig. 10.28. The distance between successive maxima represents a vary small flux (of the order of 10^{-7} gauss cm). Something like the graph of Fig. 10.29 is actually observed. Figure 10.29 differs from Fig. 10.28 in three respects:

1. the scale is different;
2. as φ increases, the oscillations die out;
3. a nonzero lower bound arises for the oscillations as φ increases.

The above experiment is analogous to the double-slit experiment in optics. In optics, if the slits have finite width, the intensity must be multiplied by the diffraction pattern of a single slit, and for large angles the oscillations die out. In our experiment, there is some magnetic field passing through the junction material, so δ and δ' are not uniform along the junctions. The spreading of δ along the junction corresponds to the "width" of the slit in the optical double-slit experiment, and it has a similar effect.

The nonzero lower bound that appears as the oscillations die out is a consequence of the fact that $\boldsymbol{v}$, the velocity of the electron pairs, is nonzero. The physical apparatus as pictured in Fig. 10.30 does not really have its left side identical to its right side, so $\oint \boldsymbol{v}\ d_s$ need not be zero.

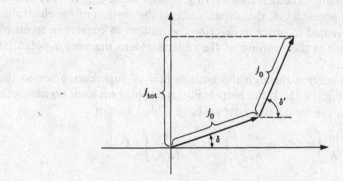

Fig. 10.27 Current j_0 in the complex plane.

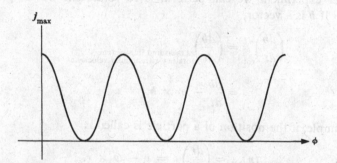

Fig. 10.28 Theoretical periodic variation of maximum current j_{max}.

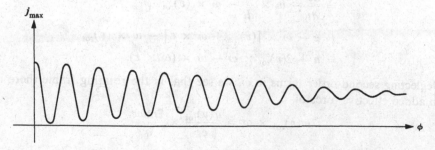

Fig. 10.29 Experimental variation of j_{max}.

If the magnetic flux through the loop is produced by a solenoid through the loop, the oscillations do not die out because no magnetic field passes through the path of the current. The fact that varying φ does cause $j_{\max}$ to vary shows that it is the vector potential, A, that directly affects the motion of the electrons. The B field was invented in order to describe the motion of objects in terms of something that exists at the position of the objects. Here the vector potential serves this purpose.

We can run a current through the bottom film of superconductor in the device pictured in Fig. 10.31. If the bottom film is doubled on itself no magnetic field is produced in the loop. But now we have added a term

$$\frac{m}{\hbar} \int v \cdot ds = \frac{m}{\hbar\rho_s} \int j \cdot ds = \frac{m}{\hbar q\rho_s} \int qj \cdot ds$$

to the phase, $\delta - \delta'$.

The integration path is along the bottom film of superconductor. By measuring the electric current needed to produce a given shift in the position of the maxima of the oscillations, we can find $m/\hbar\rho_s q$. Recall that $\Lambda = \rho_s q/m$.

Another experiment we can perform is to rotate the device pictured in Fig. 10.30. If b is a vector,

$$\left(\frac{db}{dt}\right)_{\text{rot}} = \left(\frac{db}{dt}\right)_{\substack{\text{as measured in a system} \\ \text{rotating with angular velocity } \omega}}$$

$$= \frac{db}{dt} - \omega \times b.$$

So, for example, if the position of a particle is called r,

$$(v)_{\text{rot}} = \left(\frac{dr}{dt}\right)_{\text{rot}} = v - \omega \times r.$$

$$a_{\text{rot}} = \left(\frac{dv_{\text{rot}}}{dt}\right)_{\text{rot}} = \frac{d(v)}{dt} - \omega \times (v)_{\text{rot}}$$

$$= \frac{dv}{dt} - \omega \times \frac{dr}{dt} - \omega \times (v)_{\text{rot}}$$

$$= a - \omega \times [(v)_{\text{rot}} + \omega \times r] - \omega \times (v)_{\text{rot}}$$

$$= a + 2(v)_{\text{rot}} \times \omega - \omega \times (\omega \times r).$$

Neglecting second-order terms in ω, we see that in the rotating frame there is an added effective force of

$$2m(v)_{\text{rot}} \times \omega = \frac{q(v)_{\text{rot}}}{c} \times \frac{2m\omega c}{q} \, ;$$

that is, an effective magnetic field $B = 2m\omega c/q$.

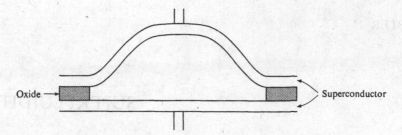

Fig. 10.30 The "double-slit" apparatus.

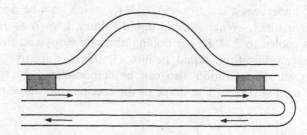

Fig. 10.31 The same, with the bottom film doubled on itself.

Going back to quantum mechanics, we see that this effective magnetic field causes an effective A to appear in the Hamiltonian. In our experiment with the rotating loop, the maximum current varies as the angular velocity is varied. By means of this experiment we can measure q/m.

Devices related to the one described above may have many uses. They react so quickly that there are rumors that one can detect radio signals by the vibrations of the magnetic fields. More likely, such devices might be useful for measuring small magnetic fields simply by counting the maxima of the current oscillations. Perhaps they could be used to maintain fields (or rather fluxes) constant to a high degree of precision. Maybe they would be useful for information storage in computers. There may be circumstances in which they could be used as a strain gauge, by keeping B fixed and detecting changes of area.

CHAPTER 11

SUPERFLUIDITY

11.1 INTRODUCTION: NATURE OF TRANSITION

Liquid helium[4] undergoes a transition at 2.18°K, which may be demonstrated in a spectacular manner. When normal liquid helium, known as HeI, at the boiling point is cooled to 2.18°K, the boiling abruptly stops and the quiescent low-temperature modification, liquid helium II, (HeII) appears. HeII does not boil. This abrupt transition also can be demonstrated by specific-heat measurements (see Fig. 11.1). The shape of the specific heat curve resembles a

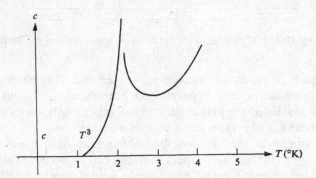

Fig. 11.1 Specific heat of helium near the transition temperature.

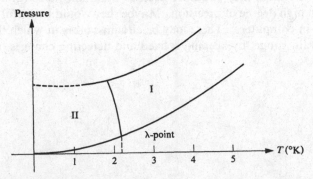

Fig. 11.2 Phase Diagram of helium near the transition temperature.

"λ" and as a result the transition temperature 2.18°K is often called the "λ-point." The specific-heat curve on both sides of the λ-point is given empirically by (see 1.9.)

$$c_v = \left.\begin{matrix} a_> \\ a_< \end{matrix}\right\} + b \ln c(|T - T_\lambda|),$$

where $a_>$ is the constant relevant for $T > T_\lambda$ (HeI) and $a_<$ for $T < T_\lambda$ (HeII). This result has not been obtained theoretically and remains one of the unsolved problems of superfluidity. At very low temperatures the specific heat is proportional to T^3, a relation that is understood.

Properties of Liquid Helium: Liquid helium below the λ-point, that is, HeII, or "superfluid" helium, exhibits several remarkable properties.

1. As has already been noted, HeII does not boil (although there is evaporation). This can be explained by assuming the heat conductivity of HeII to be essentially infinite.

2. More remarkable than the apparent infinite heat conductivity is the zero viscosity (under certain conditions) or superfluidity of HeII. It has been shown that below a critical velocity V_c, HeII flows through a thin capillary or "super-leak" with zero resistance. Furthermore, V_c increases as the capillary diameter decreases. Observing flow through a capillary is not the only way to measure viscosity, however. If a cylinder is placed in a liquid helium II bath and rotated, there is a momentum transfer from the rotating cylinder to the helium, indicating that under the conditions of this experiment, the viscosity is *not* zero! These viscosity experiments can be crudely explained if it is assumed that HeII is composed of an intimate mixture of two fluids—one fluid with zero viscosity and density ρ_s, and the other with normal viscosity and "density" ρ_n.[*] Thus it is the zero-viscosity component that flows through capillaries and the normal component that interacts with the rotating cylinder. To explain the observed data, we must assume that ρ_n/ρ_s is a function of the temperature.

3. A third remarkable property of HeII is the "thermomechanical" or "fountain effect," together with the related "mechanocaloric effect."

Consider two containers of HeII connected by a superleak (Fig. 11.3). Let the density, ρ, and temperature T be kept constant on each side. Then there will be superfluid flow through the superleak until the pressure head ΔP is equal to $\Delta P = \rho s\,\Delta T$. s is the specific entropy. This temperature difference giving rise to a pressure difference is known as the thermomechanical effect. If one container is a thin tube the pressure head will cause a fountain of liquid helium (Fig. 11.4).

---

[*] The reason for the quotation marks about "density" will become evident later.

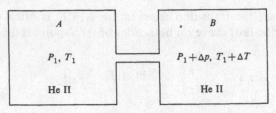

Fig. 11.3 The thermomechanical effect.

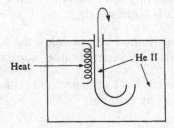

Fig. 11.4 The fountain effect.

If the containers of Fig. 11.3 are kept at a constant pressure and there is mass flow from A to B, container B will cool. This is the mechanocaloric effect. If the two-fluid concept is extended, so that now it is assumed that the superfluid component not only has zero viscosity but also carries zero entropy, the thermomechanical and mechanocaloric effects can be explained both qualitatively and quantitatively. For example, the cooling of container B in the mechanocaloric effect is exactly what one would expect if the mass transferred had zero entropy and hence zero temperature.

The superfluid flow of a zero-entropy, zero-viscosity fluid also explains the abnormal heat conductivity of HeII.

Most of the other "superfluid" properties of HeII can be explained in terms of the properties described above. For example, liquid helium II placed in a beaker will creep up the surface of the beaker and spill over the sides. This will continue until the beaker is empty (it being assumed throughout that the temperature of the helium is always below the λ-point). This rather peculiar behavior can be explained by the ordinary physics of evaporation and the infinite thermal conductivity and zero viscosity of HeII.

If any normal liquid is placed in a beaker we would expect that, due to Van der Waals attraction between liquid and beaker molecules, a layer of liquid, whose thickness decreases with height, will be formed on the beaker walls. In all liquids but HeII, however, the layer formation is inhibited by small but finite temperature, and hence vapor pressure, differences between the wall and the liquid. Thus, depending on whether the wall is warmer or colder, there will be either rapid evaporation from the wall layer or the formation of droplets that will fall back into the liquid. In HeII, however, the abnormal heat conductivity prevents the establishment of any temperature differences, and as a result the layer creeps up the side and over the top.

Landau Two-Fluid Theory: We have seen that both the viscosity experiments and the thermomechanical effect can be explained (crudely) by a two-fluid model for liquid helium II. A phenomenological theory using the two-fluid concept was introduced in 1940 by Tisza* and in another form in 1941 by Landau.† Since Landau's view is the deeper one, we shall concentrate on his formulation.

Consider HeII to consist of a perfect background fluid (which has zero entropy and viscosity), and some type of excitations, which for the moment may be regarded as phonons. This simple hypothesis explains a great many of the superfluid properties. (The model is somewhat analogous to a solid consisting of a background lattice plus phonon excitations.) The excitations according to the Landau view are the normal component.

First, the specific heat of a "phonon gas" is proportional to T^3 at low temperatures in agreement with Fig. 11.1. Second, as the perfect background fluid flows through a superleak the phonons are inhibited because they collide with the walls. Thus the perfect fluid emerges with no excitation and hence zero entropy. The two-fluid model is further strengthened by the following ingenious experiment due to Andronikashvili.‡ A pile of closely packed discs is rotated in a HeII bath. The superfluid component is unaffected, but the phonons (and any other excitations) are dragged around with the discs and have an inertial effect, which can be measured. In this way the "density" ρ_n of the inertial or normal component can be measured. (Fig. 11.5).

The two-fluid concept suggests that the two components can oscillate out of phase in such a way that the total density of HeII at a point is essentially constant but the difference or ratio of the superfluid and normal components is not. The density of excitations, however, is a function of temperature, and hence as ρ_n/ρ varies so must the temperature. This leads to a new type of wave propagation known as "second sound." Second sound is a temperature wave and will be excited by heat rather than pressure pulses.

According to the Landau view, one can think of second sound as a density wave in a phonon gas. If the phonon velocity near $T = 0$ is c (c is, of course, the

---

* Some of Tisza's ideas can be found in his following papers: *Nature*, **171**, 913 (1938); *J. Phys. et Radium*, (8) **1**, 164, 350 (1940); *Phys. Rev.* **72**, 838 (1947).

† L. D. Landau, *J. Phys. USSR* **5**, 71 (1941). See also: L. D. Landau and E. M. Lifshitz, *Fluid Mechanics*, Addison-Wesley, Reading, Mass., 1959, ch. XVI; and L. D. Landau and E. M. Lifshitz, *Statistical Physics*, Addison-Wesley, Reading, Mass., 1959, ch. VI, §§66–67.

‡ E. L. Andronikashvili, *J. Phys. USSR* **10**, 201 (1946); *Zh. Exsperim. i Teor. Fiz.* **18**, 424 (1948).

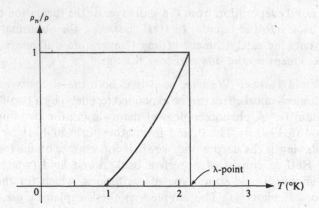

Fig. 11.5 Andronikashvili's experiment.

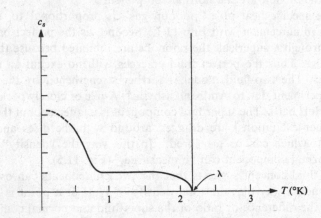

Fig. 11.6 Velocity of second sound.

velocity of first sound), the ordinary theory of sound propagation predicts that the velocity of second sound $c_s = c/\sqrt{3}$ as $T \to 0$. (See also Section 9.2.) Tisza associated second sound with the superfluid component rather than the normal (phonons), and as a result he predicted that $c \to 0$ as $T \to 0$. For a while there was a controversy on this point, but experiment soon proved Landau's supremely confident prediction that $c_s \to c/\sqrt{3}$ to be essentially correct. (Actually $c_s > c/\sqrt{3}$ at very low temperatures, less than 0.5°K.) The actual curve of second sound vs. temperature is given in Fig. 11.6.

To explain this curve between the λ-point and 1°K, it must be assumed that there are other excitations besides phonons. Landau derived the excitation curve of Fig. 11.7 empirically; it has since been developed theoretically. At the

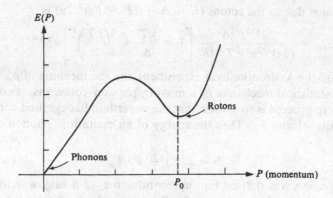

Fig. 11.7 Excitation curve for phonons and rotons.

lowest temperatures, the mean free path of the phonons becomes large; so it becomes difficult to measure the speed of second sound.

At low temperatures, $E(P) = cP$ and the excitations are phonons. In the region around P_0,

$$E(P) = \Delta + \frac{(P - P_0)^2}{2\mu},$$

where Δ is a constant and μ is some effective mass. Landau called these excitations "rotons."

The two-fluid model of Landau can be put on a more quantitative level by considering the statistical mechanics of a phonon-roton gas.

The free energy is

$$F = kT \ln (1 - e^{-E(p)/kT}) \frac{d^3 p}{(2\pi\hbar)^3} V. \qquad (11.1)$$

It is clear that the low-energy excitations (phonons) and the excitations around P_0 (rotons) contribute the most to the integral. Well below 1°K, however, the contribution due to the phonons dominates. Above 1°K, the rotons dominate.

The average energy and specific heat due to the phonons ($E = Pc$) is found as follows: The expectation value of the number of phonons of energy E is

$$\frac{\sum_{n=0}^{\infty} n e^{-B(nE)}}{\sum_{n=0}^{\infty} e^{-B(nE)}} = \frac{1}{e^{BE} - 1} = \frac{1}{e^{Pc/kT} - 1}.$$

Then

$$E_{\text{ph}} = \int_0^{\infty} \frac{Pc}{e^{Pc/kT} - 1} \frac{4\pi P^2 \, dP}{(2\pi\hbar)^3} = \frac{4\pi^5 k^4 T^4}{15h^3 c^3}; \quad C_{\text{ph}} = \frac{16\pi^5 k^4 T^3}{15h^3 c^3}. \qquad (11.2)$$

The specific heat due to the rotons ($E = \Delta + (P - P_0)^2/2\mu$) is

$$C_{rot} = \frac{2\mu^{1/2}P_0^2\Delta^2}{(2\pi)^{3/2}k^{1/2}T^{3/2}h^3}\left[1 + \frac{kT}{\Delta} + \frac{3}{4}\left(\frac{kT}{\Delta}\right)^2\right]e^{-\Delta/kT}. \qquad (11.3)$$

To understand the Andronikashvili experiment and the meaning of ρ_n, we must consider the statistical mechanics of a moving phonon–roton gas. Perhaps the simplest way to proceed is to assume that the superfluid background component is moving with velocity V_s. Then the energy of an excitation, phonon or roton, is given by

$$E = E(P) + P \cdot V_s. \qquad (11.4)$$

A similar equation was derived for superconductors. An easy way to see this relation is as follows. Suppose a gremlin floating in the fluid gives it a kick and produces an excitation of momentum P. Then the velocity of the rest of the fluid is decreased by P/M, where M is the mass of the background fluid. If the fluid winds up with velocity V_s plus the excitation, it must originally have had velocity $V_s + P/M$, along with the energy necessary to create the excitations. The total energy must therefore be $\frac{1}{2}M(V_s + P/M)^2 + E(P) \approx \frac{1}{2}MV_s^2 + P \cdot V_s + E(P)$. The energy the fluid has if it moves at velocity V_s without an excitation is $\frac{1}{2}MV_s^2$. Therefore the energy necessary to excite the fluid without changing its velocity is $E(P) + P \cdot V$. Now, applying ordinary statistical mechanics, we find that the expectation value of the number of phonons of momentum P is

$$N_P = \frac{1}{e^{(E(P) + P \cdot V_s)/kT} - 1}. \qquad (11.5)$$

The total momentum density is evidently given by $\rho V_s + \langle P \rangle$ where

$$\langle P \rangle = \int \frac{P}{e^{(E + P \cdot V_s)/kT} - 1} \frac{d^3P}{(2\pi\hbar)^3}$$

$$= -\int \frac{P(P \cdot V_s)/kT e^{E/kT}}{(e^{E/kT} - 1)^2} \frac{d^3P}{(2\pi\hbar)^3} + \text{higher order in } V_s$$

$$= -\rho_n V_s. \qquad (11.6)$$

By definition

$$\rho_n = \int \frac{(P^2/3kT)e^{E/kT}}{(e^{E/kT} - 1)^2} \frac{d^3P}{(2\pi\hbar)^3}. \qquad (11.7)$$

In other words, from the point of view just adopted, ρ_n is a derived concept and is not the density of anything. It is clear from Eq. (11.6.) that $(\rho - \rho_n)V_s = \rho_s V_s$ where ρ_s is the density of superfluid.

As in Section 9 of Chapter 10, we can consider the case of an unstable state by minimizing $E - TS$, subject to the condition that there is a fixed

expectation value of the momentum. In other words, now the excitations can
be forced to drift. The result is

$$N_p = \frac{1}{e^{\beta[E(P) - P \cdot V_n + P \cdot V_s]} - 1}$$

and $\langle P \rangle = -\rho_n (V_s - V_n)$.

11.2 SUPERFLUIDITY—AN EARLY APPROACH

We saw in the previous section that the Landau two-fluid theory and the
empirically derived excitation curve (Fig. 11.7) explain a great many of the
superfluid properties of liquid helium. Thus, a deep understanding of super-
superfluidity can be obtained by a theoretical deduction of the excitation curve
and of course, the transition. In particular, we must explain:

1. Why there are so few excitations at low energy—which is the central
   feature of superfluidity.
2. Why the excitation curve has its particular form.
3. Why there is a transition, and
4. The quantitative nature of the transition.

The Transition: *Proof that superfluidity is not explained by quantum-hydro-
dynamics.* Although the answer to why there is a transition has been known
for some time,* this problem has a rather curious history. The Einstein-Bose
condensation (Section 9 of Chapter 1) predicts that an *ideal* Bose gas at the
density of liquid helium will undergo a sharp transition at 3.2°K, which is
remarkably close to the λ-point, 2.18°K. Liquid helium[4] obeys Bose statistics,
but is of course not an ideal gas. Nevertheless, it had long been felt that the
λ-point transition is some sort of Bose-Einstein condensation, and as will be
shown later, this is indeed the case.

However, a seemingly alternative explanation of the transition was supplied
by Landau in 1941. Landau, using a procedure termed the "quantization of
hydrodynamics," developed a set of commutation relations that *seemed to suggest*
that there were an energy gap, phonons, rotons, and so forth. These were never
deduced by Landau. If, however, it is assumed that the quantization of hydrodynamics gives rise to the excitation curve, it is not unlikely that all of superfluidity including the transition can be explained by "quantum hydrodynamics."
Furthermore, since quantum hydrodynamics has nothing to do with statistics
(the way Landau developed it), the validity of Landau's approach would imply
that ^{3}He as well as ^{4}He undergoes a transition, and the Bose condensation is
not the cause of the transition.

---

* Feynman, R. P., "Application of Quantum Mechanics to Liquid Helium" in
Progress in Low Temperature Physics, Vol. 1, edited by C. J. Gorter (1955). Dingle,
R. B., "Theories of Liquid Helium II," *Advances in Physics* 1, No. 2, p. 111 (1952).

Thus, for a while there were two alternatives purporting to explain the liquid-helium transition. The refusal of ^{3}He to undergo a transition as lower and lower temperatures were attained perhaps threw doubt on the Landau hypothesis. However, I think I have proved that quantum hydrodynamics does not predict the excitation curve as had earlier been hoped by Landau and others. An outline of this proof is as follows:

We consider a fluid and adapt the Lagrangian point of view. That is, we describe the state of the fluid by giving the displacement $R(r_0, t)$ of each particle. r_0 is the position of the particle at time $t = 0$ and has the effect of labeling or distinguishing it. Each particle obeys quantum mechanics, and since the particles are distinguishable, quantum statistics need not be considered. It is then possible to find a set of commutation relations, and to show rather easily that there are a tremendous number of low-energy excitations (i.e., there is no energy gap, etc.). Now it is possible to convert the commutation relations from the Lagrangian to the Eulerian point of view. In the Eulerian point of view, the density, velocity, and so on, are functions of r and t where r is a fixed point in space. The result is that the Eulerian commutation relations so obtained are precisely those obtained by Landau!

Thus, according to Landau's quantum hydrodynamics there are a multitude of low-energy excitations, and hence this theory does not explain superfluidity. Landau's error is that by tacitly assuming the particles to be distinguishable, he neglected the effects of quantum statistics.

Theory of Superfluidity: We now want to explain superfluidity from first principles, given a collection of N He^4 atoms at density ρ_0. The two-body interatomic potential is given in Fig. 11.8. It is clear from this figure that the helium atoms resemble hard spheres of diameter 2.7 Å. Since at normal density there is 45 Å^3 per atom the atoms are not tightly squeezed.

The first question one may ask is, why is He^4 still a liquid at $T = 0$, at zero external pressure? If we calculate the potential energy of a helium lattice

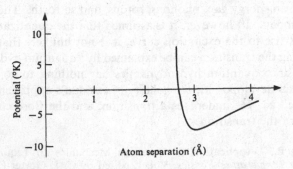

Fig. 11.8 Two-body interatomic potential.

and add the zero-point kinetic energy, we obtain the total zero-point energy, which is approximately the correct value. However, the kinetic or oscillatory energy is so large that the oscillatory amplitude is of the order of a lattice spacing. Thus the lattice "melts" due to the lattice vibration. With the exception of He^3, no other substance has such a high ratio of zero-point kinetic energy to zero-point energy. This is because helium is very light (and hence has high natural frequency) and has a saturated outer shell, causing very weak interatomic forces.

11.3 INTUITIVE DERIVATION OF WAVE FUNCTIONS: GROUND STATE

We now turn to the central feature of HeII, which is the scarcity of available low-energy excited states. Rather than look at the Hamiltonian we shall "wave our hands," use analogies with simpler systems, draw pictures, and make plausible guesses based on physical intuition to obtain a qualitative picture of the solutions (wave functions). This qualitative approach will prove singularly successful.

Consider the ground state $\varphi(R_1, R_2, \ldots, R_N)$, where R_i is the radius vector of the ith atom. Each set of $3N$ numbers (N radius vectors) will be called a "configuration" and can be represented by N points in 3-space, the N points representing the positions of the N helium atoms (Fig. 11.9). (For ease of illustration, the figure is, of course, two dimensional). For each configuration there is an amplitude or number φ, which is high for probable configurations and low for improbable ones. We may describe the difference between two configurations, A and B, by saying that the atoms "move" from configuration A to B. The word "move" does not imply any dynamics.

Now, what can be said about φ? First of all, because He^4 is a Bose-Einstein liquid, φ must be symmetric. That is, an interchange of two particles leaves φ unchanged. Secondly, in analogy with simpler Bose-system stationary states, φ is real and has no nodes. That is, φ is always positive.

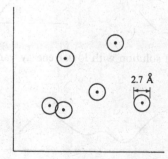

Fig. 11.9 A configuration of helium atoms in 3-space.

To see that φ can be taken to be real, consider Schrödinger's equation for φ:

$$\left[\frac{-h^2}{2m}\sum_i \nabla_i^2 + \sum_{i<j} V(r_{ij})\right]\varphi = E\varphi.$$

Clearly, φ^* satisfies the same equation; so either $i\varphi$ is real or $\varphi + \varphi^*$ is a nonzero real solution to Schrödinger's equation energy E.

If there are several degenerate eigenfunctions, first pick one solution, use it to construct a real eigenfunction, φ_1, then consider the space of degenerate eigenfunctions orthogonal to φ_1 and repeat the procedure. In this way we can obtain a set of real eigenfunctions with energy E.

To show that φ does not pass through zero if its energy is as low as possible, we will show how to find a function of lower average energy $|x\rangle$, when a wave function φ with nodes is given. Since $|x\rangle$ is the lowest possible in the ground state, it will show that φ is *not* the ground state, and therefore that the ground state has no nodes. Suppose we fix all the arguments of φ except ξ, and let $\varphi(x_1, x_2, \ldots, \xi, \ldots) = \Phi(\xi)$ look like Fig. 11.10a. Then construct a new wave function that looks like Fig. 11.10b. This wave function has the same value of

$$\left(\frac{\partial\varphi}{\partial\xi}\right)^2, \left(\frac{\partial\varphi}{\partial x_i}\right)^2, \text{ and } \varphi^2,$$

and since the energy may be expressed as

$$\langle E\rangle = \frac{\int \varphi H\varphi d^N R}{\int \varphi^2 d^N R} = \frac{\int [1/2m \sum_i (\nabla_i\varphi)^2 + \sum V\varphi^2]d^N R}{\int \varphi^2 d^N R} \tag{11.8}$$

our new wave function has the same energy as the old. We can now lower the energy by smoothing out the wave function (see Fig. 11.11), that is, we can

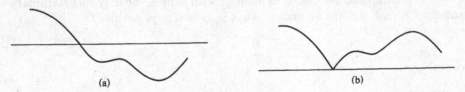

(a)   (b)

Fig. 11.10.   To show that a solution with lower energy can always be found, given a wave function with nodes.

Fig. 11.11   A smoothed version of Fig. 11.10b.

drastically decrease $(\nabla\varphi)^2$ near the node without increasing the φ^2 very much.

It follows, by the way, that the state of lowest energy is nondegenerate. If it were degenerate, we could, by appropriate choice of c_1 and c_2, make a wave function

$$\psi = c_1\varphi_1 + c_2\varphi_2$$

that has nodes, and therefore does not have the lowest possible energy. Thus, we can take the state of lowest energy to be real and everywhere nonnegative. Figure 11.8 indicates that the helium atoms are almost hard spheres. Thus any configuration (e.g., Fig. 11.9) that includes two overlapping atoms gives a very small amplitude—call it zero. From Eq. (11.8) we see that the energy is high where there are large gradients in φ. Thus we want the ground-state wave function φ to vary as slowly as possible. Now consider in Fig. 11.12, the atom $\otimes$ to "move" about, while the other atoms remain fixed. φ for configuration (b) must be very small; otherwise there would be a high gradient as $\otimes$ moves the small distance from (b) to (c). In other words, configurations where two atoms are very close are unlikely, and in the ground state the most likely configurations are those where the atoms are approximately evenly distributed (Fig. 11.12a).

Thus we have deduced that φ is symmetric, real, always positive, largest when the atoms are evenly spaced, and lowest when there is clumping. An expression that satisfies these requirements is:

$$\varphi = \exp\left[-\sum_{ij} f(r_{ij})\right] = \prod_{ij} F(r_{ij}) \tag{11.9}$$

where r_{ij} is the distance between the i and j atoms, and F is given approximately in Fig. 11.13. F is occasionally approximated by some simple function that fits Fig. 11.13, such as $F = 1 - a/r_{ij}$. Since φ is expressed in terms of the one function F, it is natural to ask why a variational principle to determine F is not used. The answer is that the integral over all the $3N$ variables cannot be expressed simply in terms of F.

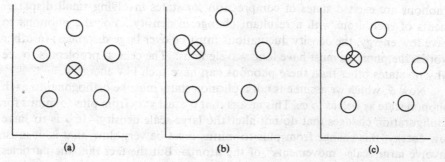

(a)   (b)   (c)

Fig. 11.12 *a, b, and* c. A configuration in which one atom "moves" about while all the others remain fixed.

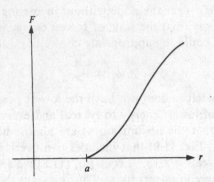

Fig. 11.13 The function F of Eq. (12.9).

Note that the nondegeneracy of the ground state implies that it is a symmetric state for *any* many-particle system, when the particles are identical. For, since the Hamiltonian is invariant under a permutation P of the identical particles, then $\varphi(PR)$ will also be a ground state when $\varphi(R)$ is. But since there is no degeneracy, we must have

$$\varphi(PR) = \alpha_p\varphi(R),$$

and since $\varphi(PR)$ is always positive and real (all φ values are) we have $\alpha = 1$. Thus φ is totally symmetric. (Remember that this function might yet not be allowed by statistics.)

Excited States: Explanation of Superfluidity: Again, in the analogy to simpler Bose systems, the first excited state wave function, ψ, must have one node and be symmetric. Since φ is always positive and $\int \varphi\psi \, d^N R = 0$, ψ must be positive for half the configurations and negative for the other half. We now come to the key argument of superfluidity: The *only* low-energy excitations are phonons. Phonons are excited states of compression, or states involving small displacements of each atom with a resultant change in density. For the phonons to have low energy, the density fluctuations must be over large distances; in other words, the phonons must have long wavelengths. The central problem is to see why no states other than these phonons can have such low energies.

Now ψ, which we assume is not a phonon state, must be orthogonal to each phonon state as well as to φ. This means that ψ must vary from plus to minus for configuration changes that do not alter the large-scale density. If ψ is to have low energy, this change from plus to minus must be very slow, that is, it must involve large-scale "movements" of the atoms. But the fact that the particles are indistinguishable and obey Bose statistics makes such a movement impossible!

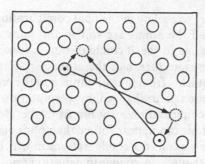

Fig. 11.14 Two configurations (solid and dotted) that result from large displacements (long arrows) of the atoms, can actually be accomplished by much smaller adjustments (short arrows) because of the identity of the atoms. (From Reference 1).

Now let us quote from the first reference cited in Section 11.2*:

"The function ψ takes on its maximum positive value for some configuration of the atoms. Let us call this configuration A, and the particular locations of the atoms α-positions. We said that the α-positions must be well spaced, so that the atoms do not overlap, and further that they are, on a large scale, at roughly uniform density. Similarly, let us call configuration B, with atomic positions β, that for which ψ has its largest negative value. Now we want B to be different as possible from A. We want it to require as much readjustment over a long distance as possible to change A to B. Otherwise ψ changes too rapidly and easily from plus to minus, our wave function has a high gradient, and the energy of the state is not as low as possible.

"Try to arrange things so that A *requires* a large displacement to be turned into B. At first you might suppose it is easy. For example (see Fig. 11.14), in A take some atoms in the left side of the box containing the liquid and move it across to the other side of the vessel, and call the resulting configuration B. One objection to this is that an atom is moved from one side to another, so a hole remains at the left and an extra atom is at the right. This represents a density variation. To avoid this we may imagine that another atom has been removed at the same time from right to left, and the various holes and tight squeezes have been ironed out by some minor adjustments of several of the neighboring atoms. This movement of two atoms each through a distance equal to the size of the vessel, one from left to right and the other from right to left, is certainly a long displacement, so B and A seem very different. But they are not.

"The atoms must be considered as identical; the amplitude must not depend on which atom is which. One cannot allow ψ to change if one simply permutes atoms. The long displacements can be accomplished in two steps. In the first step permute the atoms you wish to move to those α-positions closest to the

---

* Feynman, R. P., *Ibid.*

ultimate position they are to occupy in the final configuration B. This step does not change ψ because all the atoms are still in the same configuration of α-positions. Then the change to the B configuration is made by small readjustments, no atom moving more than half the atomic separation. In this minor motion ψ must change quickly from plus to minus and the energy cannot be low. Because the wave function is unchanged by permutation of the atoms it is impossible to get a B configuration very far from the A configuration. No very-low-energy excitations can appear (other than phonons) at all.

"In the phonon case we consider configurations in which, as ψ changes sign, the density distribution changes. A change in density cannot be accomplished by permuting atoms. That is why the Bose statistics does not affect phonon states. But it leaves them isolated as the lowest states of the system, so that the specific heat approaches zero as T approaches zero according to Debye T^3 law. This is the key argument for the understanding of the properties of liquid helium."[*]

11.4 PHONONS AND ROTONS

We have been assuming that large-scale density variations are long-wavelength phonons. That this is so can be seen as follows:

Let $\rho(R)$ be the density of the helium liquid. Then, Fourier analysis of $\rho(R)$ gives $\rho(R) = \sum_K q_K e^{-iR \cdot K}$ and

$$H = \sum_K \tfrac{1}{2}(\dot{q}_K^2 + C^2 K^2 q_K^2), \qquad (11.10)$$

which is the Hamiltonian for a collection of $3N$ independent harmonic oscillators (normal modes) each of frequency $\omega = KC$. Each mode is quantized and, by common usage, if the Kth mode is excited to energy $E_K = \hbar KC(n + \tfrac{1}{2})$, we say that there are n phonons of frequency KC. The ground-state wave function for a set of independent oscillators is

$$\chi_{gs} = \prod_K \exp\left[-\frac{1}{2}\frac{q_K^2}{KC}\right] = \exp\left[-\frac{1}{2}\sum\frac{q_K^2}{KC}\right]. \qquad (11.11)$$

The low excited state of one phonon of momentum $\hbar K$ is given by

$$\chi_K = \exp\left(-\frac{1}{2}\sum_{K' \neq K}\frac{q_{K'}^2}{K'C}\right)q_K \exp\left(-\frac{1}{2}\frac{q_K^2}{KC}\right) = q_K\chi_{gs}. \qquad (11.12)$$

However, χ_{gs} is not the true ground-state wave function in the case of liquid helium, since it does not contain enough short distance detail. For example, it doesn't tell us that two atoms cannot overlap, and so forth. Let φ be the true

[*] Feynman, R. P., *Phys. Rev.* **91**, 1291, 1301 (1953); **94**, 262 (1954). Feynman, R. P. and Cohen, M., *Phys. Rev.* **102**, 1189 (1956). Feynman, R. P., *Progress in Low Temperature Physics* (edited by C. J. Gorter), ch. II, (1955).

ground state, which takes the short distance factors into account. Then in analogy to Eq. (11.12), the true low excited state ψ_K is given by

$$\psi_K = q_K \varphi'. \tag{11.13}$$

We are talking about long-range density fluctuations, and K is small. Since the operator $\rho(R) = \sum_i \delta(R - R_i)$, we obtain

$$q_K = \int \rho(R) e^{iK \cdot R} \, d^3R = \sum_i e^{iK \cdot R_j}, \tag{11.14}$$

and from Eqs. (11.13) and (11.14)

$$\psi_{\substack{\text{Phonon} \\ \text{Small } K}} = \left(\sum_i e^{iK \cdot R_i} \right) \varphi. \tag{11.15}$$

Actually, we should write in place of Eq. (11.10)

$$\rho(R) = \sum_K (q_K \sin K \cdot R + q'_K \cos K \cdot R).$$

Then

$$H = \sum_K \left(\frac{\dot{q}_K^2}{2} + \frac{\dot{q}_K'^2}{2} \right) + C^2 K^2 (q_K^2 + q_{K'}^2).$$

The singly excited states are $\psi_K = q_K \varphi$ and $\psi'_K = q'_K \varphi$. Then the state $\psi'_K + i\psi_K$ also represents a one-phonon state, and is equal to $(\sum_i e^{iK \cdot R_i})\varphi$. Equation (11.15) helps us to understand why it is impossible to have as a low energy non-phonon excitation a single atom moving through the helium fluid. For the wave function of such an excitation is $e^{iK \cdot R_i}\varphi$, which is not symmetric. Blindly symmetrizing this function gives $\sum_i e^{iK \cdot R_i}\varphi$, which represents a phonon excitation. It is possible, however, to have a (symmetrized) wave function $\sum_i e^{iK \cdot R_i}\varphi$ if $1/K$ is of the order of the atomic spacing, a. The energy associated with such an excitation is

$$E = \hbar^2 K^2 / 2m \sim \hbar^2 / 2ma^2,$$

which is very close to the excitation energy of a roton.

Rotons: Higher Energy Excitations: We have seen that low-energy non-phonon excitations are impossible. In other words, there are *no* possible long-distance movements of the atoms that do not change the density. So now let us try to construct a wave function ψ that goes from plus to minus, without a density change, as slowly as possible, but which of necessity involves only short movements.

Again let A be the configuration for which ψ takes on its maximum positive value, and B the configuration for which ψ takes on its maximum negative values. The positions of the atoms in configuration A will be called α-positions and, in configuration B, β-positions. The change from A to B must involve only a small displacement of each atom and yet involve as large a total movement as possible.

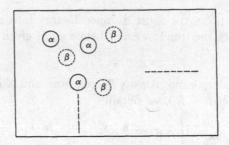

Fig. 11.15 In this configuration the β-sites are between the α-sites.

We know that the α-sites in configuration A must be evenly spaced, and the same for the β-sites in B. These conditions imply that the β-sites are between the α-sites as in Fig. 11.15.

Suppose for a moment that each atom is either on an α- or a β-site, and let N_α and N_β be the number of atoms in each type of position. Then $\psi = (N_\alpha - N_\beta)\varphi$ is maximum positive for configuration A, maximum negative for configuration B, and passes smoothly from A to B. ψ can be expressed mathematically by defining a function $f(R_i)$ to be $+1$ if R_i is at an α-site and -1 if R_i is at a β-site.

$$\psi = \left(\sum_i f(R_i) \right) \varphi. \tag{11.16}$$

For an intermediate position (that is, one for which R_i is between an α and β-site), ψ will vary smoothly if $f(R_i)$ is taken to vary in some smooth way between its extreme values of $+1$ and -1.

We now want to use the variational principle to find the best possible $f(R)$, if

$$\psi = \sum_i f(R_i)\varphi = F\varphi: \qquad F = \sum_i f(R_i). \tag{11.17}$$

The energy integral to be minimized is

$$\langle E \rangle = \frac{\int \left[\sum_i (\hbar^2/2m)|\nabla_i\psi|^2 + V|\psi|^2 \right] d^{3N}R}{\int |\psi|^2 \, d^{3N}R}. \tag{11.18}$$

When $\psi = F\varphi$, this integral can be written in an interesting way, which is quite general (though we take φ real and also assume φ vanishes at infinity). Replacing ψ by $F\varphi$ in Eq. (11.18) gives us

$$\langle E \rangle = \left\{ \sum_i \frac{\hbar^2}{2m} \left[\int |F|^2 (\nabla_i\varphi)(\nabla_i\varphi) \, d^{3N}R + \int (F^*\nabla_i\varphi)\varphi(\nabla_i F) \, d^{3N}R \right. \right.$$
$$\left. + \int (F\nabla_i\varphi)\varphi(\nabla_i F^*) \, d^{3N}R + \int \varphi^2 |\nabla_i F|^2 \, d^{3N}R \right]$$
$$\left. + \int V|F|^2\varphi^2 \, d^{3N}R \right\} \Big/ \int |F|^2\varphi^2 \, d^{3N}R. \tag{11.19}$$

Integrating the first integral on the right of Eq. (11.19) by parts and making the substitution $H\varphi = E_0\varphi$ gives finally

$$\varepsilon \equiv \langle E \rangle - E_0 = \frac{\sum_i \int (\hbar^2/2m)|\nabla_i F|^2 \rho_N \, d^{3N}R}{\int |F|^2 \rho_N \, d^{3N}R}, \tag{11.20}$$

where

$$\rho_N(R_1, \ldots, R_N) = \varphi^2(R_1, \ldots, R_N). \tag{11.21}$$

ρ_N is evidently the probability of a given configuration. ε is the energy of an excitation, and E_0 is the energy of the ground state.

Returning now to the specifics of the liquid helium problem, we have $F = \sum_i f(R_i)$ and we wish to minimize Eq. (11.20) to find the best value for f. The denominator of Eq. (11.20) becomes

$$\int |F|^2 \rho_N \, d^{3N}R = \sum_i \sum_j \int f^*(R_i) f(R_j) \left[\int \rho_N \prod_{k \neq i,j} d^3R_k \right] d^3R_i \, d^3R_j$$

$$= \rho_0 \int f^*(R_1) f(R_2) g(R_1 - R_2) \, d^3R_1 \, d^3R_2, \tag{11.22}$$

where

$$\rho_0 g(R_1 - R_2) = \sum_i \sum_j \int \delta(R_i' - R_1) \, \delta(R_j' - R_2) \rho_N \, d^{3N}R'.$$

Here φ is normalized so that $\int \rho_N \, d^{3N}R = N$ = number of atoms. ρ_0 is the liquid number density in the ground state and $g(R_1 - R_2)$ is the probability of finding an atom at R_2 per unit volume if one is known to be R_1. The numerator of Eq. (11.20) can be written as

$$\sum_i \int \frac{\hbar^2}{2m} |\nabla_i F|^2 \rho_N \, d^{3N}R = \rho_0 \frac{\hbar^2}{2m} \int |\nabla f(R)|^2 \, d^3R. \tag{11.23}$$

Equation (11.20) can now be rewritten as

$$\varepsilon = \frac{\hbar^2/2m \int |\nabla f(R)|^2 \, d^3R}{\int f^*(R_1) f(R_2) g(R_1 - R_2) \, d^3R_1 \, d^3R_2}. \tag{11.24}$$

By setting the variation $\delta\varepsilon = 0$ to minimize ε we obtain the integrodifferential equation

$$-\frac{\hbar^2}{2m} \nabla^2 f(R) = \varepsilon \int g(R - R') f(R') \, d^3R'. \tag{11.25}$$

The solution to Eq. (11.25) is $f(R) = e^{iK \cdot R'}$.

$$\psi = \left(\sum_i e^{iK \cdot R_i} \right) \varphi, \tag{11.26}$$

and

$$\varepsilon = \frac{\hbar^2 K^2}{2mS(K)} \qquad \text{where} \qquad S(K) = \int e^{i\mathbf{K} \cdot \mathbf{R}} g(R) \, d^3R. \qquad (11.27)$$

$S(K)$ is the form factor for scattering of neutrons from the liquid and can be found experimentally (Fig. 11.16). From Eq. (11.27) it is the Fourier transform of $g(R)$, which is the probability per unit volume of finding an atom at a distance R from a given atom in the liquid in the ground state.

Since states of different momenta are orthogonal, the energies of Eq. (11.27) are significant not only for K near the minimum but also in the neighborhood of this value. Although in the derivation of Eq. (11.27) it has been assumed that $K \sim 2\pi/a$ (i.e., K is large) where a is the atomic spacing, the analysis is valid for all K as long as $\psi = (\sum f(R_i))\varphi$ is a reasonably good wave function. But we see from Eq. (11.15) that for small K, $\psi = (\sum_i f(R_i))\varphi$ represents phonons, and is a very good wave function. Hence the results of Eqs (11.26) and (11.27) are valid over the entire range of K. From Eq. (11.27) and Fig. 11.16 we can plot $\varepsilon(K)$.

The minimum occurs near $K = 1/a$, and the excitations in this region are the rotons.

It is interesting to note that the excitation curve (Fig. 11.17) can now be determined directly, by measuring the energy loss suffered by a scattered neutron of incident momentum $\hbar K$. For a given angle of scattering, there will be a minimum energy loss, which is the energy $\varepsilon(K)$ of a single excitation.

11.5 ROTONS

A roton can be visualized as a group of children going down a slide. The slide is the AB part in Fig. 11.18. After sliding down, the children go around and come back to the slide again.

The returning-flow part is not included in Eq. (11.26) of Section 11.4 for the roton wave function

$$\psi = \left(\sum_i e^{i\mathbf{K} \cdot \mathbf{R}_i} \right) \varphi_{\text{ground}} \qquad (11.28)$$

but is derived as follows. When we construct a wave packet out of the rotons of Eq. (11.28), we have a localized excitation. This localized wave packet does not move, because the group velocity of the roton state is zero as is seen in Fig. 11.17 (where $d\varepsilon/dk = 0$). This localized excitation is associated with the momentum $\hbar K$. When we calculate the quantum mechanical current density j for a roton, we find it is not zero:

$$j \neq 0. \qquad (11.29)$$

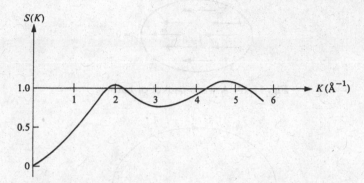

Fig. 11.16 Experimental result for the form factors for scattering neutrons from the liquid.

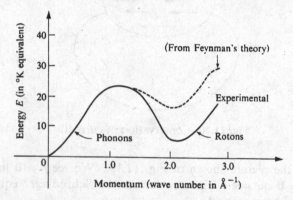

Fig. 11.17 Experimental and theoretical excitation curves. [Feynman, *Progress in Low Temperature Physics* (1955)].

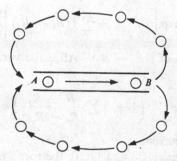

Fig. 11.18 Schematic of the movement in a roton.

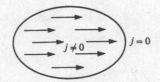

Fig. 11.19 A wave packet of rotons.

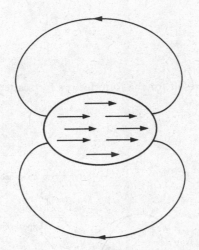

Fig. 11.20 Conservation of current.

Then we have the picture shown in Fig. 11.19. We see $j \neq 0$ in a localized region and $j = 0$ outside. This contradicts the Schrödinger equation, since the current must be conserved.

This difficulty is avoided by considering the slow return of current to the packet (Fig. 11.20). Mathematically this is achieved by improving Eq. (11.28) by writing

$$\psi = \left(\sum_i \exp\left[iK \cdot R_i + \sum_j f(R_i - R_j) \right] \right) \varphi_{\text{ground}} \qquad (11.30)$$

and finding the best form of $f(R_i - R_j)$. After the variation we find the best form is

$$\psi = \sum_i e^{iK \cdot R_i} \left(1 + \alpha \sum_j \frac{R_i - R_j}{|R_i - R_j|^2} \right) \varphi_{\text{ground}}. \qquad (11.31)$$

This form conserves the current.

As is suggested by Figs. 11.18 and 11.20, the roton is like a classical smoke ring, as shown in Fig. 11.21. There is one difference, though. The smoke ring

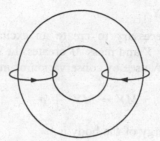

Fig. 11.21 A roton has the same shape as a perfect smoke ring.

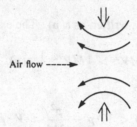

Air flow ------►

Fig. 11.22 The attraction between two opposite vortices of a smoke ring is balanced by its motion.

moves, whereas the roton does not, as has already been mentioned. Because two vortices of opposite sense attract each other (Fig. 11.22), there is another important difference: In the case of the smoke ring, this attraction is balanced by the flow of air through it, so that the ring does not shrink to a point. For the roton, since it does not move, the diameter of the ring shrinks down to the atomic scale.

When the atom in position A in Fig. 11.18 goes to position B, the state returns to itself. This leads to a condition on K in $e^{iK \cdot R_i}$. K is of the order of $1/a$ where a is the average distance between atoms. This agrees with the curve in Fig. 11.20. Thus the roton is associated with a momentum and velocity of approximately

$$p = h/a; \qquad v = h/ma. \tag{11.32}$$

When there is more than one roton, they interact. The interaction becomes important for high temperatures. Landau and Khalatnikov* calculated the attenuation of the first and second sounds from kinetic theory taking into account the cross sections of phonon–phonon, phonon–roton and roton–roton scattering.

---

* For detailed discussion and references, see I. M. Khalatnikov, *An Introduction to the Theory of Superfluidity*, W. H. Benjamin, 1965.

11.6 CRITICAL VELOCITY

Let us find the energy necessary to create an excitation of momentum p. Suppose a body of velocity V and mass M creates the excitation and winds up moving with velocity V'. We see by conservation of momentum

$$MV = MV' + p,$$

so that the new kinetic energy of the body is

$$\tfrac{1}{2}MV'^2 = \tfrac{1}{2}MV^2 - V\cdot p + \frac{p^2}{2M}. \tag{11.33}$$

The energy of the excitation is written as $\varepsilon(p)$. The excitation cannot be created unless

$$\tfrac{1}{2}MV^2 > \tfrac{1}{2}MV'^2 + \varepsilon(p),$$

so

$$\varepsilon(p) < V\cdot p - \frac{p^2}{2M} \approx V\cdot p \tag{11.34}$$

for large M.

The critical velocity necessary to create an excitation is then derived by drawing a tangent to the $\varepsilon(p)$ vs. p curve as shown in Fig. 11.23.

The critical velocity calculated from this condition is about 60 m/sec.

We might say that this is the reason why helium is a superfluid; it is not excited for small perturbations. Although this explanation seems to be qualitatively satisfactory, excitations occur for velocities of about 1 cm/sec. Now we have to reverse our position and look for the reason why liquid He can be excited with such a small velocity.

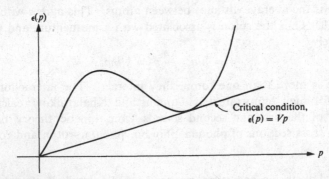

Fig. 11.23 The critical velocity is at tangent to the $\varepsilon(p)$ versus p curve.

11.7 IRROTATIONAL SUPERFLUID FLOW

When liquid helium is undergoing a mass motion of velocity V the wave function is

$$\psi = \exp\left(i\,\frac{m}{\hbar}\,\sum_i V \cdot R_i\right)\varphi_{\text{ground}}. \qquad (11.35)$$

Note the difference between Eq. (11.35) and the roton or the phonon wave function of Eq. (11.15). The exponential part in Eq. (11.35) may be interpreted as follows.

$$P \cdot R_{\text{C.G.}} = NmV \cdot \left(\sum_i R_i/N\right) = mV \cdot \sum_i R_i, \qquad (11.36)$$

where P is the momentum of the whole system, $R_{\text{C.G.}}$ the center-of-mass coordinate, and N the number of particles in the system.

As in the case of superconductivity, for a nonuniform velocity we assume

$$\psi = \exp\left[i\,\sum_i \theta(R_i)\right]\varphi_{\text{ground}}, \qquad (11.37)$$

and the velocity field $V(R)$ at R is found from $\theta(R)$ by using

$$V(R) = \frac{\hbar}{m}\,\nabla\theta(R). \qquad (11.38)$$

We see from Eq. (11.38) that

$$\nabla \times V(R) = 0,$$

so that when the system is singly connected there is no rotation. When the system is doubly connected, we can have a circulating V in the system as pictured in Fig. 11.24. We take

$$\theta = k\alpha.$$

For this choice of θ, $V(R)$ is perpendicular to the radius and has magnitude

$$|V| = \frac{\hbar k}{mr}.$$

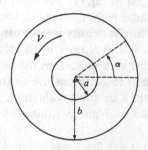

Fig. 11.24 Circulating V in a doubly connected system.

k is restricted to the integers by the requirement that the wave function must be single valued.

$$\theta = n\alpha$$

$$|V| = \frac{\hbar}{mr} n \qquad\qquad *(11.39)$$

and

$$\oint V \cdot dS = 2\pi \frac{\hbar}{m} n.$$

The unit of quantization is

$$2\pi \frac{\hbar}{m} = 2\pi \times 1.5 \times 10^{-4} \text{ cm}^2/\text{sec}.$$

What is the energy for this flow? If ρ_0 is the fluid density in atoms per cc, the kinetic energy is (for $n = 1$)

$$\text{KE} = \frac{1}{2} \int_a^b \rho_0 m \left(\frac{\hbar}{mr}\right)^2 2\pi r \, drL, \qquad (11.40)$$

where a and b are the radii as shown in Fig. 11.24, and L is the depth of the liquid. Thus:

$$\text{Line energy per unit length} = \rho_0 \pi \frac{\hbar^2}{m} \ln\left(\frac{b}{a}\right) = 10^{-8} \ln\left(\frac{b}{a}\right) \text{ ergs/cm,} \qquad (11.41)$$

where we used $\rho_0 = (3.6)^{-3} \text{ Å}^3 = \frac{1}{45} \text{ Å}^3$. If we neglect atomic structure and assume a classical continuous liquid with surface tension, a unit line makes a hole opposed by surface tension that we can calculate to be only 0.4 Å in radius. That means that there is no real hole in the liquid.

The energy associated with a vortex line can be estimated from Eq. (11.41). The lower limit a of the integral in Eq. (11.40) is considered as a length of the order of atomic spacing, because within about the atomic spacing the velocity formula is meaningless (and also the density is low near the center of the vortex). The exact determination of a would require us to solve the difficult quantum-mechanical problem. In almost all applications the ratio b/a will be very large, and the logarithm large enough to be insensitive to the exact value of a. For this reason we will not attempt a detailed evaluation, but will simply choose a to be close to the atomic spacing. We may arbitrarily take $a = 4.0$ Å.

---

* The quantization of a vortex line was first suggested by L. Onsager, *Nuovo Cimento* **6**, Suppl. 2, 246 (1949).

11.8 ROTATION OF THE SUPERFLUID*

Let us consider the state of rotating liquid helium in a can (see Fig. 11.25). Operationally this state is defined by starting from solid helium in the can (under pressure > 25 atm at $0°K$). First rotate the solid helium and then release the pressure to melt it; the liquid helium is then rotating with the angular momentum initially given to the solid. What is the final state of helium? We ask then for the lowest state of this rotating helium.

For a system of *given angular momentum*, the kinetic energy is the least if the angular velocity ω is a constant throughout the liquid. However, this motion is not rotation-free, because $\nabla \times V = 2\omega$, and a high energy (see Eq. 11.41) is required in order to set any small part of liquid helium into a rotational state. If we consider the helium to be a rigid body rotating, we get a reasonable energy for the rotation, but we cannot consider helium to be rigid. The energy to clamp the atoms into a rigid relationship with one another is very high, and must be taken into account if one wishes to calculate the energy for uniform rotation, with $\nabla \times V_s \neq 0$ everywhere. A simple way to check the validity of the above reasoning is to consider two atoms of mass m in a harmonic potential. The energy of excitation of one of the atoms is

$$h\omega = h\left(\frac{k}{m}\right)^{1/2}.$$

But if the atoms could be rigidly held together, that energy would become $h(k/2m)^{1/2}$, which apparently would mean that the presence of a second atom allows a lower energy of excitation. This paradox is resolved by noting that it takes energy to hold the atoms together. If only a limited energy is available, nearly all the parts of the fluid must be frozen out in their ground states. That is, nearly everywhere the local angular momentum is zero, so that, $\nabla \times V = 0$ except at certain discontinuities. Suppose we assume that the angular momentum is carried by excitations. That is, the liquid is not free of excitations such as rotons

Fig. 11.25 Rotating can of helium.

---

* R. P. Feynman, chapter II in vol. I *Progress in Low Temperature Physics*, edited by C. J. Gorter (1955).

and phonons even when the temperature is 0°K. The number and type of excitations is to be found by minimizing the energy for the given value of angular momentum. This energy, for vessels of centimeter dimensions turning at about one radian per second, turns out to be nearly 10^4 times the energy of a rigid body rotating at the same angular velocity. We must reject such a model.

Thus, in order to achieve a state of lower energy for the given angular momentum, we examine velocity fields that are not everywhere continuous. As we saw in Section 11.7, we know that if there is a hole at the center of the liquid, circulation can exist. Thus a solution is suggested: the liquid forms a vortex around a hole with constant circulation as analyzed in Section 11.7. The velocity varies inversely as the radius, rising to such heights near the center as to be able to maintain the hole free of liquid by centrifugal force. The energy for this state is still quite a bit higher than the kinetic energy for the rigid body case, because the velocity instead of being distributed proportionally to the radius (the solid case), actually falls as the radius increases (see Eq. 11.39). Nevertheless it is orders of magnitude below the continuous V-field with excitations as suggested above.

Once we admit the possibility that V is not necessarily continuous we can think of a state of lower energy. Suppose that the liquid has not only one vortex at the center, but several vortices. For example, suppose beside the central one there were a number distributed about the circle of radius $R/2$, half that of the vessel R, and all turning the same way as shown in Fig. 11.26. If the number of vortices at $R/2$ is large, this state is like a vortex sheet so the tangential velocity can jump as we pass from inside $R/2$ to outside as shown in Fig. 11.27. The gain in energy resulting from this improved distribution may more than compensate for the energy needed to make the additional holes (and, further, the central vortex need not now be so large and energetic).

Continuing in this way with ever more vortices we soon notice that the energy can always be reduced if more vortices form. However, there is a limit. Due to the quantization of the vortex strength, the smallest vortex has circulation $2\pi\hbar/m$. The lowest energy results if a large number of minimum-strength vortex lines (which we will call "unit lines") form throughout the fluid at nearly

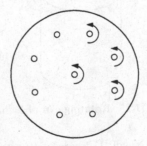

Fig. 11.26 The vessel as seen from above, with vortices all turning the same way.

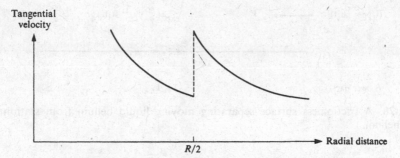

Fig. 11.27 Discontinuity in tangential velocity at $R/2$.

uniform density. The lines are all parallel to the axis of rotation. Since the curl of the velocity is the circulation per unit area, and the curl is 2ω, there will be (by Stokes's theorem)

$$2\omega \, \frac{m}{2\pi\hbar} = 2.1 \times 10^3 \omega \text{ lines/cm}^2 \tag{11.42}$$

with ω in radians/second. For $\omega = 1$ radians/second the lines are about 0.2 mm apart, so that the velocity distribution is practically uniform.

The energy associated with one vortex line is derived from Eq. (11.41). Taking $a = 4.0$ Å and $b = 0.2$ mm, and using Eq. (11.42) we find

$$\text{Total line energy per unit volume} = \rho_0 \omega \hbar \ln\left(\frac{b}{a}\right) = 14\rho_0 \omega \hbar. \tag{11.43}$$

Thus the total line energy in the rotating liquid of radius R and a unit thickness is $14\rho_0 \omega \hbar \pi R^2$. The kinetic energy for a rigid rotating body of radius R and unit thickness is $\pi m \rho_0 R^4 \omega^2/4$. The ratio of the two is of order

$$4 \, \frac{\hbar/m}{R^2 \omega} \times 14 \cong 10^{-2} \tag{11.44}$$

(the number 10^{-2} is calculated for $R = 1$ cm and $\omega = 1$ rad/sec). For macroscopic laboratory dimensions the excess energy to form the lines is small. They would form if rotating solid helium is melted by releasing the pressure, the angular velocity distribution would differ imperceptibly from uniformity, and the surface should appear parabolic.

11.9 A REASONING LEADING TO VORTEX LINES

Suppose liquid helium is separated into an upper and a lower part by a frictionless surface, and the upper part is moving with velocity V (see Fig. 11.28).

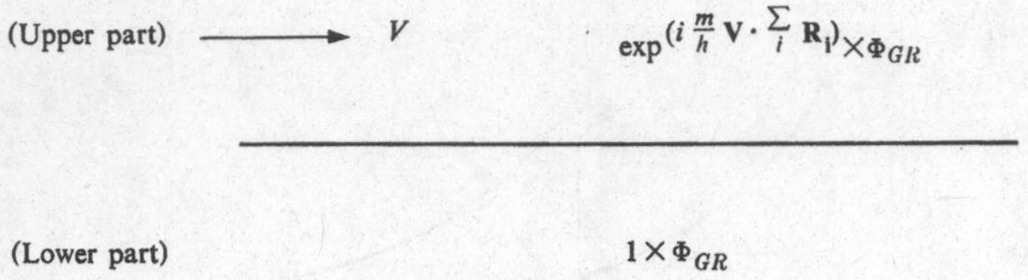

(Upper part) ——————▶ V $\exp^{(i\frac{m}{\hbar}\mathbf{V}\cdot\sum_i \mathbf{R}_i)}\times\Phi_{GR}$

(Lower part) $1\times\Phi_{GR}$

Fig. 11.28 A frictionless surface separating moving liquid helium from stationary liquid helium.

Fig. 11.29 Discontinuities at the dividing surface.

Fig. 11.30 Vortex lines.

In the upper part the wave function contains the factor

$$\exp\left(i\,\frac{m}{\hbar}\,V\cdot\sum_i R_i\right).$$

When the dividing surface is removed there appears the surface free energy or the surface tension. Because at periodic points $\exp(im/\hbar)vR = 1$, there is no surface tension at these points. The dividing surface is pictured in Fig. 11.29: around a discontinuity there is a circulation. Now you may ask, what is the best size of the slot for discontinuity? A calculation shows that the discontinuity is not the shape of a slot, but a line, so that the dividing surface would look like Fig. 11.30. Thus, this reasoning may lead to the assumption of the existence of vortex lines.

Consider what would happen if liquid is flowing out of an orifice into a reservoir of fluid at rest. If the flow is irrotational, it looks like Fig. 11.31. A very high velocity develops near the corners and there are large accelerations there. An ordinary fluid, such as water, flows in a more complicated manner. For a rough estimate, let us suppose that the tube is a long narrow slot, and suppose the fluid tries to go out in a jet, at first of the same width and velocity as in the tube. Then outside this tube we have a situation similar to that shown in Fig. 11.28 and following the same reasoning as before, we expect to see the

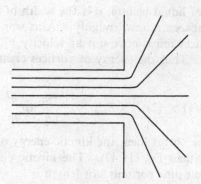

Fig. 11.31 Irrotational flow from an orifice.

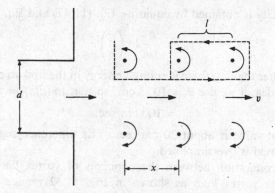

Fig. 11.32 Vortex lines outside an orifice.

vortex lines shown in Fig. 11.30 or in Fig. 11.32. Integrating over the dotted line we find

$$V_l = \oint \mathbf{V} \cdot d\mathbf{r} = n \left(\frac{2\pi\hbar}{m} \right),$$

where n is the number of vortex lines in length l. The number of vortices per unit length along the line of flow is

$$\frac{V}{(2\pi\hbar/m)} = \frac{mV}{2\pi\hbar}. \tag{11.45}$$

These vortex lines have an associated energy. From Eq. (11.41) the energy per unit length of the vortex line may be written as

$$\sim \pi\rho \, \frac{\hbar^2}{m} \ln \left(\frac{d}{a} \right), \tag{11.46}$$

where ρ is the density of liquid helium, d is the width of the jet in Fig. 11.32, and a is the spacing between atoms, roughly 4 Å as was stated at the end of Section 11.7. The vortices must move out at velocity $V/2$, the velocity at the location of the vortices. Thus the energy of vortices created per unit time per unit length of slot is

$$2 \times \frac{V}{2} \times (1) \times (2): \quad E_{\text{vort}} = \frac{V^2 m}{2\pi\hbar}\, \pi\rho\, \frac{\hbar^2}{m} \ln\left(\frac{d}{a}\right). \quad (11.47)$$

In order to form the vortex lines, the kinetic energy of the jet per unit slot length must be greater than Eq. (11.47). The kinetic energy of flow coming out of the orifice per unit time per unit slot length is

$$E_{\text{flow}} = m\, \frac{\rho V^2}{2}\, Vd. \quad (11.48)$$

The critical velocity is obtained by equating Eq. (11.47) and Eq. (11.48):

$$v_0 = \frac{\hbar}{dm} \ln\left(\frac{d}{a}\right). \quad (11.49)$$

For velocity greater than v_0, there is enough energy in the flow to create vortices. To get a rough idea, if we use $d = 10^{-5}$ cm, so that $\ln(d/a) = 6$, we see

$$v_0 = 100 \text{ cm/sec}. \quad (11.50)$$

The experimental value is about 20 cm/sec. The difference suggests that the model we have used is oversimplified.

There is a connection between the concepts of vortex lines and rotons. Starting from the vortex lines as shown in Fig. 11.32 we see that they will gradually change into rotons as shown in Fig. 11.33. A vortex ring can be broken into two smaller rings. A roton may be regarded as a small vortex ring, as discussed in Section 11.5.

There are experimental verifications of vortex lines. See, for instance, W. F. Vinen and H. E. Hall, *Proc. Roy. Soc.* **238**, 204 (1956); **238**, 215 (1956): H. E. Hall, *Proc. Roy. Soc.* **245**, 546 (1958): W. F. Vinen, *Nature* **181**, 1524 (1958).

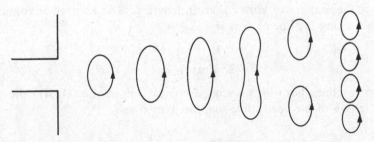

Fig. 11.33 Vortex lines changing into rotons.

11.10 THE λ TRANSITION IN LIQUID HELIUM

When we know the curve for $E(p)$, the energy versus momentum function, then the partition function and the specific heat can be calculated if we assume each excitation is independent. This calculation leads to the curve of the specific heat, which agrees with the experiment for the part drawn in solid in Fig. 11.34.

The above approach, however, does not explain the existence of the transition shown by the dotted curves in Fig. 11.34, for the interaction between excitations is not taken into account.

For the region near the transition point the $E(p)$ curve in Fig. 11.34 is not of much help. We start afresh and use a different method.

In terms of path integrals we have

$$e^{-\beta F} = \frac{1}{N!} \sum_p \int \underset{\substack{x_i(0)=x_i \\ x_i(U)=px_i}}{\iint} \exp\left(-\frac{1}{\hbar} \int_0^U \left\{ \sum_i \frac{m\dot{x}_i^2}{2} + \sum_{i<j} V[x_i - x_j] \right\} du \right)$$
$$\mathscr{D}x_1 \, \mathscr{D}x_2 \cdots \mathscr{D}x_N \, dx_1 \cdots dx_N \qquad (11.51)$$

As we concluded in Section 4.1 for high enough temperatures only the identity permutation counts, and we get approximately the classical partition function of helium. In this case, "high enough temperatures" means $(1/\hbar) \int_0^u (m\dot{x}^2/2)$ is large for $x(0) =$ position of one atom, $x(U) =$ position of another atom. If d is the average interatomic distance, $\dot{x}$ is on the average d/u and $(1/\hbar) \int_2^u (m\dot{x}^2/2)$ is of order $mkTd^2/2\hbar^2$. For $T \gg 2°K$,

$$\exp\left[\frac{mkTd^2}{2\hbar^2} \right] \gg 1,$$

and permutations cannot be important in Eq. (11.51). For low T (high U) permutations are important. Furthermore, the approximation that $V[x_i(u) - x_j(u)]$ is independent of u for important paths is true only if U is small, so that $x(u) \approx x(0)$ to have an appreciable contribution to the path

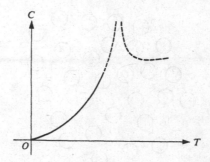

Fig. 11.34 Specific heat of liquid helium; theory and experiment agree for the part shown as a solid line.

integral. For low temperatures, $x(u)$ can vary greatly without giving rise to a large value of $\dot{x}^2(u)$, and thus $V(x_i - x_j)$ can vary over the path.

To calculate the λ transition, we must concern ourselves with what happens at low temperatures. We can visualize each path of the integral as a motion of every particle of the liquid from one position to another. "u" is thought of as time, and $x_i(u)$ is the position of the ith particle of "time" u. If, during this motion, $\dot{x}_i(u)$ becomes too large, or if V becomes too large, the path gives only a small contribution. $V[x_i(u) - x_j(u)]$ is large if at "time" u particles i and j are close to each other. Of course we are not claiming that $x_i(u)$ is really the position of some atom at an actual time. But thinking in the manner described above allows one to apply physical intuition effectively.

Consider a permutation in which a particle at position A moves to the position B, while the particle at B moves to a third position. (See Fig. 11.35.) We will consider the contribution to $(1/\hbar) \sum \int (m\dot{x}^2/2)$ caused by A.

As A moves to position B, it must somehow avoid passing through particles such as C. To get to position B in time U, particle A must move quickly enough to cover the distance. Since x is large for particle A, it is increased a large amount by increments of velocity so A does not have time to avoid C by trying to move around it. Instead, C is jostled out of the way. Suppose, now, that the distance from A to B is r, and suppose that there are n particles that have to be pushed out of the way. n is of the order of r/d, where d is the interatomic spacing. Each particle along the path has to move a distance of order d' to avoid being too close to particle A. Particle A is close to particle C for a time of order u/n, so that particle C must move with velocity $d'/(u/n)$ during that time, giving contribution

$$\frac{m}{2\hbar}\left(\frac{d'}{u/n}\right)^2 \frac{u}{n} = \frac{md'^2 n}{2\hbar u} \quad \text{to} \quad \frac{1}{\hbar}\int_0^u \sum_i \frac{m\dot{x}_i^2}{2}.$$

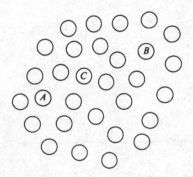

Fig. 11.35 Particle at A moves to B while particle at B moves elsewhere.

Thus the n particles give contribution

$$\frac{md'^2n^2}{2\hbar u} = \frac{m}{2\hbar^2}\left(\frac{d'}{d}\right)\frac{r^2}{\beta}.$$

Particle A itself contributes $mr^2/2\hbar^2\beta$. Thus the effect of the potential is to change $e^{-mr^2/2\hbar^2}$ to $e^{-(-\gamma mr^2/2\hbar^2\beta)}$ where $\gamma = 1 + (d'/d)^2$.

We are going to proceed as if γ were independent of r and β, and our reasoning above shows that this assumption is not too unreasonable. But remember that for displacements of order d we may not need as much adjustment, so γ may be less if r is small. Also if the velocity is especially high, it may be preferable to pass through relatively high potentials, rather than have rapid adjustments. Thus again γ would be lowered. Let us ignore such details.

It is convenient to define the effective mass m' as

$$m\gamma = m'$$

so that we may write $e^{-m'/2\hbar^2\beta r^2}$. Thus in the liquid state, so far as the path integral is concerned, a particle behaves like an ideal-gas particle* of mass m'.

We have arrived at the important conclusion that the interaction among particles is "kinetic" rather than "potential." Kinetic interaction varies as r^2/β, while potential varies as $r\beta$ (as can be seen by estimating the potential contribution for paths that do not avoid high potentials).

Therefore the partition function for low temperatures has the form

$$e^{-\beta F} = \frac{1}{N!}\sum_P \int \left(\frac{m'}{2\pi\hbar^2\beta}\right)^{3N/2}$$

$$\times \exp\left(-\frac{m'}{2\hbar^2\beta}\sum_i (R_i - PR_i)^2\right)\rho(R_1, R_2, \ldots, R_n)\, d^3R_1\, d^3R_2 \cdots.$$

$$(11.52)$$

The last factor comes from the potential-energy contribution of the initial configuration. The reason for the factor $(m'/2\pi\hbar^2\beta)^{3N/2}$ is not obvious, for it is not obvious how a change in the effective mass in the path integral will change the normalization of the integral. To be complete, we should instead write a factor $K_\beta(m'/2\pi\hbar^2\beta)^{3N/2}$. For simplicity, we will pay no attention to K_β. The factor ρ is clearly small for initial configurations (equal to the final configurations) such that atoms overlap. We can, for example, expect results that are qualitatively correct by choosing $\rho = 0$ if any two R's are closer than b (with "b" roughly 2.6 Å) and $\rho = 1$ otherwise.

A permutation among particles can be visualized as a polygon with arrows. The example shown in Fig. 11.36 consists of several cycles of length 1 (particles

---

* The above results apply only to Bose particles. Fermi particles behave differently.

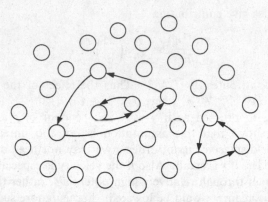

Fig. 11.36 Permutation among particles; here there is one cycle of length 2 and two of length 3.

that are not permutated), one cycle of length 2, and two cycles of length 3. Thus, the sum $\sum_p$ in Eq. (11.45) is a sum over all possible polygon patterns.

A way to estimate Eq. (11.45) for temperatures near the transition temperature will be explained here.* For such temperatures, a side (of a polygon) longer than the average interatomic distance d is not important. Thus we may assume that the sides of all polygons that contribute have length of order d.

Given a particular configuration $(R_1, R_2, \ldots, R_N)$ we restrict our permutations to be such that R_i and PR_i are close. In order to get a contribution to the integral in Eq. (11.45), ρ forces the atoms to be roughly uniformly distributed. Consider the sum

$$\sum_p \exp\left[-\frac{m'}{2\hbar^2\beta} \sum_i (R_i - PR_i)^2\right].$$

Large shifts in the R_i that preserve the uniform distribution effectively permute the atoms and thus leave the above sum unchanged, except for the change that would result from small shifts in the R_i. The effect of smaller shifts in the R_i that do not permute the atoms can be included by replacing $R_i - PR_i$ with an average interatomic distance, d. For all uniform distributions of atoms, we can therefore replace

$$\sum_p \exp\left[-\frac{m'}{2\hbar^2\beta} \sum_i (R_i - PR_i)^2\right]$$

by

$$\sum_p{}' \exp\left(-\frac{m'd^2n(p)}{2\hbar^2\beta}\right),$$

---

* For a more complete discussion, see R. P. Feynman, *Phys. Rev.* **91**, 1291 (1953).

where $\sum_p'$ is the sum over all permutations of a fixed lattice such that the resulting polygons are made up of lines joining near neighbors, and $n(p)$ is the total number of sides in the set of polygons defined by P. We can now take the above sum out of the integral in Eq. (11.45) and write

$$e^{-\beta F} = \text{constant} \times \left(\frac{m'}{2\pi\hbar^2\beta}\right)^{3N/2} \times \sum_p' e^{-m'd^2n(p)/2\hbar^2\beta}.$$

In this equation, the difficult problem is the calculation of the sum. So ignore the first two factors on the right and set $y = e^{-m'd^2/2\hbar^2\beta}$. We must compute

$$e^{-\beta F} = \sum_p' y^{n(p)}.$$

So we have the following problem: Given a lattice of points, lines can form between adjacent points to form polygons. With each line is associated an energy $(m'd^2/2\hbar^2\beta)$. We wish to calculate the free energy of this system. Let n_s be the number of polygons with s sides and let $C(n_1, n_2, \ldots)$ be the number of permutations with n_s polygons of s sides. Then

$$e^{-\beta F} = \sum_{n_1, n_2, \ldots}' C(n_1, n_2, \ldots) \, y\left[\sum_{s=2}^{\infty} sn_s\right].$$

This sum is restricted by the equation

$$\sum_{s=1}^{\infty} sn_s = N = \text{number of atoms}.$$

The restriction on the sum can be eliminated by the usual method of letting N vary while assigning to each atom a weighting factor, τ (in Chapter 1, "τ" was $e^{\beta\mu}$).

$$Q = e^{-\beta g} = \sum_{n_1, n_2, \ldots} C(n_1, n_2, \ldots) \, y\left(\sum_{s=2}^{\infty} sn_s\right) \tau \left(\sum_{s=2}^{\infty} sn_s\right) \tau^{n_1}$$

$$= \sum_N \tau^N e^{-\beta F_N}. \tag{11.53}$$

From Eq. (11.46) it is easy to see that the expectation value of N is

$$\langle N \rangle = \tau \frac{d \log Q}{d\tau} = -\tau\beta \frac{dg}{d\tau}. \tag{11.54}$$

To evaluate Eq. (11.46), we must find a way of approximating $C(n_1, n_2, \ldots)$. We will approximate C by writing

$$C(n_1, n_2, \ldots) = \prod_s \frac{R_s^{n_s}}{n_s!},$$

where R is the number of ways of drawing a polygon with s vertices on a lattice of $\langle N \rangle$ points (with all sides joining nearest neighbors). This approximation is based on the following assumptions:

a.   The number of ways of drawing a polygon with s sides is roughly independent of the polygons that have already been drawn. The effect of the competition between polygons for available atoms is taken into account by the factor "τ."

b.   If we allow overlapping of polygons there will not be much error.

c. n_s small enough so that given n_s random polygons with s sides, there probably will not be any that are on top of each other.

Then we can choose n_s polygons of length s in about $R_s^{n_s}/n_s!$ ways. By assumption (a), the total number of ways of choosing sets of polygons with n_s polygons with s sides is

$$\prod_s \frac{R_s^{n_s}}{n_s!}.$$

Equation (11.46) becomes

$$Q = \sum_{n_1, n_2, \ldots} \tau^{n_1} \frac{R_1^{n_1}}{n_1!} \prod_{s=2}^{\infty} \frac{R_s^{n_s}}{n_s!} (y\tau)^{sn_s} = \sum_{n_1} \frac{(R_1\tau)^{n_1}}{n_1!} \prod_{s=2}^{\infty} \sum_{n_s=0}^{\infty} \frac{[R_s(y\tau)^s]n_s}{n_s!}$$

$$= \exp\left[R_1\tau\right] \exp\left[\sum_{s=0}^{\infty} R_s(y\tau)^s\right]$$

or

$$g = -kT\left[R_1\tau + \sum_{s=0}^{\infty} R_s(y\tau)^s\right]. \tag{11.55a}$$

τ is chosen so that

$$\langle N \rangle = R_1\tau + \sum_{s=2}^{\infty} sR_s(y\tau)^s. \tag{11.56a}$$

We now must find R_s. If $s = 1$, we wish to find R_1, the number of ways of picking one point out of $\langle N \rangle$, so $R_1 = \langle N \rangle$. Hereafter, write $\langle N \rangle \equiv N$. In general, to find R_s we let h_s equal the number of polygons with s sides that can be drawn starting at a given atom. Then Nh_s would be the number of ways of drawing a polygon with s vertices if we started at any atom; but we have counted each polygon s times. Thus $R_s = Nh_s/s$. To avoid worrying about what happens when we get near the boundary, we simply consider the system to have periodic boundary conditions. That is, in two dimensions the system would have the topology of a torus. We will now approximate h_s for fairly large s assuming that the temperatures are such that small s does not contribute much to the sums in Eq. (11.55a) and Eq. (11.56a). Suppose l is the number of nearest neighbors of each lattice point. Consider the problem of random walk on the lattice. There are l^s equally probable random walks that start on a given atom and make s steps. In h_s, we wish to consider only those ways in which the walks terminate at the origin. The fraction of walks that end up at the origin is inversely proportional to the volume of space in which the walk is likely to end. Since the distance from the origin of the typical random walk is of the order $d\sqrt{s}$ (d = length of each step), the volume in which a typical walk ends is proportional to $s^{3/2}$. Thus h_s is proportional to $l^s/s^{3/2}$ for large s, where the proportionality constant is independent of s. Equations (11.55a) and (11.56a) become

$$g = -NKT\tau + c \sum_{s=2}^{\infty} \frac{(\tau yl)^s}{s^{5/2}}, \tag{11.55b}$$

$$1 = \tau + c \sum_{s=2}^{\infty} \frac{(\tau yl)^s}{s^{3/2}}, \tag{11.56b}$$

where c is a proportionality constant. For extremely large s, our result above for h_s is incorrect. If one chain covers a volume comparable to the size of the whole system, then $h_s = l^s/N$, where $1/N$ is the probability that a random walk will end at the original point (or any other particular point). Then a more realistic h_s might be

$$h_s \approx \left(\frac{c}{s_{3/2}} + \frac{1}{N} \right) l^s.$$

which has the correct behavior for medium and large s. In any case, the sums in Eqs. (11.55a) and (11.56a) converge for $\tau yl < 1$ and diverge for $\tau yl > 1$. At reasonably high temperatures y is small and the sum converges. Also, at such temperatures there are not too many polygons, so assumptions (a), (b), and (c) are not too unrealistic. Thus we expect our results to be qualitatively correct above the λ point. Equations (11.55b) and (11.56b) resemble the results for a Bose gas (see Section 1.8) and lead to an increase in specific heat with decreasing temperature. The shape of the specific heat curve is shown in Fig. 11.37.

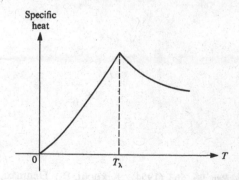

Fig. 11.37 Specific heat near the lambda point.

This analysis (which is supposed to be good for $T > T_\lambda$) thus explains why C increases as T approaches T_λ from above. This treatment does not lead to the discontinuity of the kind shown in Fig. 11.34, probably because of the approximations made between Eqs. (11.52) and (11.55a). In other words, we expect that the volume of the helium atoms described by ρ in Eq. (11.52) is the cause of the discontinuity in the specific heat. Kikuchi and others have written papers in which it is indicated that more accurate computations starting from Eq. (11.52) would lead to a discontinuity.*

Problem: How do these polygon calculations differ from the ones we did in Chapter 5 for the Onsager problem?

---

* Kikuchi, R., *Phys. Rev.* **96**, 563 (1954). Kikuchi, R., Denman, H. H., and Schreiber, C. L., *Phys. Rev.* **119**, 1823 (1960).

INDEX

Annihilation operators, *see* creation and annihilation operators

Bardeen-Cooper-Schrieffer (BCS) model, 269

BCS model, *see* Bardeen-Cooper-Schrieffer model

Blackbody radiation, 10–12

Boltzmann's constant, 1, 6

Bose-Einstein condensation, 31–33

Bose-Einstein statistics, 30–33
  classical limit of, 31
  density matrix for, 60–64
  field operator for, 173, 174–176
  free energy for, 64

Chemical potential, μ, 26–27
  in a Bose-Einstein system, 26
  in a classical system, 109
  determination from number of particles, 27, 28, 30, 34, 109
  in a Fermi-Dirac system, 26
  for the lambda transition problem, 347
  for a one-dimensional classical gas, 119, 121
  for a one-dimensional order-disorder problem, 130
  for a two-dimensional order-disorder problem, 131

Classical systems, 97–126
  condensation of, 125–126
  equation of state of, 98, 100, 102, 103, 110, 114
  Mayer cluster expansion for, 105–110
  of one-dimensional gas, 117–125
  partition function for, 97
  radial distribution function for, 111–113
  second virial coefficient for, 100–104
  thermal energy of, 98, 114

Cooper pairs, 274

Correlation energy of electrons, 249, 255–264

Creation and annihilation operators, 153
  for Bose-Einstein systems, 173, 174–176
  expression of other operators in terms of, 176–182, 188, 189, 191, 192
  for Fermi-Dirac systems, 173, 174–176
  and the harmonic oscillator, 154
  for phonons, 159, 162
  for spin waves, 217
  for a system of harmonic oscillators, 157–159
  and an unharmonic oscillator, 156

Critical temperature
  for Bose-Einstein condensation, 31
  for liquid helium, 313
  for order-disorder systems, 140
  for a superconductor, 284

Debye approximation, 19

Debye temperature, 20

Density matrix, 39–71, 40, 47
  in the classical limit, 77
  differential equation for, 48
  for free particles, 48–49, 59, 73–74
  for free Bose-Einstein particles, 60–64
  for free Fermi-Dirac particles, 64
  functions of, 45
  for an harmonic oscillator, 49–53, 80–84
  for an interacting harmonic oscillator, 81–84
  for one-dimensional motion, 74–76
  path integral formulation of, 72–96, *see* also path integrals
  perturbation expansion for, 66–67
  for polarized light, 42–43
  for a pure state, 41
  in the x-representation, 41

Printed in the United States
by Baker & Taylor Publisher Services